Springer Finance

Springer Finance

Springer Finance is a programme of books addressing students, academics and practitioners working on increasingly technical approaches to the analysis of financial markets. It aims to cover a variety of topics, not only mathematical finance but foreign exchanges, term structure, risk management, portfolio theory, equity derivatives, and financial economics.

Ammann M., Credit Risk Valuation: Methods, Models, and Application (2001)
Back K., A Course in Derivative Securities: Introduction to Theory and Computation (2005)
Barucci E., Financial Markets Theory. Equilibrium, Efficiency and Information (2003)
Bielecki T.R. and Rutkowski M., Credit Risk: Modeling, Valuation and Hedging (2002)
Bingham N.H. and Kiesel R., Risk-Neutral Valuation: Pricing and Hedging of Financial Derivatives (1998, 2nd ed. 2004)
Brigo D. and Mercurio F., Interest Rate Models: Theory and Practice (2001, 2nd ed. 2006)
Buff R., Uncertain Volatility Models – Theory and Application (2002)
Carmona R.A. and Tehranchi M.R., Interest Rate Models: An Infinite Dimensional Stochastic Analysis Perspective (2006)
Dana R.-A. and Jeanblanc M., Financial Markets in Continuous Time (2003)
Deboeck G. and Kohonen T. (Editors), Visual Explorations in Finance with Self-Organizing Maps (1998)
Delbaen F. and Schachermayer W., The Mathematics of Arbitrage (2005)
Elliott R.J. and Kopp P.E., Mathematics of Financial Markets (1999, 2nd ed. 2005)
Fengler M.R., Semiparametric Modeling of Implied Volatility (2005)
Filipović D., Term-Structure Models (2009)
Fusai G. and Roncoroni A., Implementing Models in Quantitative Finance (2008)
Geman H., Madan D., Pliska S.R. and Vorst T. (Editors), Mathematical Finance – Bachelier Congress 2000 (2001)
Gundlach M. and Lehrbass F. (Editors), CreditRisk[+] in the Banking Industry (2004)
Jeanblanc M., Yor M., Chesney M., Mathematical Methods for Financial Markets (2009 forthcoming)
Jondeau E., Financial Modeling Under Non-Gaussian Distributions (2007)
Kabanov Y.A. and Safarian M., Markets with Transaction Costs (2009 forthcoming)
Kellerhals B.P., Asset Pricing (2004)
Külpmann M., Irrational Exuberance Reconsidered (2004)
Kwok Y.-K., Mathematical Models of Financial Derivatives (1998, 2nd ed. 2008)
Malliavin P. and Thalmaier A., Stochastic Calculus of Variations in Mathematical Finance (2005)
Meucci A., Risk and Asset Allocation (2005, corr. 2nd printing 2007)
Pelsser A., Efficient Methods for Valuing Interest Rate Derivatives (2000)
Prigent J.-L., Weak Convergence of Financial Markets (2003)
Schmid B., Credit Risk Pricing Models (2004)
Shreve S.E., Stochastic Calculus for Finance I (2004)
Shreve S.E., Stochastic Calculus for Finance II (2004)
Yor M., Exponential Functionals of Brownian Motion and Related Processes (2001)
Zagst R., Interest-Rate Management (2002)
Zhu Y.-L., Wu X., Chern I.-L., Derivative Securities and Difference Methods (2004)
Ziegler A., Incomplete Information and Heterogeneous Beliefs in Continuous-time Finance (2003)
Ziegler A., A Game Theory Analysis of Options (2004)

Damir Filipović

Term-Structure Models

A Graduate Course

 Springer

Damir Filipović
University of Vienna, and Vienna University
of Economics and Business
Heiligenstädter Strasse 46-48
1190 Vienna
Austria
damir.filipovic@vif.ac.at

ISBN 978-3-642-26915-8   e-ISBN 978-3-540-68015-4
DOI 10.1007/978-3-540-68015-4
Springer Dordrecht Heidelberg London New York

Mathematics Subject Classification (2000): 60H05, 60H10, 60J60, 62P05, 91B28
JEL Classification: E43, G12, G13

©Springer-Verlag Berlin Heidelberg 2009
Softcover reprint of the hardcover 1st edition 2009

Printed on acid-free paper

Springer is part of Springer Science+Business Media (www.springer.com)

for Susanne and Elena Christina

Preface

Changing interest rates constitute one of the major risk sources for banks, insurance companies, and other financial institutions. Modeling the term-structure movements of interest rates is a challenging task. One simple reason lies in the high dimensionality of this object, which is often assumed to be infinite. This creates a demand for mathematical models which differ from the standard stock market models. The origin of the term-structure models treated in this book can be traced back more than thirty years to the seminal work of Vasiček [160]. Since that time, the volume of traded interest rate sensitive derivatives has grown enormously.

This book gives an introduction to the mathematics of term-structure models in continuous time. It is suitable for a one-semester graduate course in mathematics, financial engineering, or quantitative finance. The focus is on a mathematically straightforward but rigorous development of the theory, which is illustrated with examples whenever possible. Each chapter ends with a set of exercises that provides a source for homework and exam questions. Readers are expected to be familiar with elementary Itô calculus, and analysis and probability theory on the level of e.g. Rudin [138] and Williams [161], respectively.

This book has emerged in several stages. I wrote the first version as lecture notes for a one-semester graduate course on fixed-income models in the fall term 2002/03 at the Department of Operations Research and Financial Engineering at Princeton University. The text has been gradually improved in subsequent lectures held at the Mathematics Institute at the University of Munich, at the Vienna Graduate School of Finance (VGSF), and at the Executive Academy of the Vienna University of Economics and Business Administration (WU). In the winter term of 2008/09 I completed the book by substantial revision and extension of the text, the inclusion of exercise and notes sections, and the addition of a completely new chapter on affine processes.

The number of books on term-structure models is rapidly growing, yet it is difficult, with a few exceptions, to find a convenient textbook for a one-semester graduate course on term-structure models for mathematicians and financial engineers. There are several reasons for this:

- Until recently, many textbooks on mathematical finance have treated stochastic interest rates as an appendix to the elementary arbitrage pricing theory, which usually requires constant (zero) interest rates.
- Interest rate theory is not as standardized as the arbitrage pricing models for stocks, such as the fundamental Black–Scholes model.
- The very nature of fixed-income instruments causes difficulties, other than for stock derivatives, in implementing and calibrating models. These issues should therefore not be left out.

Being aware that I must have overlooked important other contributions, I mention the following incomplete list of related books in alphabetic order: Björk [13] (introduction to mathematical finance, with a part on interest rate models), Brigo and Mercurio [27] (interest rate and credit risk models, practical implementation and calibration of selected models), Cairns [33] (a graduate course book on interest rate models), Carmona and Tehranchi [35] (mathematically advanced text on an infinite-dimensional analysis approach to interest rate models), James and Webber [100] (comprehensive resource on interest rate models, includes some historic account), Jarrow [103] (discrete-time introduction to interest rates), Musiela and Rutkowski [127] (comprehensive introduction to mathematical finance, with a large part on interest rate modeling and market pricing practice), Pelsser [131] (introduction to interest rate models and their efficient implementation), Rebonato [134] (emphasis on market practice for pricing and handling interest rate derivatives), Shreve [149] (introduction to mathematical finance with a chapter on term-structure models), and Zagst [163] (introduction to mathematical finance, interest rate modeling and risk management). In particular, more term-structure-related exercises, besides the set provided in this book, can be found in Björk [13], Cairns [33], and Shreve [149].

What distinguishes this book from others in particular is its comprehensive chapter on affine diffusion processes, which are among the most widely used factor models in finance. Another feature of this book is its section on the interplay between curve-fitting methods and factor models for the term-structure of interest rates.

I owe a lot of thanks for their helpful comments and contributions to Francesca Biagini, Rama Cont, Christa Cuchiero, Jason Chung, Zehra Eksi, Luiz Paulo Feijó Fichtner, Nikolaos Georgiopoulos, Paul Glasserman, Georg Grafendorfer, Michael Kupper, Eberhard Mayerhofer, Antoon Pelsser, Daniel Rost, Mykhaylo Shkolnikov, Gregor Svindland, Stefan Tappe, Takahiro Tsuchiya, Nicolas Vogelpoth, Mario Wüthrich, and Vilimir Yordanov. Financial support during the final writing of this book from WWTF (Vienna Science and Technology Fund) is gratefully acknowledged. Moreover, I am grateful to Catriona Byrne and the Editorial Assistants at Springer-Verlag for their valuable support. I also thank Jef Boys for thoroughly copy-editing the manuscript.

Most I owe to my wife Susanne for her loving support and patience during the time-consuming writing of this book.

Vienna   *Damir Filipović*

Contents

1	Introduction	1

2 Interest Rates and Related Contracts 5

2.1	Zero-Coupon Bonds	5
2.2	Interest Rates	6
	2.2.1 Market Example: LIBOR	7
	2.2.2 Simple vs. Continuous Compounding	8
	2.2.3 Forward vs. Future Rates	9
2.3	Money-Market Account and Short Rates	9
	2.3.1 Proxies for the Short Rate	10
2.4	Coupon Bonds, Swaps and Yields	11
	2.4.1 Fixed Coupon Bonds	11
	2.4.2 Floating Rate Notes	12
	2.4.3 Interest Rate Swaps	12
	2.4.4 Yield and Duration	15
2.5	Market Conventions	17
	2.5.1 Day-Count Conventions	17
	2.5.2 Coupon Bonds	18
	2.5.3 Accrued Interest, Clean Price and Dirty Price	18
	2.5.4 Yield-to-Maturity	19
2.6	Caps and Floors	19
	2.6.1 Caps	20
	2.6.2 Floors	20
	2.6.3 Caps, Floors and Swaps	21
	2.6.4 Black's Formula	21
2.7	Swaptions	22
	2.7.1 Black's Formula	24
2.8	Exercises	24
2.9	Notes	27

3 Estimating the Term-Structure 29

3.1	A Bootstrapping Example	29
3.2	Non-parametric Estimation Methods	34
	3.2.1 Bond Markets	35
	3.2.2 Money Markets	36
	3.2.3 Problems	38
3.3	Parametric Estimation Methods	38
	3.3.1 Estimating the Discount Function with Cubic B-splines	38

    3.3.2   Smoothing Splines   43
    3.3.3   Exponential–Polynomial Families   49
  3.4   Principal Component Analysis   51
    3.4.1   Principal Components of a Random Vector   51
    3.4.2   Sample Principle Components   52
    3.4.3   PCA of the Forward Curve   53
    3.4.4   Correlation   55
  3.5   Exercises   56
  3.6   Notes   57

4   Arbitrage Theory   59
  4.1   Stochastic Calculus   59
    4.1.1   Stochastic Integration   60
    4.1.2   Quadratic Variation and Covariation   61
    4.1.3   Itô's Formula   62
    4.1.4   Stochastic Differential Equations   63
    4.1.5   Stochastic Exponential   64
  4.2   Financial Market   65
    4.2.1   Self-Financing Portfolios   65
    4.2.2   Numeraires   66
  4.3   Arbitrage and Martingale Measures   67
    4.3.1   Martingale Measures   68
    4.3.2   Market Price of Risk   69
    4.3.3   Admissible Strategies   70
    4.3.4   The First Fundamental Theorem of Asset Pricing   70
  4.4   Hedging and Pricing   71
    4.4.1   Complete Markets   71
    4.4.2   Arbitrage Pricing   74
  4.5   Exercises   75
  4.6   Notes   77

5   Short-Rate Models   79
  5.1   Generalities   79
  5.2   Diffusion Short-Rate Models   80
    5.2.1   Examples   82
    5.2.2   Inverting the Forward Curve   83
  5.3   Affine Term-Structures   84
  5.4   Some Standard Models   85
    5.4.1   Vasiček Model   85
    5.4.2   CIR Model   87
    5.4.3   Dothan Model   88
    5.4.4   Ho–Lee Model   89
    5.4.5   Hull–White Model   90
  5.5   Exercises   91
  5.6   Notes   92

6   Heath–Jarrow–Morton (HJM) Methodology 93
  6.1   Forward Curve Movements 93
  6.2   Absence of Arbitrage 95
  6.3   Short-Rate Dynamics 96
  6.4   HJM Models 97
      6.4.1   Proportional Volatility 98
  6.5   Fubini's Theorem 99
  6.6   Exercises 102
  6.7   Notes 103

7   Forward Measures 105
  7.1 T-Bond as Numeraire 105
  7.2   Bond Option Pricing 109
      7.2.1   Example: Vasiček Short-Rate Model 110
  7.3   Black–Scholes Model with Gaussian Interest Rates 110
      7.3.1   Example: Black–Scholes–Vasiček Model 113
  7.4   Exercises 114
  7.5   Notes 116

8   Forwards and Futures 117
  8.1   Forward Contracts 117
  8.2   Futures Contracts 118
      8.2.1   Interest Rate Futures 119
  8.3   Forward vs. Futures in a Gaussian Setup 120
  8.4   Exercises 121
  8.5   Notes 122

9   Consistent Term-Structure Parametrizations 123
  9.1   Multi-factor Models 123
  9.2   Consistency Condition 125
  9.3   Affine Term-Structures 127
  9.4   Polynomial Term-Structures 128
      9.4.1   Special Case: $m = 1$ 129
      9.4.2   General Case: $m \geq 1$ 131
  9.5   Exponential–Polynomial Families 134
      9.5.1   Nelson–Siegel Family 134
      9.5.2   Svensson Family 135
  9.6   Exercises 138
  9.7   Notes 140

10   Affine Processes 143
  10.1   Definition and Characterization of Affine Processes 143
  10.2   Canonical State Space 146
  10.3   Discounting and Pricing in Affine Models 151
      10.3.1   Examples of Fourier Decompositions 157
      10.3.2   Bond Option Pricing in Affine Models 161

        10.3.3 Heston Stochastic Volatility Model 166
    10.4 Affine Transformations and Canonical Representation 168
    10.5 Existence and Uniqueness of Affine Processes 171
    10.6 On the Regularity of Characteristic Functions 173
    10.7 Auxiliary Results for Differential Equations 177
        10.7.1 Some Invariance Results 177
        10.7.2 Some Results on Riccati Equations 180
        10.7.3 Proof of Theorem 10.3 185
    10.8 Exercises 186
    10.9 Notes 194

11 Market Models 197
    11.1 Heuristic Derivation 197
    11.2 LIBOR Market Model 199
        11.2.1 LIBOR Dynamics Under Different Measures 201
    11.3 Implied Bond Market 201
    11.4 Implied Money-Market Account 204
    11.5 Swaption Pricing 206
        11.5.1 Forward Swap Measure 207
        11.5.2 Analytic Approximations 209
    11.6 Monte Carlo Simulation of the LIBOR Market Model 210
    11.7 Volatility Structure and Calibration 212
        11.7.1 Principal Component Analysis 212
        11.7.2 Calibration to Market Quotes 213
    11.8 Continuous-Tenor Case 219
    11.9 Exercises 221
    11.10 Notes 223

12 Default Risk 225
    12.1 Default and Transition Probabilities 225
    12.2 Structural Approach 227
    12.3 Intensity-Based Approach 229
        12.3.1 Construction of Doubly Stochastic Intensity-Based Models 235
        12.3.2 Computation of Default Probabilities 236
        12.3.3 Pricing Default Risk 236
        12.3.4 Measure Change 240
    12.4 Exercises 242
    12.5 Notes 243

References 245

Index 253

Chapter 1
Introduction

A term-structure is a function that relates a certain financial variable or parameter to its maturity. Prototypical examples are the term-structure of interest rates or zero-coupon bond prices. But there are also term-structures of option implied volatilities, credit spreads, variance swaps,[1] etc. Term-structures are high-dimensional objects, which often are not directly observable. On the empirical side this requires estimation methods that are flexible enough to capture the entire market information. But flexibility often comes at the cost of irregular term-structure shapes and a great number of factors. Principal component analysis and parametric estimation methods can put things right. On the modeling side we find several challenging tasks. Bonds and other forward contracts expire at maturity where they have to satisfy a formally predetermined terminal condition. For example, a zero-coupon bond has value one at maturity, a European-style option has a predetermined payoff contingent on some underlying instrument, etc. Under the absence of arbitrage this has non-trivial implications for any dynamic term-structure model. As a consequence, various approaches to modeling the term-structure of interest rates have been proposed in the last decades, starting with the seminal work of Vasiček [160]. By arbitrage we mean an investment strategy that yields no negative cash flow in any future state of the world and a positive cash flow in at least one state; in simple terms, a risk-free profit. The assumption of no arbitrage is justified by market efficiency as a consequence of which prices tend to converge to arbitrage-free prices due to demand and supply effects.

The goal of this book is to give a self-contained and rigorous introduction to the mathematics of term-structure models in continuous time. After an elementary introduction to bond and interest rate markets, we review some of the related term-structure estimation methods in use, which will eventually be tested for consistency with arbitrage-free stochastic models. Before that we gradually introduce the mathematical tools and principles of arbitrage pricing needed to analyze the stochastic models most widely used in the industry and academia. This includes short-rate models, the Heath–Jarrow–Morton framework for term-structure movements, and the LIBOR and swap market models. A special feature of this book is a thorough chapter on affine diffusions, which are among the most widely used models in finance. Their main applications lie in the theory of the term-structure of interest rates, stochastic volatility option pricing and the modeling of credit risk, hence ranging well beyond interest rates. An outline of the most common approaches to default risk modeling completes this book.

---

[1]Given a stock index S, a variance swap exchanges the payment of realized variance of the log-returns of S against a previously agreed strike price. See Bühler [31].

D. Filipović, *Term-Structure Models*,
Springer Finance,
DOI 10.1007/978-3-540-68015-4_1, © Springer-Verlag Berlin Heidelberg 2009

In what follows, we give a quick overview of the contents of each chapter.

Chapter 2 introduces the basic notions of bond and interest rate markets: zero-coupon bonds, spot and forward rates, yields, short rates and the money-market account. Market practice such as day-count conventions will be briefly discussed. We then look at the prototypical interest rate derivatives, such as swaps, caps, floors and swaptions, and we learn how traders price them using Black's formula.

Chapter 3 reviews some of the most common term-structure estimation methods. We start with a bootstrapping example, and then consider more general aspects of non-parametric and parametric estimation methods by means of illustrating examples. In the last section we perform a principal component analysis for the term-structure movements, which is the best-known dimension reduction technique in multivariate data analysis.

Chapter 4 briefly recalls the fundamental arbitrage principles in a Brownian-motion-driven financial market. The basics of stochastic calculus are introduced, including the stochastic integral, stochastic differential equations and the stochastic exponential function. The main pillars are Itô's formula, Girsanov's change of measure theorem, and the martingale representation theorem. Based on these, the first and second fundamental theorems of asset pricing are established, which form the basis for hedging and pricing in a financial market model.

Chapter 5 gives an introduction to diffusion short-rate models, which are the earliest arbitrage-free interest rate models. Particular focus is on affine term-structures. We survey the most common standard models, including the Vasiček and Cox–Ingersoll–Ross models. It turns out that short-rate models are not always flexible enough to calibrate them to the observed initial term-structure.

Chapter 6 provides the essentials of the Heath–Jarrow–Morton (HJM) framework for modeling the entire forward curve directly. This is a very general setup, which includes all models presented in this book. The only substantive economic restrictions are the continuous sample paths assumption for the forward rate process, and the finite number of driving Brownian motions. The absence of arbitrage leads to the celebrated HJM drift condition, which implies that all bond option prices do only depend on the volatility curve. For the sake of completeness, we provide here a full proof of a version of Fubini's theorem for stochastic integrals, which is needed in the derivation of the HJM drift condition.

In Chap. 7 we replace the risk-free numeraire by another traded asset, such as the T-bond. This change of numeraire technique proves most useful for option pricing and provides the basis for the market models studied below. We derive explicit option price formulas for Gaussian HJM models. This includes the Vasiček short-rate model and some extension of the Black–Scholes stock model with stochastic interest rates.

Chapter 8 introduces two common types of term contracts: forwards and futures. The latter are actively traded on many exchanges. In particular, we discuss interest rate futures and futures rates, and relate them to forward rates in the Gaussian HJM model.

Chapter 9 brings together the aforementioned parametric estimation methods and arbitrage-free factor models for the term-structure of interest rates. We provide the appropriate consistency conditions and explore some important examples. This includes affine, quadratic, and more general polynomial term-structures, as well as the Nelson–Siegel and Svensson curve families, which are commonly used by central banks.

Chapter 10 is a special feature of this book. It provides a comprehensive discussion of affine diffusions and their applications in finance. We give a complete characterization, establishing existence and uniqueness, of multivariate affine diffusions on the common state space $\mathbb{R}_+^m \times \mathbb{R}^n$. A general pricing formula is derived from Fourier transform methods. We find explicit expressions for European call and put options, exchange options, and spread options. These are further illustrated for specific models, including the Vasiček and Cox–Ingersoll–Ross short-rate models, as well as Heston's stochastic volatility model.

In Chap. 11 we introduce the lognormal LIBOR and swap market models. The principal idea of these approaches builds on the above change of numeraire technique, and is to choose a different numeraire than the risk-free account. Both approaches lead to Black's formula for either caps (LIBOR models) or swaptions (swap rate models). Because of this they are usually referred to as "market models". We discuss the Monte Carlo simulation and calibration of the LIBOR market model in some detail.

Chapter 12 reviews the two most common approaches to credit risk modeling: the structural and the intensity-based approach. The structural approach models the value of a firm's assets. Default is when this value hits a certain lower bound. In the intensity-based approach, default is specified exogenously by a stopping time with given intensity process. The scope is on single name risk only. That said, it provides the ground for further studies in the very active research area of default risk.

Each chapter ends with a set of exercises that provides a source for homework and exam questions, and a notes section. The notes sections provide background information, further reading and references to data and text sources used in the main text.

Finally a word on notation. We write ab for the matrix product and $a^\top$ for the transpose of matrices a, b. The ith standard basis vector in $\mathbb{R}^n$ is denoted by e_i. We write $\mathbb{R}_+ = [0, \infty)$ for the nonnegative real numbers and $\mathbb{R}_- = (-\infty, 0]$ for the nonpositive real numbers, and accordingly $\mathbb{R}_+^m$ and $\mathbb{R}_-^m$ in higher dimension. The real and imaginary part of a complex number or vector z is denoted by $\Re(z)$ and $\Im(z)$, respectively. We define $\mathbb{C}_+^m$ as the set of $z \in \mathbb{C}^m$ with $\Re(z) \in \mathbb{R}_+^m$, and analogously $\mathbb{C}_-^m$ as the set of $z \in \mathbb{C}^m$ with $\Re(z) \in \mathbb{R}_-^m$. With $x \wedge y$ and $x \vee y$ we denote the minimum and the maximum of x and y, respectively. All other notation is standard or explained in the text.

Chapter 2
Interest Rates and Related Contracts

A bond is a securitized form of a loan. Bonds are the primary financial instruments in the market where the time value of money is traded. This chapter provides the basis concepts of interest rates and bond markets. We start with zero-coupon bonds and define a number of related interest rates. We then look at market conventions and learn how caps, floors and swaptions are priced by market practice.

2.1 Zero-Coupon Bonds

A dollar today is worth more than a dollar tomorrow. The time t value of a dollar at time $T \geq t$ is expressed by the *zero-coupon bond* with maturity T, $P(t, T)$, briefly also T-*bond*. This is a contract which guarantees the holder one dollar to be paid at the maturity date T, see Fig. 2.1.

In theory we will assume that:

- There exists a frictionless market for T-bonds for every $T > 0$.
- $P(T, T) = 1$ for all T.
- $P(t, T)$ is differentiable in T.

In reality these assumptions are not always satisfied: zero-coupon bonds are not traded for all maturities, and $P(T, T)$ might be less than one if the issuer of the T-bond defaults. Yet, this is a good starting point for doing the mathematics. More realistic models will be introduced and discussed in the sequel.

The third condition is purely technical and implies that the *term-structure of zero-coupon bond prices*[1] $T \mapsto P(t, T)$ is a smooth curve, see Fig. 2.2 for an example.

Note that $t \mapsto P(t, T)$ is a stochastic process since bond prices $P(t, T)$ are not known with certainty before t, see Fig. 2.3.

A reasonable assumption would also be that $T \mapsto P(t, T) \leq 1$ is a nonincreasing curve (which is equivalent to nonnegativity of interest rates). However, already classical interest rate models imply zero-coupon bond prices greater than 1. Therefore we leave aside this requirement.

Fig. 2.1 Cash flow of a T-bond

---

[1] $T \mapsto P(t, T)$ is also called the discount curve.

D. Filipović, *Term-Structure Models,*
Springer Finance,
DOI 10.1007/978-3-540-68015-4_2, © Springer-Verlag Berlin Heidelberg 2009

Fig. 2.2 Term-structure
$T \mapsto P(t, T)$

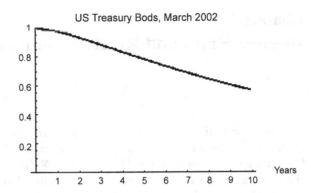

Fig. 2.3 T-bond price
process $t \mapsto P(t, T)$

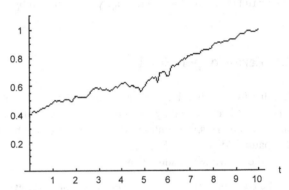

2.2 Interest Rates

The term-structure of zero-coupon bond prices does not contain much visual information (strictly speaking it does). A better measure is given by the implied interest rates. There is a variety of them.

A prototypical *forward rate agreement* (*FRA*) is a contract involving three time instants $t < T < S$: the current time t, the expiry time $T > t$, and the maturity time $S > T$.

- At t: sell one T-bond and buy $\frac{P(t,T)}{P(t,S)}$ S-bonds. This results in a zero net investment.
- At T: pay one dollar.
- At S: receive $\frac{P(t,T)}{P(t,S)}$ dollars.

The net effect is a forward investment of one dollar at time T yielding $\frac{P(t,T)}{P(t,S)}$ dollars at S with certainty.

We are led to the following definitions.

- The *simple* (or, *simply compounded*) *forward rate* for $[T, S]$ prevailing at t is given by

$$F(t; T, S) = \frac{1}{S - T} \left(\frac{P(t, T)}{P(t, S)} - 1 \right),$$

which is equivalent to

$$1 + (S - T)F(t; T, S) = \frac{P(t, T)}{P(t, S)}.$$

- The *simple spot rate* for $[t, T]$ is

$$F(t, T) = F(t; t, T) = \frac{1}{T - t}\left(\frac{1}{P(t, T)} - 1\right).$$

- The *continuously compounded forward rate* for $[T, S]$ prevailing at t is given by

$$R(t; T, S) = -\frac{\log P(t, S) - \log P(t, T)}{S - T},$$

which is equivalent to

$$e^{R(t;T,S)(S-T)} = \frac{P(t, T)}{P(t, S)}.$$

- The *continuously compounded spot rate* for $[t, T]$ is

$$R(t, T) = R(t; t, T) = -\frac{\log P(t, T)}{T - t}.$$

- As we let S tend to T, we arrive at the instantaneous *forward rate* with maturity T prevailing at time t, which is defined as

$$f(t, T) = \lim_{S \downarrow T} R(t; T, S) = -\frac{\partial \log P(t, T)}{\partial T}. \qquad (2.1)$$

The function $T \mapsto f(t, T)$ is called the *forward curve* at time t.
- The instantaneous *short rate* at time t is defined by

$$r(t) = f(t, t) = \lim_{T \downarrow t} R(t, T).$$

Notice that (2.1) together with the requirement $P(T, T) = 1$ is equivalent to

$$P(t, T) = e^{-\int_t^T f(t, u)\, du}.$$

2.2.1 Market Example: LIBOR

"Interbank rates" are rates at which deposits between banks are exchanged, and at which swap transactions (see below) between banks occur. The most important interbank rate usually considered as a reference for fixed-income contracts is the

LIBOR (*London Interbank Offered Rate*)[2] for a series of possible maturities, ranging from *overnight* to 12 months. These rates are quoted on a simple compounding basis. For example, the three-months forward LIBOR for the period $[T, T + 1/4]$ at time t is given by

$$L(t, T) = F(t; T, T + 1/4).$$

While individual banks may calculate their own LIBOR rates, the British Bankers' Association's (BBA) LIBOR serves as the primary benchmark globally. It is used as the basis for settlement of interest rate contracts on many of the world's major futures and options exchanges as well as most over the counter (OTC) and lending transactions. BBA LIBOR rates are published on www.bba.org.uk.

Throughout this book, we consider LIBOR as risk-free. In reality, LIBOR rates may reflect liquidity and credit risk. If the LIBOR rates are significantly above the interest rates as set by the central bank (such as the US Federal Reserve and the European Central Bank) it indicates that lenders are more worried about defaults on loans. This has been observed, for instance, after the dislocation of the credit markets in August 2007 and the following dry-up of the funding markets.

2.2.2 Simple vs. Continuous Compounding

One dollar invested for one year at an interest rate of R per annum grows to $1 + R$. If the rate is compounded twice per year the terminal value is $(1 + R/2)^2$, etc. It is a mathematical fact that

$$\left(1 + \frac{R}{m}\right)^m \to e^R \quad \text{as } m \to \infty.$$

On the other hand, we know that

$$e^R = 1 + R + o(R) \quad \text{for } R \text{ small}$$

where $o(R)/R \to 0$ for $R \to 0$. For example $e^{0.04} = 1.04081$.

Since the exponential function has nicer analytic properties than power functions, we often consider continuously compounded interest rates. Note, however, that in practice differences of the order of basis points,[3] such as $e^{0.04} - 1.04081 = 8.1$ bp, do matter as this can be scaled by the appropriate nominal investment!

---

[2]To be more precise: this is the rate at which high-credit financial institutions can *borrow* in the interbank market.

[3]Recall that a basis point (bp) equals $1/100$ percent: $1\,\text{bp} = 0.01\%$.

2.2.3 Forward vs. Future Rates

Can forward rates predict the future spot rates? Let us first consider a hypothetical deterministic world. If markets are free of arbitrage (that is, there is no risk-free profit) we have necessarily

$$P(t, S) = P(t, T)P(T, S), \quad t \leq T \leq S. \tag{2.2}$$

Proof Suppose that $P(t, S) > P(t, T)P(T, S)$ for some $t \leq T \leq S$. Then we follow the strategy (where do we use the assumption of a deterministic world?):

- At t: sell $\frac{P(T,S)P(t,T)}{P(t,S)}$ S-bonds, and buy $P(T, S)$ T-bonds. This results in a zero net investment.
- At T: receive $P(T, S)$ dollars and buy one S-bond.
- At S: pay $\frac{P(T,S)P(t,T)}{P(t,S)}$ dollars, receive one dollar.

The net profit of $1 - \frac{P(T,S)P(t,T)}{P(t,S)} > 0$ is risk-free. This is an arbitrage opportunity, which contradicts the assumption.

If $P(t, S) < P(t, T)P(T, S)$ the same profit can be realized by changing sign in the strategy ($\rightarrow$ Exercise 2.2), whence (2.2) is proved. □

Taking logarithms in (2.2) yields

$$\int_T^S f(t, u) \, du = \int_T^S f(T, u) \, du, \quad t \leq T \leq S.$$

This is equivalent to

$$f(t, S) = f(T, S) = r(S), \quad t \leq T \leq S.$$

As time goes by we walk along the forward curve: the forward curve is shifted. In this case, the forward rate with maturity S prevailing at time $t \leq S$ is exactly the future short rate at S.

The real world is not deterministic though. We will see in Sect. 7.1 below that the forward rate $f(t, T)$ is the conditional expectation of the short rate $r(T)$ under a particular probability measure, the T-forward measure, depending on T. Hence the forward rate is a biased estimator for the future short rate. Forecasts of future short rates by forward rates have in fact little or no predictive power.

2.3 Money-Market Account and Short Rates

The return of a one dollar investment today ($t = 0$) over the period $[0, \Delta t]$ is given by

$$\frac{1}{P(0, \Delta t)} = e^{\int_0^{\Delta t} f(0,u) \, du} = 1 + r(0)\Delta t + o(\Delta t)$$

where $o(\Delta t)/\Delta t \to 0$ for $\Delta t \to 0$. Instantaneous reinvestment in $2\Delta t$-bonds yields

$$\frac{1}{P(0, \Delta t)} \frac{1}{P(\Delta t, 2\Delta t)} = (1 + r(0)\Delta t)(1 + r(\Delta t)\Delta t) + o(\Delta t)$$

at time $2\Delta t$, etc. This strategy of "rolling over"[4] just-maturing bonds leads in the limit to the *money-market account*[5] $B(t)$. Hence $B(t)$ is the asset which grows at time t instantaneously at short rate $r(t)$

$$B(t + \Delta t) = B(t)(1 + r(t)\Delta t) + o(\Delta t).$$

For $\Delta t \to 0$ this converges to

$$dB(t) = r(t)B(t)dt$$

and with $B(0) = 1$ we obtain

$$B(t) = e^{\int_0^t r(s)\,ds}.$$

B is a *risk-free* asset insofar as its future value at time $t + \Delta t$ is known (up to order Δt) at time t. For the same reason we speak of $r(t)$ as the *risk-free rate of return* over the infinitesimal period $[t, t + dt]$.

B is important for relating amounts of currencies available at different times: in order to have one dollar in the money-market account at time T we need to have

$$\frac{B(t)}{B(T)} = e^{-\int_t^T r(s)\,ds}$$

dollars in the money-market account at time $t \leq T$. This *discount factor* is stochastic: it is not known with certainty at time t. There is a close connection to the discount factor given by $P(t, T)$, which is known at time t. Indeed, we will see that the latter is the conditional expectation of the former under the risk-neutral probability measure.

2.3.1 Proxies for the Short Rate

The short rate $r(t)$ is a key interest rate in all models and fundamental to no-arbitrage pricing. But it cannot be directly observed.

The overnight interest rate is not usually considered to be a good proxy for the short rate, because the motives and needs driving overnight borrowers are very different from those of borrowers who want money for a month or more. Moreover,

---

[4]This limiting process is made rigorous in [16].

[5]The money-market account is also called savings account, bank account, or money account.

microstructure effects, such as the second Wednesday settlement effect in the US Federal Funds market,[6] may create systematic spikes in the raw data that have to be smoothed. To avoid this problem, Aït-Sahalia [2] selects a slightly longer rate: the seven-day Eurodollar rate.

In general, also other daily interest rate series have to be used to check the robustness of the short-rate estimation results: one- or three-month spot LIBOR rates are considered as best available proxies since they are very liquid.

More information on estimating the term-structure of interest rates is provided in Chap. 3 below, see also [100, Chap. 3.5].

2.4 Coupon Bonds, Swaps and Yields

In most bond markets, there is only a relatively small number of zero-coupon bonds traded. Most bonds include coupons.

2.4.1 Fixed Coupon Bonds

A (*fixed*) *coupon bond* is a contract specified by:

- a number of future dates $T_1 < \cdots < T_n$ (the coupon dates)
  (T_n is the maturity of the bond),
- a sequence of (deterministic) coupons $c_1, \ldots, c_n$,
- a nominal value N,

such that the owner receives c_i at time T_i, for $i = 1, \ldots, n$, and N at terminal time T_n. The price $p(t)$ at time $t \le T_1$ of this coupon bond is given by the sum of discounted cash flows

$$p(t) = \sum_{i=1}^{n} P(t, T_i)c_i + P(t, T_n)N.$$

Typically, it holds that $T_{i+1} - T_i \equiv \delta$, and the coupons are given as a fixed percentage of the nominal value: $c_i \equiv K\delta N$, for some fixed interest rate K. The above formula reduces to

$$p(t) = \left(K\delta \sum_{i=1}^{n} P(t, T_i) + P(t, T_n) \right) N.$$

---

[6]Most depository institutions in the United States are subject to the Federal Reserves statutory reserve requirements. These rules require that, on every second Wednesday, a banks total actual reserves over the two-week period equal or exceed its total required reserves for that two-week period. This may create pressure in the federal funds market and cause spikes in interest rate changes and volatility on settlement Wednesdays, which are the settlement effects. See also [87].

2.4.2 Floating Rate Notes

There are versions of coupon bonds for which the value of the coupon is not fixed at the time the bond is issued, but rather reset for every coupon period. Most often the resetting is determined by some market interest rate (e.g. LIBOR).

A *floating rate note* is specified by:

- a number of future dates $T_0 < T_1 < \cdots < T_n$,
- a nominal value N.

The deterministic coupon payments for the fixed coupon bond are now replaced by

$$c_i = (T_i - T_{i-1})F(T_{i-1}, T_i)N,$$

where $F(T_{i-1}, T_i)$ is the prevailing simple market interest rate, and we note that $F(T_{i-1}, T_i)$ is determined already at time T_{i-1} (this is why here we have T_0 in addition to the coupon dates $T_1, \ldots, T_n$), but that the cash flow c_i is at time T_i.

The value $p(t)$ of this note at time $t \leq T_0$ is obtained as follows. Without loss of generality we set $N = 1$. By definition of $F(T_{i-1}, T_i)$ we then have

$$c_i = \frac{1}{P(T_{i-1}, T_i)} - 1.$$

The time t value of -1 paid out at T_i is $-P(t, T_i)$. The time t value of $\frac{1}{P(T_{i-1}, T_i)}$ paid out at T_i is $P(t, T_{i-1})$:

- At t: buy a T_{i-1}-bond. Cost: $P(t, T_{i-1})$.
- At T_{i-1}: receive one dollar and buy $\frac{1}{P(T_{i-1}, T_i)}$ T_i-bonds. This is a zero net investment.
- At T_i: receive $\frac{1}{P(T_{i-1}, T_i)}$ dollars.

The time t value of c_i therefore is

$$P(t, T_{i-1}) - P(t, T_i). \tag{2.3}$$

Summing up we obtain the (surprisingly simple) formula

$$p(t) = P(t, T_n) + \sum_{i=1}^{n} (P(t, T_{i-1}) - P(t, T_i)) = P(t, T_0).$$

In particular, for $t = T_0$, we obtain $p(T_0) = 1$.

2.4.3 Interest Rate Swaps

An interest rate swap is a scheme where you exchange a payment stream at a *fixed* rate of interest for a payment stream at a *floating* rate (typically LIBOR).

There are many versions of interest rate swaps. A *payer interest rate swap* settled in arrears is specified by:

- a number of future dates $T_0 < T_1 < \cdots < T_n$ with $T_i - T_{i-1} \equiv \delta$
  (T_n is the maturity of the swap),
- a fixed rate K,
- a nominal value N.

Of course, the equidistance hypothesis is only for convenience of notation and can easily be relaxed. Cash flows take place only at the coupon dates $T_1, \ldots, T_n$. At T_i, the holder of the contract:

- pays fixed $K\delta N$,
- and receives floating $F(T_{i-1}, T_i)\delta N$.

The net cash flow at T_i is thus

$$(F(T_{i-1}, T_i) - K)\delta N, \tag{2.4}$$

and using the previous results, see (2.3), we can compute the value at $t \le T_0$ of this cash flow as

$$N(P(t, T_{i-1}) - P(t, T_i) - K\delta P(t, T_i)).$$

The total value $\Pi_p(t)$ of the swap at time $t \le T_0$ is thus

$$\Pi_p(t) = N\left(P(t, T_0) - P(t, T_n) - K\delta \sum_{i=1}^{n} P(t, T_i)\right).$$

A *receiver interest rate swap* settled in arrears is obtained by changing the sign of the cash flows at times $T_1, \ldots, T_n$. Its value at time $t \le T_0$ is thus

$$\Pi_r(t) = -\Pi_p(t).$$

The remaining question is how the "fair" fixed rate K is determined. The *forward swap rate* (also called par swap rate) $R_{swap}(t)$ at time $t \le T_0$ is the fixed rate K above which gives $\Pi_p(t) = \Pi_r(t) = 0$. Hence

$$R_{swap}(t) = \frac{P(t, T_0) - P(t, T_n)}{\delta \sum_{i=1}^{n} P(t, T_i)}.$$

The following alternative representation of $R_{swap}(t)$ is sometimes useful. The value at time $t \le T_0$ of the cash flow (2.4) can directly be written as

$$N\delta P(t, T_i)(F(t; T_{i-1}, T_i) - K).$$

Summing up yields

$$\Pi_p(t) = N\delta \sum_{i=1}^{n} P(t, T_i)(F(t; T_{i-1}, T_i) - K),$$

and thus we can write the swap rate as weighted average of simple forward rates

$$R_{swap}(t) = \sum_{i=1}^{n} w_i(t) F(t; T_{i-1}, T_i),$$

with weights

$$w_i(t) = \frac{P(t, T_i)}{\sum_{j=1}^{n} P(t, T_j)}.$$

These weights follow stochastic processes, but there seems to be empirical evidence that the variability of $w_i(t)$ is small compared to that of $F(t; T_{i-1}, T_i)$. This will be used in Chap. 11 below for the approximation of swaption price formulas in LIBOR market models: the swap rate volatility is written as a linear combination of the forward LIBOR volatilities.

Swaps were developed because different companies could borrow at fixed or at floating rates in different markets. Here is an example: consider two companies A and B, and suppose that:

- company A is borrowing fixed for five years at $5\frac{1}{2}\%$, but could borrow floating at LIBOR plus $\frac{1}{2}\%$;
- company B is borrowing floating at LIBOR plus 1%, but could borrow fixed for five years at $6\frac{1}{2}\%$.

By agreeing to swap streams of cash flows both companies could be better off, and a mediating institution would also make money:

- company A pays LIBOR to the intermediary in exchange for fixed at $5\frac{3}{16}\%$ (receiver swap);
- company B pays the intermediary fixed at $5\frac{5}{16}\%$ in exchange for LIBOR (payer swap).

This is visualized in Fig. 2.4. The net payments are as follows:

- company A is now paying LIBOR plus $\frac{5}{16}\%$ instead of LIBOR plus $\frac{1}{2}\%$;
- company B is paying fixed at $6\frac{5}{16}\%$ instead of $6\frac{1}{2}\%$;
- the intermediary receives fixed at $\frac{1}{8}\%$.

Everyone seems to be better off. But there is implicit credit risk; this is why company B had higher borrowing rates in the first place. This risk has been partly taken up by the intermediary, in return for the money it makes on the spread.

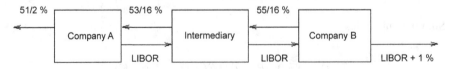

Fig. 2.4 A swap with mediating institution

Fig. 2.5 Yield curve
$T \mapsto R(t,T)$

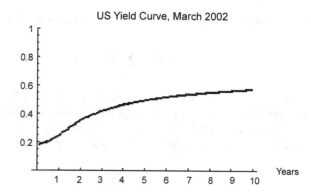

Interest rate swap markets[7] are extremely liquid, and thus swaps can be used to hedge interest rate risk at a low cost. Maturities from 1 to 30 years are standard, swap rate quotes are available up to 60 years. This gives market participants, such as life insurers, the opportunity to create synthetically long-dated investments.

2.4.4 Yield and Duration

For a zero-coupon bond $P(t,T)$ the *zero-coupon yield* is simply the continuously compounded spot rate $R(t,T)$. That is,

$$P(t,T) = e^{-R(t,T)(T-t)}.$$

Accordingly, the function $T \mapsto R(t,T)$ is referred to as the *(zero-coupon) yield curve*, see Fig. 2.5 for an example.

The term "yield curve" is ambiguous. There is a variety of other terminologies, such as "zero-rate curve" (Zagst [163]), or zero-coupon curve (Brigo and Mercurio [27]). On the other hand, in [27] the "yield curve" is a combination of simple spot rates (for maturities up to 1 year) and annually compounded spot rates (for maturities greater than 1 year), etc. In this book, by yield curve we mean the above term-structure $T \mapsto R(t,T)$ of continuously compounded spot rates.

2.4.4.1 Yield-to-Maturity

Now let $p(t)$ be the time t market value of a fixed coupon bond with coupon dates $T_1 < \cdots < T_n$, coupon payments $c_1, \ldots, c_n$ and nominal value N (see Sect. 2.4.1).

---

[7]These markets are over the counter, there exists no swap exchange. But swap contracts exist in standardized form, e.g. by the ISDA (International Swaps and Derivatives Association, Inc.).

For simplicity we suppose that c_n already contains N, that is,

$$p(t) = \sum_{i=1}^{n} P(t, T_i)c_i, \quad t \leq T_1.$$

Again we ask for the bond's "internal rate of interest"; that is, the constant (over the period $[t, T_n]$) continuously compounded rate which generates the market value of the coupon bond: the (*continuously compounded*) *yield-to-maturity* $y(t)$ of this bond at time $t \leq T_1$ is defined as the unique solution to

$$p(t) = \sum_{i=1}^{n} c_i e^{-y(t)(T_i - t)}.$$

How does the bond price change as function of $y(t)$? To simplify the notation we assume now that $t = 0$, and write $p = p(0)$, $y = y(0)$, etc. The *Macaulay duration* of the coupon bond is defined as

$$D_{Mac} = \frac{\sum_{i=1}^{n} T_i c_i e^{-yT_i}}{p}.$$

The duration is thus a weighted average of the coupon dates $T_1, \ldots, T_n$, and it provides us in a certain sense with the "mean time to coupon payment". As such it is an important concept for interest rate risk management: it acts as a measure of the first-order sensitivity of the bond price w.r.t. changes in the yield-to-maturity. This is shown by the obvious formula

$$\frac{dp}{dy} = \frac{d}{dy}\left(\sum_{i=1}^{n} c_i e^{-yT_i}\right) = -D_{Mac}p.$$

However, it is argued by Schaefer [141] that the yield-to-maturity is an inadequate statistic for the bond market:

- coupon payments occurring at the *same point in time* are discounted by *different discount factors*, but
- coupon payments at *different points in time* from the same bond are *discounted by the same rate*.

In reality, one would wish to do exactly the opposite. So let us stick to the zero-coupon yields!

2.4.4.2 Duration and Convexity

A first-order sensitivity measure of the bond price w.r.t. *parallel shifts* of the entire zero-coupon yield curve $T \mapsto R(0, T)$ is given by the *duration* of the bond

$$D = \frac{\sum_{i=1}^{n} T_i c_i e^{-y_i T_i}}{p} = \sum_{i=1}^{n} \frac{c_i P(0, T_i)}{p} T_i,$$

with $y_i = R(0, T_i)$. In fact, we have

$$\frac{d}{ds}\left(\sum_{i=1}^{n} c_i e^{-(y_i+s)T_i}\right)\bigg|_{s=0} = -Dp.$$

Hence duration is essentially for bonds (w.r.t. parallel shift of the yield curve) what delta is for stock options. The bond equivalent of the gamma is *convexity*:

$$C = \frac{d^2}{ds^2}\left(\sum_{i=1}^{n} c_i e^{-(y_i+s)T_i}\right)\bigg|_{s=0} = \sum_{i=1}^{n} c_i e^{-y_i T_i}(T_i)^2.$$

We thus obtain the second-order approximation for the change Δp of the bond price with respect to a parallel shift by Δy of the zero-coupon yield curve:

$$\Delta p \approx -Dp\Delta y + \frac{1}{2}C(\Delta y)^2.$$

2.5 Market Conventions

In this intermediary section, we shall see how our theoretical continuous time framework with an infinite maturity time span can actually be brought into connection with the real market conventions. We consider day-count conventions, range of available bonds and their price quotes.

2.5.1 Day-Count Conventions

By convention, we measure time in units of years. But if t and T denote two dates expressed as day/month/year, it is not clear what $T - t$ should be. The market evaluates the year fraction between t and T in different ways.

The *day-count convention* decides upon the time measurement between two dates t and T. Here are three examples of day-count conventions:

- Actual/365: a year has 365 days, and the day-count convention for $T - t$ is given by

$$\frac{\text{actual number of days between } t \text{ and } T}{365}.$$

- Actual/360: as above but the year counts 360 days.
- 30/360: months count 30 and years 360 days. Let $t = d_1/m_1/y_1$ and $T = d_2/m_2/y_2$. The day-count convention for $T - t$ is given by

$$\frac{\min(d_2, 30) + (30 - d_1)^+}{360} + \frac{(m_2 - m_1 - 1)^+}{12} + y_2 - y_1.$$

Example: The time between $t = 4$ January 2000 and $T = 4$ July 2002 is given by

$$\frac{4 + (30 - 4)}{360} + \frac{7 - 1 - 1}{12} + 2002 - 2000 = 2.5.$$

When extracting information on interest rates from data, it is important to realize for which day-count convention a specific interest rate is quoted.

2.5.2 Coupon Bonds

Coupon bonds issued in the American (European) markets typically have semiannual (annual) coupon payments.

Debt securities issued by the US Treasury are divided into three classes:

- *Bills:* zero-coupon bonds with time to maturity less than one year.
- *Notes:* coupon bonds (semiannual) with time to maturity between 2 and 10 years.
- *Bonds:* coupon bonds (semiannual) with time to maturity between 10 and 30 years.[8]

In addition to bills, notes and bonds, Treasury securities called *STRIPS* (separate trading of registered interest and principal of securities) have traded since August 1985. These are the coupons or principal (= nominal) amounts of Treasury bonds trading separately through the Federal Reserve's book-entry system. They are synthetically created zero-coupon bonds of longer maturities than a year. They were created in response to investor demands.

2.5.3 Accrued Interest, Clean Price and Dirty Price

Remember that we had for the price of a coupon bond with coupon dates $T_1, \ldots, T_n$ and payments $c_1, \ldots, c_n$ the price formula

$$p(t) = \sum_{i=1}^{n} c_i P(t, T_i), \quad t \leq T_1.$$

For $t \in (T_1, T_2]$ we have

$$p(t) = \sum_{i=2}^{n} c_i P(t, T_i),$$

etc. Hence there are systematic discontinuities of the price trajectory at $t = T_1, \ldots, T_n$ which is due to the coupon payments. This is why prices are differently quoted at the exchange.

---

[8] 30-year Treasury bonds were not offered from 2002 to 2005.

The *accrued interest* at time $t \in (T_{i-1}, T_i]$ is defined by

$$AI(i; t) = c_i \frac{t - T_{i-1}}{T_i - T_{i-1}}$$

(where now time differences are taken according to the day-count convention). The quoted price, or *clean price*, of the coupon bond at time t is

$$p_{clean}(t) = p(t) - AI(i; t), \quad t \in (T_{i-1}, T_i].$$

That is, whenever we buy a coupon bond quoted at a clean price of $p_{clean}(t)$ at time $t \in (T_{i-1}, T_i]$, the cash price, or *dirty price*, we have to pay is

$$p(t) = p_{clean}(t) + AI(i; t).$$

2.5.4 Yield-to-Maturity

The quoted (*annual*) *yield-to-maturity* $\hat{y}(t)$ on a Treasury bond at time $t = T_i$ is defined by the relationship

$$p_{clean}(T_i) = \sum_{j=i+1}^{n} \frac{r_c N/2}{(1 + \hat{y}(T_i)/2)^{j-i}} + \frac{N}{(1 + \hat{y}(T_i)/2)^{n-i}},$$

and at $t \in [T_i, T_{i+1})$

$$p_{clean}(t) = \sum_{j=i+1}^{n} \frac{r_c N/2}{(1 + \hat{y}(t)/2)^{j-i-1+\tau}} + \frac{N}{(1 + \hat{y}(t)/2)^{n-i-1+\tau}},$$

where r_c is the (annualized) coupon rate, N the nominal amount and

$$\tau = \frac{T_{i+1} - t}{T_{i+1} - T_i}$$

is again given by the day-count convention, and we assume here that

$$T_{i+1} - T_i \equiv 1/2 \quad \text{(semiannual coupons)}.$$

2.6 Caps and Floors

In the following last two sections of this chapter we introduce the two main derivative products in the interest rate market, caps/floors and swaptions. We learn how traders price them with Black's formula. In Chap. 11 on market models below, we will come back to these formulas from a stochastic model point of view.

2.6.1 Caps

A *caplet* with reset date T and settlement date $T + \delta$ pays the holder the difference between a simple market rate $F(T, T + \delta)$ (e.g. LIBOR) and the strike rate κ. Its cash flow at time $T + \delta$ is

$$\delta(F(T, T + \delta) - \kappa)^+.$$

A *cap* is a strip of caplets. It thus consists of:

- a number of future dates $T_0 < T_1 < \cdots < T_n$ with $T_i - T_{i-1} \equiv \delta$
  (T_n is the maturity of the cap),
- a *cap rate* κ.

Cash flows take place at the dates $T_1, \ldots, T_n$. At T_i the holder of the cap receives

$$\delta(F(T_{i-1}, T_i) - \kappa)^+. \tag{2.5}$$

Let $t \leq T_0$. We write

$$Cpl(t; T_{i-1}, T_i), \quad i = 1, \ldots, n,$$

for the time t price of the ith caplet with reset date T_{i-1} and settlement date T_i, and

$$Cp(t) = \sum_{i=1}^{n} Cpl(t; T_{i-1}, T_i)$$

for the time t price of the cap.

A cap gives the holder a protection against rising interest rates. It guarantees that the interest to be paid on a floating rate loan never exceeds the predetermined cap rate κ.

It can be shown ($\rightarrow$ Exercise 2.7) that the cash flow (2.5) at time T_i is the equivalent to $(1 + \delta\kappa)$ times the cash flow at date T_{i-1} of a put option on a T_i-bond with strike price $1/(1 + \delta\kappa)$ and maturity T_{i-1}, that is,

$$(1 + \delta\kappa)\left(\frac{1}{1 + \delta\kappa} - P(T_{i-1}, T_i)\right)^+.$$

This is an important fact because many interest rate models have explicit formulae for bond option values, which means that caps can be priced very easily in those models.

2.6.2 Floors

A *floor* is the converse to a cap. It protects against low rates. A floor is a strip of *floorlets*, the cash flow of which is – with the same notation as above – at time T_i

$$\delta(\kappa - F(T_{i-1}, T_i))^+.$$

Write $Fll(t; T_{i-1}, T_i)$ for the price of the ith floorlet and

$$Fl(t) = \sum_{i=1}^{n} Fll(t; T_{i-1}, T_i)$$

for the price of the floor.

2.6.3 Caps, Floors and Swaps

Caps and floors are strongly related to swaps. Indeed, one can show the parity relation ($\rightarrow$ Exercise 2.7)

$$Cp(t) - Fl(t) = \Pi_p(t),$$

where $\Pi_p(t)$ is the value at t of a payer swap with rate κ, nominal one and the same tenor structure as the cap and floor.

Let $t = 0$. The cap/floor is said to be *at-the-money (ATM)* if

$$\kappa = R_{swap}(0) = \frac{P(0, T_0) - P(0, T_n)}{\delta \sum_{i=1}^{n} P(0, T_i)},$$

the forward swap rate. The cap (floor) is *in-the-money (ITM)* if $\kappa < R_{swap}(0)$ ($\kappa > R_{swap}(0)$), and *out-of-the-money (OTM)* if $\kappa > R_{swap}(0)$ ($\kappa < R_{swap}(0)$).

2.6.4 Black's Formula

It is market practice to price a cap/floor according to *Black's formula*, see Exercise 2.5. Let $t \leq T_0$. Black's formula for the value of the ith caplet is

$$Cpl(t; T_{i-1}, T_i) = \delta P(t, T_i)(F(t; T_{i-1}, T_i)\Phi(d_1(i; t)) - \kappa \Phi(d_2(i; t))), \quad (2.6)$$

where

$$d_{1,2}(i; t) = \frac{\log\left(\frac{F(t; T_{i-1}, T_i)}{\kappa}\right) \pm \frac{1}{2}\sigma(t)^2(T_{i-1} - t)}{\sigma(t)\sqrt{T_{i-1} - t}},$$

Φ stands for the standard Gaussian cumulative distribution function, and $\sigma(t)$ is the *cap (implied) volatility* (it is the same for all caplets belonging to a cap).

Correspondingly, Black's formula for the value of the ith floorlet is

$$Fll(t; T_{i-1}, T_i) = \delta P(t, T_i)(\kappa \Phi(-d_2(i; t)) - F(t; T_{i-1}, T_i)\Phi(-d_1(i; t))).$$

Cap/floor prices are quoted in the market in terms of their implied volatilities. Typically, we have $t = 0$, $T_0 = \delta$ and $\delta = T_i - T_{i-1}$ being equal to three months (US

Table 2.1 US dollar ATM
cap volatilities, 23 July 1999

Maturity (in years)	ATM vols (in %)
1	14.1
2	17.4
3	18.5
4	18.8
5	18.9
6	18.7
7	18.4
8	18.2
10	17.7
12	17.0
15	16.5
20	14.7
30	12.4

Fig. 2.6 US dollar ATM cap
volatilities, 23 July 1999

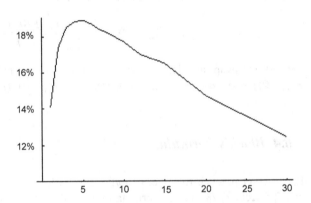

market) or half a year (euro market). An example of a US dollar ATM market cap
volatility curve is shown in Table 2.1 and Fig. 2.6.

It is a challenge for any market realistic interest rate model to match the given
volatility curve.

2.7 Swaptions

A European *payer (receiver) swaption* with strike rate K is an option giving the
right to enter a payer (receiver) swap with fixed rate K at a given future date, the
swaption maturity. Usually, the swaption maturity coincides with the first reset date
of the underlying swap. The underlying swap length $T_n - T_0$ is called the *tenor* of
the swaption.

Recall that the value of a payer swap with fixed rate K at its first reset date, T_0, is

$$\Pi_p(T_0, K) = N \sum_{i=1}^{n} P(T_0, T_i)\delta(F(T_0; T_{i-1}, T_i) - K).$$

Hence the payoff of the swaption with strike rate K at maturity T_0 is

$$N\left(\sum_{i=1}^{n} P(T_0, T_i)\delta(F(T_0; T_{i-1}, T_i) - K)\right)^{+}. \tag{2.7}$$

Notice that, contrary to the cap case, this payoff cannot be decomposed into more elementary payoffs. This is a fundamental difference between caps/floors and swaptions. Here the stochastic dependence between different forward rates will enter the valuation procedure.

Since $\Pi_p(T_0, R_{swap}(T_0)) = 0$, one can show ($\rightarrow$ Exercise 2.3) that the payoff (2.7) of the payer swaption at time T_0 can also be written as

$$N\delta(R_{swap}(T_0) - K)^{+} \sum_{i=1}^{n} P(T_0, T_i), \tag{2.8}$$

and for the receiver swaption

$$N\delta(K - R_{swap}(T_0))^{+} \sum_{i=1}^{n} P(T_0, T_i).$$

Accordingly, at time $t \leq T_0$, the payer (receiver) swaption with strike rate K is said to be *ATM, ITM, OTM*, if

$$K = R_{swap}(t), \qquad K < (>)R_{swap}(t), \qquad K > (<)R_{swap}(t),$$

respectively. A (payer/receiver) swaption with maturity in x years and whose underlying swap is y years long is briefly called a (payer/receiver) $x \times y$-*swaption*.

Swaptions can be used to synthetically create callable bonds as the following example illustrates.

Example 2.1 Suppose a company has issued a bond maturing in 10 years with annual coupons of 4%, and wants to add the option to call the bond at par after 5 years. This option means that the company has the right to prepay the nominal N of the bond and stop paying coupons after 5 years. If the company cannot change the original bond, they could buy a 5×5 receiver swaption with strike rate 4%. This works as follows: suppose after 5 years the company decides to call the bond, that is, to exercise the swaption. Clearly, the fixed coupon leg of the swap will then cancel the fixed coupon payments of the bond. On the other hand, paying the floating rate leg of the swap and the nominal N at maturity $T = 10$ is equivalent to paying the nominal N at $T = 5$, as desired ($\rightarrow$ Exercise 2.6).

Table 2.2 Black's implied volatilities (in %) of ATM swaptions on May 16, 2000. Maturities are 1, 2, 3, 4, 5, 7, 10 years, swaps lengths from 1 to 10 years

	1y	2y	3y	4y	5y	6y	7y	8y	9y	10y
1y	16.4	15.8	14.6	13.8	13.3	12.9	12.6	12.3	12.0	11.7
2y	17.7	15.6	14.1	13.1	12.7	12.4	12.2	11.9	11.7	11.4
3y	17.6	15.5	13.9	12.7	12.3	12.1	11.9	11.7	11.5	11.3
4y	16.9	14.6	12.9	11.9	11.6	11.4	11.3	11.1	11.0	10.8
5y	15.8	13.9	12.4	11.5	11.1	10.9	10.8	10.7	10.5	10.4
7y	14.5	12.9	11.6	10.8	10.4	10.3	10.1	9.9	9.8	9.6
10y	13.5	11.5	10.4	9.8	9.4	9.3	9.1	8.8	8.6	8.4

2.7.1 Black's Formula

Black's formula, see Exercise 2.5, for the price at time $t \leq T_0$ of the payer ($Swpt_p(t)$) and receiver ($Swpt_r(t)$) swaption is

$$Swpt_p(t) = N\delta \left(R_{swap}(t)\Phi(d_1(t)) - K\Phi(d_2(t))\right) \sum_{i=1}^{n} P(t, T_i),$$

$$\tag{2.9}$$

$$Swpt_r(t) = N\delta \left(K\Phi(-d_2(t)) - R_{swap}(t)\Phi(-d_1(t))\right) \sum_{i=1}^{n} P(t, T_i),$$

with

$$d_{1,2}(t) = \frac{\log\left(\frac{R_{swap}(t)}{K}\right) \pm \frac{1}{2}\sigma(t)^2(T_0 - t)}{\sigma(t)\sqrt{T_0 - t}},$$

and $\sigma(t)$ is the prevailing Black's swaption volatility.

Swaption prices are quoted in terms of implied volatilities in matrix form. Note that the accrual period $\delta = T_i - T_{i-1}$ for the underlying swap can be different from the prevailing δ for caps within the same market region.[9]

A typical example of implied swaption volatilities is shown in Table 2.2 and Fig. 2.7.

An interest rate model for swaptions valuation must fit the given today's volatility surface.

2.8 Exercises

Exercise 2.1 Consider a forward rate agreement (FRA) with current, expiry and maturity time $t < T < S$, respectively, and cash flow to the lender:

---

[9]For instance, in the euro zone caps are written on semiannual LIBOR ($\delta = 1/2$), while swaps pay annual coupons ($\delta = 1$).

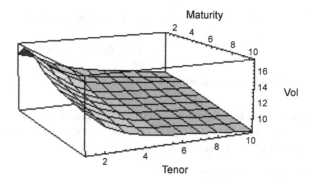

Fig. 2.7 Black's implied volatilities (in %) of ATM swaptions on May 16, 2000

- At time T: $-K$,
- At time S: $Ke^{R^*(S-T)}$,

for some predetermined principal K and interest rate R^*.

(a) Compute the value $\Pi(t)$ at time t of the cash flow above in terms of zero-coupon bond prices.

(b) Show that in order for the value of the FRA to equal zero at t, the rate R^* has to equal the forward rate $R(t; S, T)$.

Exercise 2.2 Finish the proof of (2.2) for the case $P(t, S) < P(t, T)P(T, S)$.

Exercise 2.3 Consider a swap with reset and cash flow dates

$$0 < T_0 < T_1 < \cdots < T_n$$

(T_0 is the first reset date) such that $T_i - T_{i-1} \equiv \delta$, and coupons $c_i = K\delta N$, for some fixed rate K and nominal N.

(a) Let $t \leq T_0$. Show that the time t value of the payer swap equals

$$\Pi_p(t) = N\delta(R_{swap}(t) - K)\sum_{i=1}^{n} P(t, T_i).$$

We now consider a numerical example: today is $t = 0$, first reset date is $T_0 = 1/4$, cash flow dates are $T_i = T_{i-1} + 1/4$, maturity is $T_7 = 2$. The forward curve is given by

$$\begin{pmatrix} F(0; 0, 1/4) \\ \vdots \\ \vdots \\ F(0; 7/4, 2) \end{pmatrix} = \begin{pmatrix} 0.06 \\ 0.09 \\ 0.1 \\ 0.1 \\ 0.1 \\ 0.09 \\ 0.09 \\ 0.09 \end{pmatrix}. \tag{2.10}$$

(b) Find the corresponding term-structure of bond prices

$$P(0, T_0), \ldots, P(0, T_7).$$

(c) Find the corresponding swap rate $R_{swap}(0)$.

Exercise 2.4 Duration and Convexity: we take the forward curve (2.10) from Exercise 2.3. Today is $t = 0$. Consider a coupon bond with maturity in two years, semiannual coupons $c = 5$ and nominal 100.

(a) What is its price p?
(b) What is its continuously compounded yield-to-maturity y?
(c) Compute the yield curve $y_i = R(0, i/4)$, $i = 1, \ldots, 8$.
(d) Compute the Macaulay duration D_{Mac}, the duration D and convexity C of the bond.
(e) Consider a parallel shift of the yield curve

$$y_i \to \tilde{y}_i = y_i + s, \quad i = 1, \ldots, 8$$

by $s = 0.0001$ (one basis point) and $s = 0.01$. How does the bond price change (same maturity, coupons and nominal)?
(f) Compare the first- and second-order approximations

$$p - D_{Mac} ps, \qquad p - Dps, \qquad p - Dps + \frac{1}{2}Cs^2$$

for both values of s. Are there any differences?

Exercise 2.5 Under the assumption that $\log F(T_{i-1}, T_i)$ is Gaussian distributed with mean $\log F(0; T_{i-1}, T_i) - \frac{\sigma^2(0)}{2}T_{i-1}$ and variance $\sigma^2(0)T_{i-1}$, show that Black's formula (2.6) for the ith caplet price at $t = 0$ equals

$$\delta P(0, T_i)\mathbb{E}[(F(T_{i-1}, T_i) - \kappa)^+].$$

It will become clear in Chap. 11 below, under which measure this equality holds.

Exercise 2.6 Show that the receiver swaption cash flow has the desired call effect for the bond in Example 2.1.

Exercise 2.7 The following swap, cap and floor are determined by the sequence of reset/cash flow dates

$$0 < T_0 < T_1 < \cdots < T_n$$

(T_0 is the first reset date and maturity for the swaption, cap and floor) such that $T_i - T_{i-1} \equiv \delta$, a fixed rate $\kappa > 0$, and a nominal value N. Let $t \leq T_0$.

(a) Show that the cash flow of the ith caplet

$$\delta(F(T_{i-1}, T_i) - \kappa)^+$$

at time T_i is equivalent to the cash flow

$$(1 + \delta\kappa)\left(\frac{1}{1 + \delta\kappa} - P(T_{i-1}, T_i)\right)^+$$

at maturity T_{i-1} of a put option on a T_i-bond.
(b) Show that a payer swaption price is always dominated by the corresponding cap price.
(c) Prove the parity relations

$$Cp(t) - Fl(t) = \Pi_p(t), \qquad Swpt_p(t) - Swpt_r(t) = \Pi_p(t). \tag{2.11}$$

(d) Let $Cpl_{Black}(t; T_{i-1}, T_i)$, $Fll_{Black}(t; T_{i-1}, T_i)$, $Cp_{Black}(t)$, $Fl_{Black}(t)$ be the (caplet etc.) prices according to Black's formula. First, show that

$$Cpl_{Black}(t; T_{i-1}, T_i) - Fll_{Black}(t; T_{i-1}, T_i) = \delta P(t, T_i)(F(t; T_{i-1}, T_i) - \kappa),$$

and this equality holds if and only if Black's volatility is the same for the cap and floor. Now argue that Black's formula for caps and floors is consistent with the parity relation (2.11), and that therefore caps and floors with the same underlying tenor and strike always imply the same Black's volatility.
(e) Similarly for swaps: let $Swpt_p^{Black}(t)$, $Swpt_r^{Black}(t)$ denote the prices according to Black's formula, and show that they satisfy the parity relation (2.11).
(f) Now suppose the time points T_i are not equidistant: $T_i - T_{i-1} \neq T_j - T_{j-1}$ for $i \neq j$. Derive the formula for the swap rate $R_{swap}(t)$ in this case.

Exercise 2.8 We take the forward curve (2.10) from Exercise 2.3. Today's ($t = 0$) price of the ATM cap with reset date $T_0 = 1/4$ and maturity in two years is 0.01.

(a) What is its implied volatility?
(b) Conversely, suppose the implied volatility is 14.1%. What is the corresponding price?

2.9 Notes

There is a vast literature where interest rates are introduced. The first part of this chapter follows partly the outline in Björk [13, Sect. 20]. The example at the end of Sect. 2.4.3 is taken from James and Webber [100, p. 11]. Duration and "greeks" based hedging of bond portfolios is thoroughly discussed in Zagst [163, Chaps. 6 and 7]. More information on market conventions can be found in e.g. Brigo and Mercurio [27, Chap. 1], Carmona and Tehranchi [35, Sect. 1], Jarrow [103, Chap. 2],

Musiela and Rutkowski [126, Chap. 9], Zagst [163, Chap. 5], and many more. Caps and swaptions are standard topics in the interest rate literature. Here will follow the exposition in [27, Sect. 1.6] and [163, Sect. 5.6]. The cap data from Table 2.1 are taken from James and Webber [100, p. 49]. Example 2.1 was brought to the author's attention by Antoon Pelsser. The swaption data from Table 2.2 are taken from Brigo and Mercurio [27, Sect. 6.17].

Chapter 3
Estimating the Term-Structure

In our theoretical framework we often assume a term-structure for the continuum of maturities T. In other words, we assume that the forward or zero-coupon yield curve is given by a function of the continuous variable T. This should be seen as approximation of the reality, which comes along with finitely many (possibly noisy) market quote observations. In Chap. 11 we will model the term-structure of interest rates by choosing finitely many maturities. This is appropriate if we want to price a predetermined finite set of derivatives, such as caps and swaptions. However, as soon as more exotic derivatives be priced whose cash flow dates possibly do not match the predetermined finite time grid, one has to interpolate the term-structure. In this chapter, we learn some term-structure estimation methods. We start with a bootstrapping example, which is the most used method among the trading desks. We then consider more general aspects of non-parametric and parametric term-structure estimation methods. In the last part we perform a principal component analysis for the term-structure movements, which is the best-known dimension reduction technique in multivariate data analysis.

3.1 A Bootstrapping Example

We present in this section an iterative extraction procedure for fitting to a money-market term-structure. It is commonly called the bootstrapping method, albeit the term "bootstrapping" has a different meaning in statistics. The idea is to build up the term structure from shorter maturities to longer maturities.

We take yen data from 9 January 1996, as shown in Table 3.1. The spot date t_0 is 11 January, 1996. The day-count convention is actual/360:

$$\delta(T, S) = \frac{\text{actual number of days between } T \text{ and } S}{360}.$$

The first column contains the LIBOR ($=$ simple spot rates) $F(t_0, S_i)$ for maturities

$$\{S_1, \ldots, S_5\} = \{12/1/96, 18/1/96, 13/2/96, 11/3/96, 11/4/96\}$$

hence for 1, 7, 33, 60 and 91 days to maturity, respectively. The zero-coupon bonds are

$$P(t_0, S_i) = \frac{1}{1 + \delta(t_0, S_i)F(t_0, S_i)}.$$

D. Filipović, *Term-Structure Models*,
Springer Finance,
DOI 10.1007/978-3-540-68015-4_3, © Springer-Verlag Berlin Heidelberg 2009

Table 3.1 Yen data, 9 January 1996

LIBOR (%)		Futures		Swaps (%)	
o/n	0.49	20 Mar 96	99.34	2y	1.14
1w	0.50	19 Jun 96	99.25	3y	1.60
1m	0.53	18 Sep 96	99.10	4y	2.04
2m	0.55	18 Dec 96	98.90	5y	2.43
3m	0.56			7y	3.01
				10y	3.36

The futures[1] in the second column are quoted as

$$\text{futures price for settlement day } T_i = 100(1 - F_F(t_0; T_i, T_{i+1})),$$

where $F_F(t_0; T_i, T_{i+1})$ is the futures rate for period $[T_i, T_{i+1}]$ prevailing at t_0, and

$$\{T_1, \ldots, T_5\} = \{20/3/96, 19/6/96, 18/9/96, 18/12/96, 19/3/97\},$$

hence $\delta(T_i, T_{i+1}) \equiv 91/360$. We treat futures rates as if they were simple forward rates, that is, we set

$$F(t_0; T_i, T_{i+1}) = F_F(t_0; T_i, T_{i+1}).$$

To calculate zero-coupon bond from futures prices we need $P(t_0, T_1)$. Note that $S_4 < T_1 < S_5$. We thus use geometric interpolation

$$P(t_0, T_1) = P(t_0, S_4)^q \, P(t_0, S_5)^{1-q},$$

which is equivalent to using linear interpolation of continuously compounded spot rates

$$R(t_0, T_1) = q \, R(t_0, S_4) + (1 - q) \, R(t_0, S_5),$$

where

$$q = \frac{\delta(T_1, S_5)}{\delta(S_4, S_5)} = \frac{22}{31} = 0.709677.$$

Then we use the relation

$$P(t_0, T_{i+1}) = \frac{P(t_0, T_i)}{1 + \delta(T_i, T_{i+1}) \, F(t_0; T_i, T_{i+1})}$$

to derive $P(t_0, T_2), \ldots, P(t_0, T_5)$.

---

[1]Interest rate futures will be discussed more thoroughly in Sect. 8.2.1 below.

The yen swaps in the third column have semiannual cash flows at dates

$$\{U_1, \ldots, U_{20}\} = \left\{ \begin{array}{l} 11/7/96, \ 13/1/97, \\ 11/7/97, \ 12/1/98, \\ 13/7/98, \ 11/1/99, \\ 12/7/99, \ 11/1/00, \\ 11/7/00, \ 11/1/01, \\ 11/7/01, \ 11/1/02, \\ 11/7/02, \ 13/1/03, \\ 11/7/03, \ 12/1/04, \\ 12/7/04, \ 11/1,\ 05, \\ 11/7/05, \ 11/1/06 \end{array} \right\}.$$

Recall that for a swap with maturity U_n the swap rate at t_0 is given by

$$R_{swap}(t_0, U_n) = \frac{1 - P(t_0, U_n)}{\sum_{i=1}^{n} \delta(U_{i-1}, U_i) P(t_0, U_i)} \quad \text{(set } U_0 = t_0). \quad (3.1)$$

From the data we have $R_{swap}(t_0, U_i)$ for $i = 4, 6, 8, 10, 14, 20$. Note the overlapping time intervals: $T_2 < U_1 < T_3$ and $T_4 < U_2 < T_5$. As above, we thus obtain $P(t_0, U_1)$, $P(t_0, U_2)$ (and hence $R_{swap}(t_0, U_1)$, $R_{swap}(t_0, U_2)$) by linear interpolation of the continuously compounded spot rates:

$$R(t_0, U_1) = \frac{69}{91} R(t_0, T_2) + \frac{22}{91} R(t_0, T_3),$$

$$R(t_0, U_2) = \frac{65}{91} R(t_0, T_4) + \frac{26}{91} R(t_0, T_5).$$

All remaining swap rates are derived by linear interpolation. For maturity U_3 this is

$$R_{swap}(t_0, U_3) = \frac{1}{2} (R_{swap}(t_0, U_2) + R_{swap}(t_0, U_4)).$$

We then solve (3.1) for $P(t_0, U_n)$ and obtain

$$P(t_0, U_n) = \frac{1 - R_{swap}(t_0, U_n) \sum_{i=1}^{n-1} \delta(U_{i-1}, U_i) P(t_0, U_i)}{1 + R_{swap}(t_0, U_n) \delta(U_{n-1}, U_n)}.$$

This gives $P(t_0, U_n)$ for $n = 3, \ldots, 20$.

Eventually, we set $P(t_0, t_0) = 1$, and we have constructed the term structure of zero-coupon bond prices $P(t_0, t_i)$ for 30 maturity points in increasing order:

$$t_i = t_0, S_1, \ldots, S_4, T_1, S_5, T_2, U_1, T_3, T_4, U_2, T_5, U_3, \ldots, U_{20}.$$

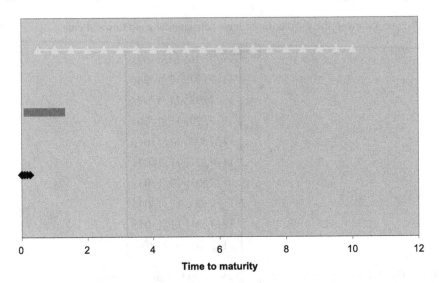

Fig. 3.1 Overlapping maturity segments (from bottom up) of LIBOR, futures and swap markets

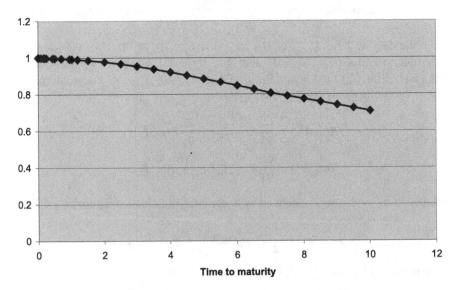

Fig. 3.2 Zero-coupon bond curve

The segments of LIBOR, futures and swap markets overlap, as is illustrated in Fig. 3.1. Figure 3.2 shows the implied zero-coupon bond price curve.

The spot and forward rate curves are in Fig. 3.3, and in Fig. 3.4 on a larger time scale, where spot and forward rates are continuously compounded:

$$R(t_0, t_i) = -\frac{\log P(t_0, t_i)}{\delta(t_0, t_i)}, \quad i = 1, \ldots, 30, \quad \text{and}$$

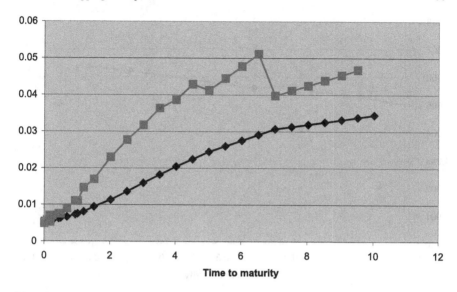

Fig. 3.3 Spot rates (lower curve), forward rates (upper curve with a "sawtooth")

$$R(t_0, t_i, t_{i+1}) = -\frac{\log P(t_0, t_{i+1}) - \log P(t_0, t_i)}{\delta(t_i, t_{i+1})}, \quad i = 0, \ldots, 29.$$

We observe that the forward curve, reflecting the derivative $-\partial_T \log P(t_0, T)$, is very irregular and sensitive to slight variations (or, errors) in bond prices.

The "sawtooth" in Fig. 3.3 indicates that the linear interpolation of swap rates is inappropriate for implied forward rates. Note, however, in some markets intermediate swaps are indeed priced as if their prices were found by linear interpolation.

The "sawtooth" in Fig. 3.4 is due to some systematic inconsistency of our use of LIBOR and futures rates data. Indeed, we have treated futures rates as forward rates. In reality futures rates are often greater than forward rates. The amount by which they differ is called the convexity adjustment, which is model dependent. An example is

$$\text{forward rate} = \text{futures rate} - \frac{1}{2}\sigma^2\tau^2,$$

where τ is the time to maturity of the futures contract, and σ is the corresponding volatility parameter. We will derive a more general formula in Sect. 8.2.1 below.

It thus becomes evident that the three curves resulting from LIBOR, futures and swaps, respectively, are not coincident to a common underlying curve. Our naive method made no attempt to meld the three curves together.

In summary, we have constructed the entire term-structure from relatively few instruments. The method exactly reconstructs market prices, which is often desirable for interest rate option traders who have to benchmark their positions to current market prices (marking to market). But it produces an unstable, irregular forward curve.

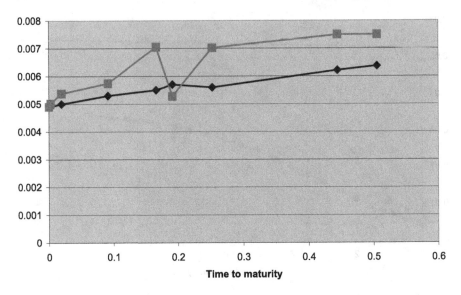

Fig. 3.4 Spot rates (lower curve), forward rates (upper curve with a "sawtooth")

This bootstrapping example can be classified as non-parametric estimation method. This class is discussed in more generality in the following section.

3.2 Non-parametric Estimation Methods

The general problem of finding today's (t_0) term-structure of zero-coupon bond prices (or the discount curve)

$$x \mapsto D(x) = P(t_0, t_0 + x)$$

can be formulated as

$$p = C\, d + \varepsilon,$$

where p is a column vector of n market prices, C the related cash flow matrix, and $d = (D(x_1), \ldots, D(x_N))^\top$ with cash flow dates $t_0 < T_1 < \cdots < T_N$,

$$T_i - t_0 = x_i,$$

and ε a vector of pricing errors, which is subject to being minimized. Including errors is reasonable since prices are never exact simultaneously quoted, and there are usually bid ask spreads. Moreover, it allows for smoothing.

Next, we shall see how to bring data from bond and money markets into the above format.

Table 3.2 Market prices for UK gilts, 4/9/96

	Coupon (%)	Next coupon	Maturity date	Dirty price (p_i)
Bond 1	10	15/11/96	15/11/96	103.82
Bond 2	9.75	19/01/97	19/01/98	106.04
Bond 3	12.25	26/09/96	26/03/99	118.44
Bond 4	9	03/03/97	03/03/00	106.28
Bond 5	7	06/11/96	06/11/01	101.15
Bond 6	9.75	27/02/97	27/08/02	111.06
Bond 7	8.5	07/12/96	07/12/05	106.24
Bond 8	7.75	08/03/97	08/09/06	98.49
Bond 9	9	13/10/96	13/10/08	110.87

3.2.1 Bond Markets

Here the basic instruments are coupon bonds. We thus can formalize the available market data as follows:

- a vector of quoted market bond prices $p = (p_1, \ldots, p_n)^\top$,
- the dates of all cash flows $t_0 < T_1 < \cdots < T_N$,
- bond $i = 1, \ldots, n$ with cash flows (coupon and principal payments) $c_{i,j}$ at time T_j (may be zero), forming the $n \times N$ cash flow matrix

$$C = (c_{i,j})_{\substack{1 \leq i \leq n \\ 1 \leq j \leq N}}.$$

As an example, we consider data from the UK government bond (gilt) market on 4 September 1996: a selection of nine gilts shown in Table 3.2. The coupon payments are semiannual. The spot date is 4/9/96, and the day-count convention is actual/365.

Hence $n = 9$ and $N = 1 + 3 + 6 + 7 + 11 + 12 + 19 + 20 + 25 = 104$,

$$T_1 = 26/09/96, \qquad T_2 = 13/10/96, \qquad T_3 = 06/11/97, \ldots.$$

Note that there are no bonds that have cash flows at the same date, whence N is so large. The 9×104 cash flow matrix is

$$C = \begin{pmatrix} 0 & 0 & 0 & 105 & 0 & 0 & 0 & 0 & 0 & 0 & \cdots \\ 0 & 0 & 0 & 0 & 0 & 4.875 & 0 & 0 & 0 & 0 & \cdots \\ 6.125 & 0 & 0 & 0 & 0 & 0 & 0 & 0 & 0 & 6.125 & \cdots \\ 0 & 0 & 0 & 0 & 0 & 0 & 0 & 4.5 & 0 & 0 & \cdots \\ 0 & 0 & 3.5 & 0 & 0 & 0 & 0 & 0 & 0 & 0 & \cdots \\ 0 & 0 & 0 & 0 & 0 & 0 & 4.875 & 0 & 0 & 0 & \cdots \\ 0 & 0 & 0 & 0 & 4.25 & 0 & 0 & 0 & 0 & 0 & \cdots \\ 0 & 0 & 0 & 0 & 0 & 0 & 0 & 0 & 3.875 & 0 & \cdots \\ 0 & 4.5 & 0 & 0 & 0 & 0 & 0 & 0 & 0 & 0 & \cdots \end{pmatrix}.$$

3.2.2 Money Markets

In the money market, the term-structure of interest rates is derived from the prices of a variety of different types of instruments, such as LIBOR rates, forward rate agreements (FRA), and swaps. On a stylized level, money-market data can be put into the same price/cash flow form as for bond markets:

- LIBOR (rate L, maturity T): $p = 1$ and $c = 1 + (T - t_0)L$ at T.
- FRA (forward rate F for $[T, S]$): $p = 0$, $c_1 = -1$ at $T_1 = T$, $c_2 = 1 + (S - T)F$ at $T_2 = S$.
- Swap (receiver, swap rate K, tenor $t_0 \le T_0 < \cdots < T_n$, $T_i - T_{i-1} \equiv \delta$): since the swap rate was defined to make floating equal to fixed leg in value:

$$0 = -D(T_0 - t_0) + \delta K \sum_{j=1}^{n-1} D(T_j - t_0) + (1 + \delta K)D(T_n - t_0),$$

we can choose
- if $T_0 = t_0$: $p = 1$, $c_1 = \cdots = c_{n-1} = \delta K$, $c_n = 1 + \delta K$,
- if $T_0 > t_0$: $p = 0$, $c_0 = -1$, $c_1 = \cdots = c_{n-1} = \delta K$, $c_n = 1 + \delta K$.

Hence, at t_0, LIBOR and swaps have notional price 1, FRAs and forward swaps have notional price 0.

As an example, we consider data from the US money market on 6 October 1997, as shown in Table 3.3. The day-count convention is actual/360. The spot date t_0 is

Table 3.3 US money market, 6 October 1997

	Period	Rate	Maturity date
LIBOR	o/n	5.59375	9/10/97
	1m	5.625	10/11/97
	3m	5.71875	8/1/98
Futures	Oct 97	94.27	15/10/97
	Nov 97	94.26	19/11/97
	Dec 97	94.24	17/12/97
	Mar 98	94.23	18/3/98
	Jun 98	94.18	17/6/98
	Sep 98	94.12	16/9/98
	Dec 98	94	16/12/98
Swaps	2	6.01253	
	3	6.10823	
	4	6.16	
	5	6.22	
	7	6.32	
	10	6.42	
	15	6.56	
	20	6.56	
	30	6.56	

8/10/97. LIBOR is for o/n (1/360), 1m (33/360), and 3m (92/360). Futures are three-month rates ($\delta = 91/360$). We take them as forward rates. That is, the quote of the futures contract with maturity date (settlement day) T is

$$100(1 - F(t_0; T, T + \delta)).$$

Swaps are annual ($\delta = 1$). The first payment date is 8/10/98.

Here $n = 3 + 7 + 9 = 19$, $N = 3 + (14 - 4) + 30 = 43$, $T_1 = 9/10/97$, $T_2 = 15/10/97$ (first future), $T_3 = 10/11/97, \ldots$. The first 14 columns of the 19×47 cash flow matrix C are

c_{11}	0	0	0	0	0	0	0	0	0	0	0	0	0
0	0	c_{23}	0	0	0	0	0	0	0	0	0	0	0
0	0	0	0	0	c_{36}	0	0	0	0	0	0	0	0
0	-1	0	0	0	0	c_{47}	0	0	0	0	0	0	0
0	0	0	-1	0	0	0	c_{58}	0	0	0	0	0	0
0	0	0	0	-1	0	0	0	c_{69}	0	0	0	0	0
0	0	0	0	0	0	0	0	-1	$c_{7,10}$	0	0	0	0
0	0	0	0	0	0	0	0	0	-1	$c_{8,11}$	0	0	0
0	0	0	0	0	0	0	0	0	0	-1	0	$c_{9,13}$	0
0	0	0	0	0	0	0	0	0	0	0	0	-1	$c_{10,14}$
0	0	0	0	0	0	0	0	0	0	0	$c_{11,12}$	0	0
0	0	0	0	0	0	0	0	0	0	0	$c_{12,12}$	0	0
0	0	0	0	0	0	0	0	0	0	0	$c_{13,12}$	0	0
0	0	0	0	0	0	0	0	0	0	0	$c_{14,12}$	0	0
0	0	0	0	0	0	0	0	0	0	0	$c_{15,12}$	0	0
0	0	0	0	0	0	0	0	0	0	0	$c_{16,12}$	0	0
0	0	0	0	0	0	0	0	0	0	0	$c_{17,12}$	0	0
0	0	0	0	0	0	0	0	0	0	0	$c_{18,12}$	0	0
0	0	0	0	0	0	0	0	0	0	0	$c_{19,12}$	0	0

with

$$c_{11} = 1.00016, \quad c_{23} = 1.00516, \quad c_{36} = 1.01461,$$

$$c_{47} = 1.01448, \quad c_{58} = 1.01451, \quad c_{69} = 1.01456, \quad c_{7,10} = 1.01459,$$

$$c_{8,11} = 1.01471, \quad c_{9,13} = 1.01486, \quad c_{10,14} = 1.01517$$

$$c_{11,12} = 0.060125, \quad c_{12,12} = 0.061082, \quad c_{13,12} = 0.0616,$$

$$c_{14,12} = 0.0622, \quad c_{15,12} = 0.0632, \quad c_{16,12} = 0.0642,$$

$$c_{17,12} = c_{18,12} = c_{19,12} = 0.0656.$$

3.2.3 Problems

As seen in both examples above, we typically have $n \ll N$. Moreover, many entries of C are zero, which is due to the many different cash flow dates. This makes the linear optimization problem

$$\min_{d \in \mathbb{R}^N} \| p - C d \|^2$$

ill-posed. Indeed, any solution $\hat{d}$ is characterized by the first-order condition

$$C^\top (p - C\hat{d}) = 0, \quad \text{and thus} \quad C^\top C\hat{d} = C^\top p.$$

But $\dim \ker(C^\top C) = \dim \ker(C) \geq N - n$. Hence the solution space is at least $N - n$-dimensional.

One could choose the data set such that cash flows are at same points in time (say four dates each year) and the cash flow matrix C is not entirely full of zeros, such as in Carleton and Cooper [34]. Still the regression method has big problems. There are as many parameters as there are cash flow dates, and there is nothing to regularize the discount curve found from the regression. As a result, the discount factors of similar maturity can be very different, which leads to a ragged spot rate (yield) curve, and even worse for forward rates.

An alternative and better method would be to estimate a smooth yield curve parametrically from the market rates. This approach is taken up in the following section.

3.3 Parametric Estimation Methods

Reduction of parameters and smooth term-structure of interest rates can be achieved by using parameterized families of smooth curves. A particular case is the class of linear families, where we fix a set of basis functions, preferably with compact support. As a first example consider B-splines.

3.3.1 Estimating the Discount Function with Cubic B-splines

A cubic spline is a piecewise cubic polynomial that is everywhere twice differentiable. It interpolates values at $q + 1$ knot points $\xi_0 < \cdots < \xi_q$. Its general form is

$$\sigma(x) = \sum_{i=0}^{3} a_i x^i + \sum_{j=1}^{q-1} b_j (x - \xi_j)_+^3,$$

hence it has $q + 3$ parameters $\{a_0, \ldots, a_3, b_1, \ldots, b_{q-1}\}$ (a kth-degree spline has $q + k$ parameters). The spline is uniquely characterized by specification of σ' or σ'' at ξ_0 and ξ_q.

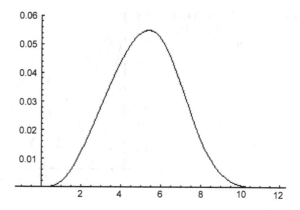

Fig. 3.5 B-spline with knot points $\{0, 1, 6, 8, 11\}$

If we introduce six extra knot points

$$\xi_{-3} < \xi_{-2} < \xi_{-1} < \xi_0 < \cdots < \xi_q < \xi_{q+1} < \xi_{q+2} < \xi_{q+3},$$

we obtain a basis for the cubic splines on $[\xi_0, \xi_q]$ given by the $q + 3$ *B-splines*

$$\psi_k(x) = \sum_{j=k}^{k+4} \left(\prod_{i=k, i \neq j}^{k+4} \frac{1}{\xi_i - \xi_j} \right) (x - \xi_j)_+^3, \quad k = -3, \ldots, q - 1.$$

The B-spline ψ_k is zero outside $[\xi_k, \xi_{k+4}]$. See Fig. 3.5 for an example.

We now use B-splines to estimate the discount curve:

$$D(x; z) = z_1 \psi_1(x) + \cdots + z_m \psi_m(x),$$

such as done by Steeley [155]. With

$$d(z) = \begin{pmatrix} D(x_1; z) \\ \vdots \\ D(x_N; z) \end{pmatrix} = \begin{pmatrix} \psi_1(x_1) & \cdots & \psi_m(x_1) \\ \vdots & & \vdots \\ \psi_1(x_N) & \cdots & \psi_m(x_N) \end{pmatrix} \begin{pmatrix} z_1 \\ \vdots \\ z_m \end{pmatrix} =: \Psi z$$

this leads to the linear optimization problem

$$\min_{z \in \mathbb{R}^m} \| p - C\Psi z \|^2.$$

If the $n \times m$ matrix $A = C\Psi$ has full rank m, the unique unconstrained solution is

$$z^* = (A^\top A)^{-1} A^\top p.$$

A reasonable constraint would be

$$D(0; z) = \psi_1(0) z_1 + \cdots + \psi_m(0) z_m = 1.$$

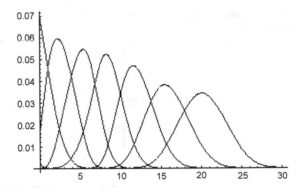

Fig. 3.6 B-splines with knots $\{-20, -5, -2, 0, 1, 6, 8, 11, 15, 20, 25, 30\}$

As an example, we take the UK government bond market data from Table 3.2 in the last section. The maximum time to maturity, x_{104}, is 12.11 [years]. Notice that the first bond is a zero-coupon bond. Its exact yield is

$$y = -\frac{365}{72} \log \frac{103.822}{105} = -\frac{1}{0.197} \log 0.989 = 0.0572.$$

As a basis we use the 8 (resp. first 7) B-splines with the 12 knot points

$$\{-20, -5, -2, 0, 1, 6, 8, 11, 15, 20, 25, 30\}$$

shown in Fig. 3.6.

The estimation with all 8 B-splines leads to

$$\min_{z \in \mathbb{R}^8} \| p - C\Psi z \| = \| p - C\Psi z^* \| = 0.23$$

with

$$z^* = \begin{pmatrix} 13.8641 \\ 11.4665 \\ 8.49629 \\ 7.69741 \\ 6.98066 \\ 6.23383 \\ -4.9717 \\ 855.074 \end{pmatrix},$$

and the discount function, yield curve (cont. comp. spot rates), and forward curve (cont. comp. 3-monthly forward rates) shown in Fig. 3.8.

The estimation with only the first 7 B-splines leads to

$$\min_{z \in \mathbb{R}^7} \| p - C\Psi z \| = \| p - C\Psi z^* \| = 0.32$$

Fig. 3.7 Five B-splines with knot points $\{-10, -5, -2, 0, 4, 15, 20, 25, 30\}$

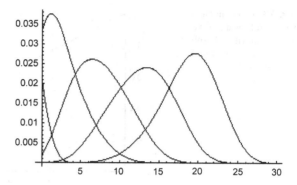

with

$$z^* = \begin{pmatrix} 17.8019 \\ 11.3603 \\ 8.57992 \\ 7.56562 \\ 7.28853 \\ 5.38766 \\ 4.9919 \end{pmatrix},$$

and the discount curve, yield curve (cont. comp. spot rates), and forward curve (cont. comp. 3-month forward rates) shown in Fig. 3.9.

Next we use only 5 B-splines with the 9 knot points

$$\{-10, -5, -2, 0, 4, 15, 20, 25, 30\}$$

shown in Fig. 3.7.

The estimation with this 5 B-splines leads to

$$\min_{z\in\mathbb{R}^5} \| p - C\Psi z\| = \| p - C\Psi z^*\| = 0.39$$

with

$$z^* = \begin{pmatrix} 15.652 \\ 19.4385 \\ 12.9886 \\ 7.40296 \\ 6.23152 \end{pmatrix},$$

and the discount curve, yield curve (cont. comp. spot rates), and forward curve (cont. comp. 3-monthly forward rates) shown in Fig. 3.10.

We thus find there is an obvious trade-off between the quality (or regularity) and the correctness of the fit. The curves in Figs. 3.9 and 3.10 are more regular than those in Fig. 3.8, but their correctness criteria (0.32 and 0.39) are worse than for the fit with 8 B-splines (0.23). We also see from these figures that estimating the discount curve leads to unstable and irregular yield and forward curves. The problems are

Fig. 3.8 Discount curve,
yield and forward curves for
estimation with 8 B-splines.
The dot is the exact yield of
the first bond

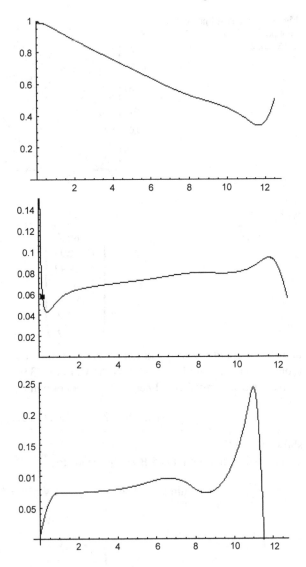

most prominent at the short- and long-term maturities. Obviously splines are not useful for extrapolating to long-term maturities. It can further be shown that the B-spline fits are extremely sensitive to the number and location of the knot points.

In sum, from this example we may conclude that splines can produce bad fits in general. We learn that we need criteria asserting smooth yield and forward curves that do not fluctuate too much and flatten towards the long end. Indeed it is advisable to directly estimate the yield or forward curve. Ideally, the number and location of the knot points for the splines are optimally adjusted to the data. As we shall see in the next section, all these criteria can be achieved by smoothing splines.

Fig. 3.9 Discount curve, yield and forward curves for estimation with 7 B-splines. The dot is the exact yield of the first bond

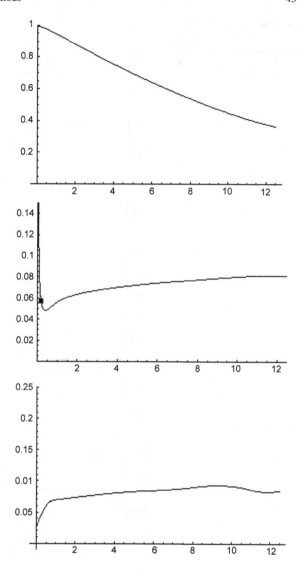

3.3.2 Smoothing Splines

Smoothing splines combine the objectives of a good data fit and curve regularity. In other words, the least-squares criterion

$$\min_z \| p - C\, d(z) \|^2$$

has to be extended by criterions for the smoothness of the yield or forward curve. We exemplify this idea with the spline method developed by Sabine Lorimier in her

Fig. 3.10 Discount curve, yield and forward curves for estimation with 5 B-splines. The dot is the exact yield of the first bond

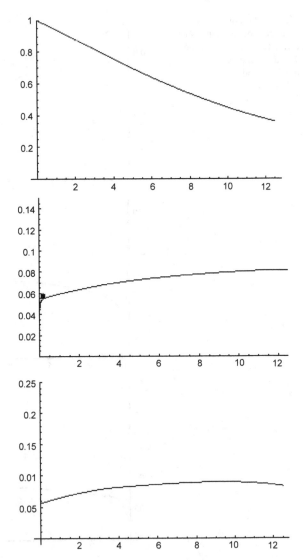

Ph.D. thesis [119], where the number and location of the knots are determined by the observed data itself.

For ease of notation we set $t_0 = 0$ (today). The data is given by N observed zero-coupon bonds $P(0, T_1), \ldots, P(0, T_N)$ at $0 < T_1 < \cdots < T_N \equiv T$, and consequently the N yields

$$Y_1, \ldots, Y_N, \quad P(0, T_i) = e^{-T_i Y_i}.$$

Let $f(u)$ denote the forward curve. The fitting requirement now is for the forward curve

$$\int_0^{T_i} f(u)\,du + \varepsilon_i/\sqrt{\alpha} = T_i Y_i, \qquad (3.2)$$

with an arbitrary constant $\alpha > 0$. The aim is to minimize $\|\varepsilon\|^2$ as well as the smoothness criterion

$$\int_0^T (f'(u))^2\,du. \qquad (3.3)$$

Recall that a function $g : [0, T] \to \mathbb{R}$ is absolutely continuous if and only if there exists some Lebesgue integrable function g' such that

$$g(x) = g(0) + \int_0^x g'(u)\,du \quad \text{for all } x \in [0, T].$$

The set H of absolutely continuous functions $g : [0, T] \to \mathbb{R}$ with

$$\int_0^T (g'(u))^2\,du < \infty$$

endowed with the scalar product

$$\langle g, h \rangle_H = g(0)h(0) + \int_0^T g'(u)h'(u)\,du,$$

becomes a Hilbert space.[2]

Define the nonlinear functional on H

$$F(f) = \int_0^T (f'(u))^2\,du + \alpha \sum_{i=1}^N \left(Y_i T_i - \int_0^{T_i} f(u)\,du \right)^2.$$

The optimization problem is then

$$\min_{f \in H} F(f). \qquad (3.4)$$

The parameter α tunes the trade-off between smoothness and correctness of the fit.

Theorem 3.1 *Problem (3.4) has a unique solution f, which is a second-order spline characterized by*

$$f(u) = f(0) + \sum_{k=1}^N a_k h_k(u), \qquad (3.5)$$

---

[2]This particular Hilbert space is known as a Sobolev space, see e.g. Brezis [26].

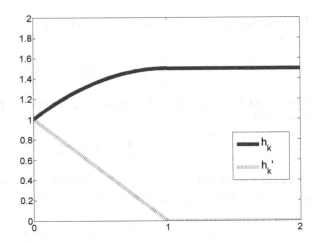

where $h_k \in C^1[0, T]$ *is a second-order polynomial on* $[0, T_k]$ *with*

$$h'_k(u) = (T_k - u)^+, \quad h_k(0) = T_k, \quad k = 1, \ldots, N, \tag{3.6}$$

see Fig. 3.11, and $f(0)$ *and* a_k *solve the linear system of equations*

$$\sum_{k=1}^{N} a_k T_k = 0, \tag{3.7}$$

$$\alpha \left(Y_k T_k - f(0) T_k - \sum_{l=1}^{N} a_l \langle h_l, h_k \rangle_H \right) = a_k, \quad k = 1, \ldots, N. \tag{3.8}$$

Proof Integration by parts yields

$$\int_0^{T_k} g(u) \, du = T_k g(T_k) - \int_0^{T_k} u g'(u) \, du$$

$$= T_k g(0) + T_k \int_0^{T_k} g'(u) \, du - \int_0^{T_k} u g'(u) \, du$$

$$= T_k g(0) + \int_0^{T} (T_k - u)^+ g'(u) \, du = \langle h_k, g \rangle_H,$$

for all $g \in H$. In particular,

$$\int_0^{T_k} h_l \, du = \langle h_l, h_k \rangle_H.$$

A local minimizer f of F satisfies, for any $g \in H$, the first-order condition

$$\frac{d}{d\varepsilon} F(f + \varepsilon g)|_{\varepsilon=0} = 0$$

or equivalently

$$\int_0^T f'g' \, du = \alpha \sum_{k=1}^{N} \left(Y_k T_k - \int_0^{T_k} f \, du \right) \int_0^{T_k} g \, du. \tag{3.9}$$

In particular, for all $g \in H$ with $\langle g, h_k \rangle_H = 0$ we obtain

$$\langle f - f(0), g \rangle_H = \int_0^T f'(u)g'(u) \, du = 0.$$

Hence

$$f - f(0) \in \text{span}\{h_1, \ldots, h_N\},$$

which proves (3.5), (3.6) and (3.7) (set $u = 0$). Hence we have

$$\int_0^T f'g' \, du = \sum_{k=1}^{N} \alpha_k \int_0^T (T_k - u)^+ g'(u) \, du$$

$$= \sum_{k=1}^{N} a_k \left(-T_k g(0) + \int_0^{T_k} g(u) \, du \right) = \sum_{k=1}^{N} a_k \int_0^{T_k} g(u) \, du,$$

and (3.9) can be rewritten as

$$\sum_{k=1}^{N} \left(a_k - \alpha \left(Y_k T_k - f(0) T_k - \sum_{l=1}^{N} a_l \langle h_l, h_k \rangle_H \right) \right) \int_0^{T_k} g(u) \, du = 0$$

for all $g \in H$. This is true if and only if (3.8) holds.

Thus we have shown that (3.9) is equivalent to (3.5)–(3.8).

Next we show that (3.9) is a sufficient condition for f to be a global minimizer of F. Let $g \in H$, then

$$F(g) = \int_0^T ((g' - f') + f')^2 \, du + \alpha \sum_{k=1}^{N} \left(Y_k T_k - \int_0^{T_k} g \, du \right)^2$$

$$\stackrel{(3.9)}{=} F(f) + \int_0^T (g' - f')^2 \, du + \alpha \sum_{k=1}^{N} \left(\int_0^{T_k} f \, du - \int_0^{T_k} g \, du \right)^2$$

$$\geq F(f),$$

where we used (3.9) with g replaced by $g - f$.

It remains to show that f exists and is unique; that is, the linear system (3.7)–(3.8) has a unique solution $(f(0), a_1, \ldots, a_N)^{\top}$. The corresponding $(N + 1) \times$

$(N + 1)$ matrix is

$$
A = \begin{pmatrix}
0 & T_1 & T_2 & \cdots & T_N \\
\alpha T_1 & \alpha \langle h_1, h_1 \rangle_H + 1 & \alpha \langle h_1, h_2 \rangle_H & \cdots & \alpha \langle h_1, h_N \rangle_H \\
\vdots & \vdots & \ddots & \ddots & \vdots \\
\alpha T_N & \alpha \langle h_N, h_1 \rangle_H & \alpha \langle h_N, h_2 \rangle_H & \cdots & \alpha \langle h_N, h_N \rangle_H + 1
\end{pmatrix}. \quad (3.10)
$$

That is, the system (3.7)–(3.8) reads

$$
A \begin{pmatrix} f(0) \\ a \end{pmatrix} = \begin{pmatrix} 0 \\ Z \end{pmatrix}, \quad (3.11)
$$

where $a = (a_1, \ldots, a_N)^\top$ and $Z = \alpha (Y_1 T_1, \ldots, Y_N T_N)^\top$. Let $\lambda = (\lambda_0, \ldots, \lambda_N)^\top \in \mathbb{R}^{N+1}$ such that $A\lambda = 0$, that is,

$$
\sum_{k=1}^{N} T_k \lambda_k = 0,
$$

$$
\alpha T_k \lambda_0 + \alpha \sum_{l=1}^{N} \langle h_k, h_l \rangle_H \lambda_l + \lambda_k = 0, \quad k = 1, \ldots, N.
$$

Multiplying the latter equation with λ_k and summing up over k yields

$$
\alpha \left\| \sum_{k=1}^{N} \lambda_k h_k \right\|_H^2 + \sum_{k=1}^{N} \lambda_k^2 = 0,
$$

where we write $\|g\|_H = \sqrt{\langle g, g \rangle_H}$ for the corresponding norm on H. Hence $\lambda = 0$, whence A is non-singular, and the theorem is proved. $\qquad \square$

The parameter α tunes the trade-off between smoothness and correctness of the fit as follows:

- If $\alpha \to 0$ then by (3.5) and (3.8) we have $f(u) \equiv f(0)$, a constant function. That is, we achieve maximal regularity

$$
\int_0^T (f'(u))^2 \, du = 0
$$

but obviously no fitting of the data, see (3.2).
- If $\alpha \to \infty$ then (3.9) implies that

$$
\int_0^{T_k} f(u) \, du = Y_k T_k, \quad k = 1, \ldots, N, \quad (3.12)
$$

which means a perfect fit. That is, f minimizes (3.3) subject to the constraints (3.12).

To estimate the forward curve from N zero-coupon bonds—that is, yields $Y_1, \ldots,$ Y_N—one has to solve the linear system (3.11).

Of course, if coupon bond prices are given, then the above method has to be modified and becomes nonlinear. With $p \in \mathbb{R}^n$ denoting the market price vector and c_{kl} the cash flows at dates T_l, $k = 1, \ldots, n$, $l = 1, \ldots, N$, this reads

$$\min_{f \in H} \left\{ \int_0^T (f')^2 \, du + \alpha \sum_{k=1}^n \left(\log p_k - \log \left[\sum_{l=1}^N c_{kl} e^{-\int_0^{T_l} f \, du} \right] \right)^2 \right\}.$$

If the coupon payments are small compared to the nominal ($=1$), then this problem has a unique solution. This and much more is carried out in Lorimier's thesis [119].

3.3.3 Exponential–Polynomial Families

Let us now introduce the parametric curve families which are used by most central banks for term-structure estimation. As in the preceding section the forward curve

$$\mathbb{R}_+ \ni x \mapsto f(t_0, t_0 + x) = \phi(x) = \phi(x; z)$$

is estimated. The implied discount curve is

$$D(x) = D(x; z) = e^{-\int_0^x \phi(u; z) \, du}, \quad z \in \mathcal{Z}.$$

If we calibrate to bond prices, we are led to a nonlinear optimization problem

$$\min_{z \in \mathcal{Z}} \| p - C \, d(z) \|,$$

with

$$d_i(z) = e^{-\int_0^{x_i} \phi(u; z) \, du}$$

for some payment tenor $0 < x_1 < \cdots < x_N$. This criterion can be modified in an obvious way to fit implied yields.

The first example is the Nelson–Siegel family [128], where we have four parameters $z_1, \ldots, z_4$. It is defined by

$$\phi_{NS}(x; z) = z_1 + (z_2 + z_3 x) e^{-z_4 x}.$$

The typical shape of these functions has one hump, see Fig. 3.12.

To improve the curve flexibility, Svensson [157] proposed an extension of Nelson–Siegel's family by including six parameters $z_1, \ldots, z_6$. The Svensson family is then defined by

$$\phi_S(x; z) = z_1 + (z_2 + z_3 x) e^{-z_5 x} + z_4 x e^{-z_6 x}.$$

Fig. 3.12 Nelson–Siegel curves for $z_1 = 7.69$, $z_2 = -4.13$, $z_4 = 0.5$ and 7 different values for $z_3 = 1.76, 0.77, -0.22, -1.21, -2.2, -3.19, -4.18$

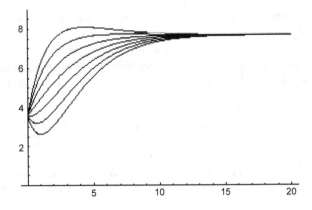

Table 3.4 Overview of estimation procedures by several central banks. BIS 1999 [11]. NS is for Nelson–Siegel, S for Svensson, wp for weighted prices

Central bank	Method	Minimized error
Belgium	S or NS	wp
Canada	S	sp
Finland	NS	wp
France	S or NS	wp
Germany	S	yields
Italy	NS	wp
Japan	smoothing splines	prices
Norway	S	yields
Spain	S	wp
Sweden	S	yields
UK	S	yields
USA	smoothing splines	bills: wp bonds: prices

Obviously, both the Nelson–Siegel and Svensson families belong to the family of general exponential–polynomial functions

$$p_1(x)e^{-\alpha_1 x} + \cdots + p_n(x)e^{-\alpha_n x},$$

where p_i denote polynomials of degree n_i.

Table 3.4 is taken from a document of the Bank for International Settlements (BIS) 1999 [11]. It illustrates the role the Nelson–Siegel and Svensson families play in the world of monetary regulation.

Let us end this section with a list of desirable features for curve families to be suitable for the estimation of the term-structure:

- Flexible: the curves shall fit a wide range of term structures.

- Parsimonious: the number of factors shall not be too large (curse of dimensionality).
- Regular: we prefer smooth yield or forward curves that flatten out towards the long end.
- Consistent: the curve families shall be compatible with dynamic interest rate models! This point will be explained and exploited in more detail in Chap. 9 below.

Let us recall that one of the main problems in the term-structure estimation stems from its high dimensionality, which makes it difficult to have a good intuition about its behavior. If we could learn from observations of the data which basis shapes are the main determinants of the zero-coupon yield curve (or its increments), we could in fact reduce the dimension of this problem. It turns out that this is a standard problem in multivariate data analysis. It goes under the name of principal component analysis, which will be exploited in the following section.

3.4 Principal Component Analysis

Principal component analysis (PCA) is a dimension reduction technique in multivariate analysis. It can be used for constructing the components of the stochastic term-structure movements that account for most of the variability, in some appropriately defined sense.

The key mathematical principle behind PCA is the spectral decomposition theorem of linear algebra, which states that any real symmetric $n \times n$ matrix Q can be written as

$$Q = ALA^\top, \tag{3.13}$$

where:

- $L = \mathrm{diag}(\lambda_1, \ldots, \lambda_n)$ is the diagonal matrix of eigenvalues of Q with $\lambda_1 \geq \lambda_2 \geq \cdots \geq \lambda_n$;
- A is an orthogonal matrix (that is, $A^{-1} = A^\top$) whose columns $a_1, \ldots, a_n$ are the normalized eigenvectors of Q (that is, $Qa_i = \lambda_i a_i$), which form an orthonormal basis of $\mathbb{R}^n$.

Recall that $A^\top$ denotes the transpose of A.

3.4.1 Principal Components of a Random Vector

Consider an $\mathbb{R}^n$-valued square-integrable random vector X with mean $\mu = \mathbb{E}[X]$ and covariance matrix $Q = \mathrm{cov}[X]$. Since Q is symmetric and positive semidefinite, the above decomposition (3.13) applies with $\lambda_i \geq 0$ for all i. The *principal components transform* of X is defined as

$$Y = A^\top (X - \mu),$$

which can be seen as a recentering and rotation of X. Note that

$$Y_i = a_i^\top (X - \mu)$$

is the projection of $X - \mu$ onto the ith eigenvector a_i of Q. One calls Y_i the ith *principal component*, and a_i the ith vector of *loadings*, of X. We thus obtain the decomposition

$$X = \mu + AY = \mu + \sum_{i=1}^{n} Y_i a_i.$$

Simple calculations reveal that

$$\mathbb{E}[Y] = 0 \quad \text{and} \quad \text{Cov}[Y] = A^\top Q A = A^\top A L A^\top A = L.$$

Hence the principal components of X are uncorrelated and have variances $\text{Var}[Y_i] = \lambda_i$, which are ordered from largest, λ_1, to smallest, λ_n.

Moreover, it can be shown ($\rightarrow$ Exercise 3.4) that the first principal component, Y_1, has maximal variance among all standardized linear combinations of X. That is,

$$\text{Var}[a_1^\top X] = \max \left\{ \text{Var}[b^\top X] \mid b^\top b = 1 \right\}. \tag{3.14}$$

For $i = 2, \ldots, n$, the ith principal component, Y_i, can be shown to have maximal variance among all such linear combinations that are orthogonal to the first $i - 1$ linear combinations.

Next, we observe that

$$\sum_{i=1}^{n} \text{Var}[X_i] = \text{trace}(Q) = \sum_{i=1}^{n} \lambda_i = \sum_{i=1}^{n} \text{Var}[Y_i].$$

Hence

$$\frac{\sum_{i=1}^{k} \lambda_i}{\sum_{i=1}^{n} \lambda_i}$$

represents the amount of variability in X explained by the first k principal components $Y_1, \ldots, Y_k$.

We may think of X as a high-dimensional stationary model for (daily changes of) the forward curve. Suppose that the first k principal components $Y_1, \ldots, Y_k$ explain a significant amount (e.g. 99%) of the variability in X. It is then most useful to approximate X by $X \approx \mu + \sum_{i=1}^{k} Y_i a_i$. That is, the loadings $a_1, \ldots, a_k$ are the main components of the stochastic forward curve movements.

3.4.2 Sample Principle Components

Now assume that we have multivariate data observations

$$x = [x(1), \ldots, x(N)],$$

where each column $x(t) = (x_1(t), \ldots, x_n(t))^\top$ is a sample realization of a random vector $X(t)$ which is identically distributed as X with mean $\mu = \mathbb{E}[X]$ and covariance matrix $Q = \text{cov}[X]$. We consider the empirical $n \times n$ covariance matrix

$$\hat{Q}_{ij} = \text{Cov}[x_i, x_j] = \frac{1}{N} \sum_{t=1}^{N} (x_i(t) - \hat{\mu}_i)(x_j(t) - \hat{\mu}_j),$$

where

$$\hat{\mu} = \frac{1}{N} \sum_{t=1}^{N} x(t)$$

denotes the empirical mean. Then $\hat{Q}$ is positive semi-definite and the above PCA applies by analogy ($\rightarrow$ Exercise 3.3). We thus obtain the empirical principal components $y = \hat{A}^\top (x - \hat{\mu})$ with loadings $\hat{a}_i$ given as column vectors of $\hat{A}$, where $\hat{Q} = \hat{A}\hat{L}\hat{A}^\top$. That is,

$$x = \hat{\mu} + \sum_{i=1}^{n} y_i \hat{a}_i \quad \text{and}$$

$$\text{Cov}[y_i, y_j] = \frac{1}{N} \sum_{t=1}^{N} y_i(t) y_j(t) = \begin{cases} \hat{\lambda}_i, & \text{if } i = j, \\ 0, & \text{else.} \end{cases}$$

(3.15)

The empirical mean $\hat{\mu}$ and covariance matrix $\hat{Q}$ are standard estimators for the true parameters μ and Q, if the observations $X(t)$ are either independent or at least serially uncorrelated (i.e. $\text{Cov}[X(t), X(t+h)] = 0$ for all $h \neq 0$, see [122, Chap. 3] for a brief account of multivariate statistical analysis and further references). When this kind of stationarity of the time series $X(t)$ is in doubt, the standard practice is to differentiate the series and to consider the increments $\Delta X(t) = X(t) - X(t-1)$ instead. This approach is illustrated in the following section for the forward curve of interest rates. See also Sect. 11.7.1 for a more specific model context.

3.4.3 PCA of the Forward Curve

Now let $x(t) = (x_1(t), \ldots, x_n(t))^\top$ denote the increments of the forward curve, say

$$x_i(t) = R(t + \Delta t; t + \Delta t + \tau_{i-1}, t + \Delta t + \tau_i) - R(t; t + \tau_{i-1}, t + \tau_i),$$

for some maturity spectrum $0 = \tau_0 < \cdots < \tau_n$. Here τ_i denotes time *to* maturity. Therefore we have to adjust the maturity arguments for $R(t + \Delta t; \cdot)$ by adding Δt.

PCA typically leads to the following picture, which is taken from Rebonato [134, Sect. 3.1]. The analysis is based on UK market data from the years 1989–1992, where the original maturity spectrum has been divided into eight distinct buckets, i.e. $n = 8$. The first three vectors of loadings are

$$a_1 = \begin{pmatrix} 0.329 \\ 0.354 \\ 0.365 \\ 0.367 \\ 0.364 \\ 0.361 \\ 0.358 \\ 0.352 \end{pmatrix}, \quad a_2 = \begin{pmatrix} -0.722 \\ -0.368 \\ -0.121 \\ 0.044 \\ 0.161 \\ 0.291 \\ 0.316 \\ 0.343 \end{pmatrix}, \quad a_3 = \begin{pmatrix} 0.490 \\ -0.204 \\ -0.455 \\ -0.461 \\ -0.176 \\ 0.176 \\ 0.268 \\ 0.404 \end{pmatrix}.$$

Figure 3.13 gives the plots of the first three loadings. We observe that:

- the first loading is roughly flat, causing parallel shifts of the forward curve (affects the average rate);
- the second loading is upward sloping, this is tilting of the forward curve (affects the slope);
- the third loading hump-shaped, causing a flex (affecting the curvature).

Moreover, Table 3.5 shows that the first three principal components explain more than 99% of the variance of x. This suggests that any of the forward curves from

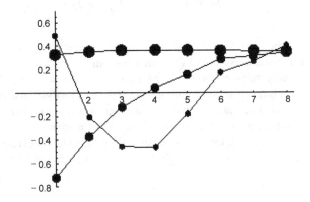

Fig. 3.13 First three forward curve loadings

Table 3.5 Explained variance of the principal components

PC	Explained variance (%)
1	92.17
2	6.93
3	0.61
4	0.24
5	0.03
6–8	0.01

this period can be approximated by a linear combination of the first three loadings, the relative error being small.

These features are very typical, and should be expected in most PCA of the forward curve or its increments. See also the findings of Carmona and Tehranchi [35, Sect. 1.7]. PCA of the forward curve goes back to the seminal paper by Litterman and Scheinkman [117].

3.4.4 Correlation

Let us finally have a look at some stylized fact about the correlation of the original forward curve increments. A typical example of correlation among forward rates is provided by Brown and Schaefer [28]. The data is from the US Treasury term structure 1987–1994. The following matrix,

$$\begin{pmatrix} 1 & 0.87 & 0.74 & 0.69 & 0.64 & 0.6 \\ & 1 & 0.96 & 0.93 & 0.9 & 0.85 \\ & & 1 & 0.99 & 0.95 & 0.92 \\ & & & 1 & 0.97 & 0.93 \\ & & & & 1 & 0.95 \\ & & & & & 1 \end{pmatrix}$$

shows the correlation for changes of forward rates of maturities

$$0,\ 0.5,\ 1,\ 1.5,\ 2,\ 3 \text{ years.}$$

Figure 3.14 illustrates the first row of this correlation matrix. In a stylized fashion we note that de-correlation occurs quickly, so that an exponentially decaying correlation structure is plausible.

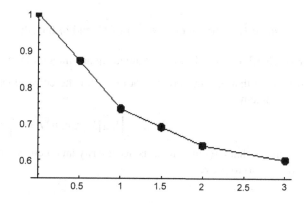

Fig. 3.14 Correlation between the short rate and instantaneous forward rates for the US Treasury curve 1987–1994

3.5 Exercises

Exercise 3.1 Compute the first 14 columns of the cash flow matrix of the US money-market data (6 October 1997) given in Sect. 3.2.2.

Exercise 3.2 Consider the yield data

T_k	2	3	4	5	7	10
Y_k	0.948	1.129	1.300	1.454	1.704	1.955

and find the optimal forward curve of the form (3.5) for

$$\alpha = 0.01, \quad \alpha = 0.1 \quad \text{and} \quad \alpha = 1.$$

That is,

(a) Compute the respective coefficients $f(0), a_1, \ldots, a_N$.
(b) Plot the forward and yield curves, respectively, and the data points for comparison of smoothness and quality of fit.

Exercise 3.3 Let $x(1), \ldots, x(N)$ be a sample of a random vector $X = (X_1, \ldots, X_n)^\top$, and let $\hat{\mu}$ and $\hat{Q}$ denote the empirical mean and covariance matrix of x, respectively. Prove the following:

(a) $\hat{Q}$ is symmetric and positive semi-definite.
(b) The PCA decomposition (3.15) holds.
(c) If $\hat{Q}$ is degenerate then one can express some x_i as a linear function of the other components x_j, $j \neq i$.
(d) Assume that $\hat{Q}$ is non-degenerate. Find a sample of vectors

$$w(t) = (w_1(t), \ldots, w_n(t))^\top,$$

for $t = 1, \ldots, N$, such that

$$x = \hat{\mu} + \sum_{i=1}^{n} \hat{a}_i \sqrt{\hat{\lambda}_i} w_i \quad \text{and} \quad \text{Cov}[w_i, w_j] = \begin{cases} 1, & \text{if } i = j, \\ 0, & \text{else,} \end{cases}$$

where $\hat{\lambda}_i$ is the ith eigenvector of $\hat{Q}$, and $\hat{a}_i$ the ith vector of loadings of x.

Exercise 3.4 Let X be an $\mathbb{R}^n$-valued square-integrable random vector.

(a) Show that the first-order conditions for the constrained quadratic optimization problem

$$\max \left\{ \text{Var}[b^\top X] \mid b^\top b = 1 \right\}$$

imply that any maximizer is necessarily an eigenvector of $\text{Cov}[X]$.
(b) Use (a) to prove (3.14).

Exercise 3.5 The Excel file

www.snb.ch/ext/stats/statmon/xls/en/statmon_E3_M1_M.xls

of the Monthly Statistical Bulletin from the Swiss National Bank contains monthly spot interest rates (that is, yields $R(t, T)$) for Swiss Confederation bonds for a time to maturity $(T - t)$ spectrum of 2, 3, 4, 5, 7, 10, 20 and 30 years. Perform a principal component analysis of the monthly yield curve changes as shown in Sect. 3.4.3 for the forward curve from the last ten years. In particular, determine:

(a) the empirical covariance matrix;
(b) its eigenvectors and eigenvalues in decreasing order;
(c) the explained variances of the principal components.

3.6 Notes

The bootstrapping example in Sect. 3.1, including the data, is from James and Webber [100, Sect. 5.4]. Sections 3.2 and 3.3.1, including the data, are from [100, Sects. 15.1–15.3]. Section 3.4 is from [122, Sect. 3.4.4] and [35, Sect. 1.7]. The figures and data on the PCA and correlation of the forward curve increments in Sects. 3.4.3 and 3.4.4 are taken from Rebonato [134, Sect. 3.1]. Other sources for PCA of the yield curve include Rebonato [134], Carmona and Tehranchi [35], the seminal paper by Litterman and Scheinkman [117], or the paper by Bouchaud et al. [20].

Chapter 4
Arbitrage Theory

This chapter briefly recalls the fundamental arbitrage principles in a Brownian-motion-driven financial market. The basics of stochastic calculus are provided without proofs. Standard terminology is employed without further explanation. Readers are requested to consult one of the many text books on stochastic calculus. References are given in the notes section. The main pillars for financial applications are Itô's formula, Girsanov's change of measure theorem, and the martingale representation theorem.

4.1 Stochastic Calculus

The stochastic basis is a filtered probability space $(\Omega, \mathcal{F}, (\mathcal{F}_t)_{t \geq 0}, \mathbb{P})$ satisfying the usual conditions[1] and carrying a d-dimensional $(\mathcal{F}_t)$-adapted Brownian motion $W = (W_1, \ldots, W_d)^\top$. We shall assume that $\mathcal{F} = \mathcal{F}_\infty = \bigvee_{t \geq 0} \mathcal{F}_t$, and do not a priori fix a finite time horizon. This is not a restriction since always one can stop a stochastic process at a finite time T if this were the ultimate time horizon, such as in the Black–Scholes model ($\rightarrow$ Exercise 4.7).

For random variables X, Y, it is always understood that "$X = Y$" means "$X = Y$ a.s." that is, $\mathbb{P}[X = Y] = 1$. The same applies to inequalities "$X \geq Y$", etc. We write $\mathcal{B}[0, t]$, $\mathcal{B}(\mathbb{R}_+)$ or simply $\mathcal{B}$, if there is no ambiguity, for the respective Borel σ-algebras. A stochastic process $X = X(\omega, t)$ is called:

- adapted if $\Omega \ni \omega \mapsto X(\omega, t)$ is $\mathcal{F}_t$-measurable for all $t \geq 0$,
- progressively measurable (or simply progressive) if $\Omega \times [0, t] \ni (\omega, s) \mapsto X(\omega, s)$ is $\mathcal{F}_t \otimes \mathcal{B}[0, t]$-measurable for all $t \geq 0$.

A progressive process is obviously adapted. Progressive measurability of X is needed in order that composed processes such as $\int_0^t X(s) \, ds$ and $X(t \wedge \tau)$, for any stopping time[2] τ, are adapted.

We denote by Prog the progressive σ-algebra, generated by all progressive processes, on $\Omega \times \mathbb{R}_+$. Progressive and Prog-measurability are equivalent[3] ($\rightarrow$ Exercise 6.1).

---

[1] The usual conditions are (1) completeness: $\mathcal{F}_0$ contains all of the null sets, and (2) right-continuity: $\mathcal{F}_t = \bigcap_{s > t} \mathcal{F}_s$ for all $t \geq 0$.

[2] A $[0, \infty]$-valued random variable τ is a stopping time of the filtration $(\mathcal{F}_t)$ if the event $\{\tau \leq t\}$ belongs to $\mathcal{F}_t$, for every $t \geq 0$.

[3] See Proposition 1.41 in [130].

D. Filipović, *Term-Structure Models*,
Springer Finance,
DOI 10.1007/978-3-540-68015-4_4, © Springer-Verlag Berlin Heidelberg 2009

4.1.1 Stochastic Integration

We now define $\mathcal{L}^2$ and $\mathcal{L}$ as the sets of $\mathbb{R}^d$-valued progressive processes $h = (h_1, \ldots, h_d)$ that satisfy

$$\mathbb{E}\left[\int_0^\infty \|h(s)\|^2 \, ds\right] < \infty$$

and

$$\int_0^t \|h(s)\|^2 \, ds < \infty \quad \text{for all } t > 0,$$

respectively. The inclusion $\mathcal{L}^2 \subset \mathcal{L}$ is obvious.

Theorem 4.1 (Stochastic Integral) *For every $h \in \mathcal{L}$ one can define the stochastic integral*

$$(h \bullet W)_t = \int_0^t h(s) \, dW(s) = \sum_{j=1}^d \int_0^t h_j(s) \, dW_j(s)$$

with the following properties:

(a) *The process $h \bullet W$ is a continuous local martingale.*
(b) *Linearity*: $(\lambda g + h) \bullet W = \lambda(g \bullet W) + h \bullet W$, *for $g, h \in \mathcal{L}$ and $\lambda \in \mathbb{R}$.*
(c) *For any stopping time τ, the stopped integral equals*

$$\int_0^{t \wedge \tau} h(s) \, dW(s) = \int_0^t 1_{\{s \leq \tau\}} h(s) \, dW(s) \quad \text{for all } t > 0.$$

(d) *If $h \in \mathcal{L}^2$ then $h \bullet W$ is a martingale and the Itô isometry holds*:

$$\mathbb{E}\left[\left(\int_0^\infty h(s) \, dW(s)\right)^2\right] = \mathbb{E}\left[\int_0^\infty \|h(s)\|^2 \, ds\right].$$

(e) *Dominated convergence: if $(h_n) \subset \mathcal{L}$ is a sequence with $\lim_n h_n = 0$ pointwise and such that $|h_n| \leq k$ for some finite constant k then $\lim_n \sup_{s \leq t} |(h_n \bullet W)_s| = 0$ in probability for all $t > 0$.*

Proof See [135, Sect. 2, Chap. IV]. □

Remark 4.1 Note that the stochastic integrands are row vectors, and the integrator Brownian motion is a column vector, by convention. This is a convenient way to avoid writing the transpose $\cdot^\top$ in the stochastic integral every time.

An Itô process is the sum of an absolutely continuous drift plus a continuous local martingale of the form

$$X(t) = X(0) + \int_0^t a(s)\,ds + \int_0^t \rho(s)\,dW(s), \tag{4.1}$$

where $\rho \in \mathcal{L}$ and a is a progressive process satisfying $\int_0^t |a(s)|\,ds < \infty$ for all $t > 0$, such that the above integrals are defined. Here comes an important identification result:

Lemma 4.1 *The decomposition (4.1) of X is unique in the sense that*

$$X(t) = X(0) + \int_0^t a'(s)\,ds + \int_0^t \rho'(s)\,ds$$

implies $a' = a$ and $\rho' = \rho \; d\mathbb{P} \otimes dt$-a.s.

Proof This follows from Proposition 1.2 in [135, Chap. IV]. $\qquad\square$

We also write

$$dX(t) = a(t)\,dt + \rho(t)\,dW(t) \quad \text{or, shorter,} \quad dX = a\,dt + \rho\,dW,$$

and define

$$\mathcal{L}^2(X) = \left\{ h \text{ progressive} \;\middle|\; \mathbb{E}\left[\int_0^\infty |h(s)a(s)|^2\,ds\right] < \infty \text{ and } h\rho \in \mathcal{L}^2 \right\},$$

$$\mathcal{L}(X) = \left\{ h \text{ progressive} \;\middle|\; \int_0^t |h(s)a(s)|\,ds < \infty \text{ for all } t > 0 \text{ and } h\rho \in \mathcal{L} \right\}.$$

For $h \in \mathcal{L}(X)$ we can define the stochastic integral with respect to X as

$$\int_0^t h(s)\,dX(s) = \int_0^t h(s)a(s)\,ds + \int_0^t h(s)\rho(s)\,dW(s).$$

4.1.2 Quadratic Variation and Covariation

Now let

$$Y(t) = Y(0) + \int_0^t b(s)\,ds + \int_0^t \sigma(s)\,dW(s)$$

be another Itô process. The covariation process of X and Y is defined as

$$\langle X, Y \rangle_t = \int_0^t \rho(s)\sigma(s)^\top\,ds,$$

and $\langle X, X \rangle$ is called the quadratic variation process of X. It can be shown[4] that

$$\langle X, Y \rangle_t = \lim \sum_{i=0}^{m} (X_{t_{i+1}} - X_{t_i})(Y_{t_{i+1}} - Y_{t_i}) \quad \text{in probability,}$$

for any sequence of partitions $0 = t_0 < t_1 < \cdots < t_m = t$ with $\max_i |t_{i+1} - t_i| \to 0$. For the Brownian motion W we obtain[5] $\langle W_i, W_j \rangle_t = \delta_{ij} t$. In fact, this property distinguishes Brownian motion among all continuous local martingales, which is the content of the following important theorem.

Theorem 4.2 (Lévy's Characterization Theorem) *An $\mathbb{R}^d$-valued continuous local martingale X vanishing at $t = 0$ is a Brownian motion if and only if $\langle X_i, X_j \rangle_t = \delta_{ij} t$ for every $1 \leq i, j \leq d$.*

Proof See Theorem 3.6 in [135, Chap. IV]. □

4.1.3 Itô's Formula

We call $X = (X_1, \ldots, X_n)^\top$ an n-dimensional Itô process if every component X_i is an Itô process. We denote by $\mathcal{L}^2(X)$ ($\mathcal{L}(X)$) the set of progressive processes $h = (h_1, \ldots, h_n)$ such that h_i is in $\mathcal{L}^2(X_i)$ ($\mathcal{L}(X_i)$) for all i. In this sense, $\mathcal{L}^2 = \mathcal{L}^2(W)$ and $\mathcal{L} = \mathcal{L}(W)$. The stochastic integral of $h \in \mathcal{L}(X)$ with respect to X is defined coordinate-wise as

$$(h \bullet X)_t = \int_0^t h(s)\, dX(s) = \sum_{i=1}^{n} \int_0^t h_i(s)\, dX_i(s).$$

Next we consider the core formula of stochastic calculus.

Theorem 4.3 (Itô's Formula) *Let $f \in C^2(\mathbb{R}^n)$. Then $f(X)$ is an Itô process and*

$$f(X(t)) = f(X(0)) + \sum_{i=1}^{n} \int_0^t \frac{\partial f(X(s))}{\partial x_i}\, dX_i(s)$$

$$+ \frac{1}{2} \sum_{i,j=1}^{n} \int_0^t \frac{\partial^2 f(X(s))}{\partial x_i \partial x_j}\, d\langle X_i, X_j \rangle_s.$$

Proof See Theorem 3.3 in [135, Chap. IV]. □

---

[4] See Theorem 1.8 and Definition 1.20 in [135, Chap. IV].
[5] See [154, Sect. 6.4] or Exercise 1.27 in [135, Chap. IV].

As a corollary ($\rightarrow$ Exercise 4.3), for $f(x, y) = xy$, we obtain the integration by parts formula for two real Itô processes X and Y:

$$X(t)Y(t) = X(0)Y(0) + \int_0^t X(s)\,dY(s) + \int_0^t Y(s)\,dX(s) + \langle X, Y \rangle_t. \qquad (4.2)$$

4.1.4 Stochastic Differential Equations

Let $b : \Omega \times \mathbb{R}_+ \times \mathbb{R}^n \rightarrow \mathbb{R}^n$ and $\sigma : \Omega \times \mathbb{R}_+ \times \mathbb{R}^n \rightarrow \mathbb{R}^{n \times d}$ be $\text{Prog} \otimes \mathcal{B}(\mathbb{R}^n)$-measurable functions. Let ξ be some $\mathcal{F}_0$-measurable initial value. A process X is said to be a solution[6] of the stochastic differential equation

$$dX(t) = b(t, X(t))\,dt + \sigma(t, X(t))\,dW(t),$$
$$X(0) = \xi \qquad (4.3)$$

if X is an Itô process satisfying

$$X(t) = \xi + \int_0^t b(s, X(s))\,ds + \int_0^t \sigma(s, X(s))\,dW(s).$$

We say that X is unique if any other solution X' of (4.3) is indistinguishable from X, that is, $X(t) = X'(t)$ for all $t \geq 0$ a.s.

If $b(\omega, t, x) = b(t, x)$ and $\sigma(\omega, t, x) = \sigma(t, x)$ are deterministic functions, a solution X of (4.3) is also called a (time-inhomogeneous) diffusion with diffusion matrix $a(t, x) = \sigma(t, x)\sigma(t, x)^\top$ and drift $b(t, x)$.

Here is a basic existence and uniqueness theorem for diffusions.

Theorem 4.4 *Suppose $b(t, x)$ and $\sigma(t, x)$ satisfy the Lipschitz and linear growth conditions*

$$\|b(t, x) - b(t, y)\| + \|\sigma(t, x) - \sigma(t, y)\| \leq K\|x - y\|,$$

$$\|b(t, x)\|^2 + \|\sigma(t, x)\|^2 \leq K^2(1 + \|x\|^2),$$

for all $t \geq 0$ and $x, y \in \mathbb{R}^n$, where K is some finite constant. Then, for every time–space initial point $(t_0, x_0) \in \mathbb{R}_+ \times \mathbb{R}^n$, there exists a unique solution $X = X^{(t_0, x_0)}$ of the stochastic differential equation

$$dX(t) = b(t_0 + t, X(t))\,dt + \sigma(t_0 + t, X(t))\,dW(t),$$
$$X(0) = x_0. \qquad (4.4)$$

---

[6]By a solution we mean in this book what is also called a strong solution in other texts, such as in Karatzas and Shreve [106, Chap. 5].

Proof See [106, Theorem 5.2.9]. □

We note that existence and uniqueness for (4.3) can be established, in special cases, without the Lipschitz condition on $\sigma(t, x)$. See Lemma 10.6 below.

The next theorem recalls that diffusion processes have the Markov property.

Theorem 4.5 *Suppose* $b(t, x)$ *and* $a(t, x) = \sigma(t, x)\sigma(t, x)^\top$ *are continuous in* (t, x), *and assume that for every time–space initial point* $(t_0, x_0) \in \mathbb{R}_+ \times \mathbb{R}^n$, *there exists a unique solution* $X = X^{(t_0, x_0)}$ *of the stochastic differential equation* (4.4). *Then* X *has the Markov property. That is, for every bounded measurable function* f *on* $\mathbb{R}^n$, *there exists a measurable function* F *on* $\mathbb{R}_+ \times \mathbb{R}_+ \times \mathbb{R}^n$ *such that*

$$\mathbb{E}[f(X(T)) \mid \mathcal{F}_t] = F(t, T, X(t)), \quad t \leq T.$$

In words, the $\mathcal{F}_t$-conditional distribution[7] of $X(T)$ is a function of t, T and $X(t)$ only.

Proof Follows from [106, Theorem 4.20]. □

Remark 4.2 The reason why we impose the continuity assumption on the diffusion matrix $a(t, x)$ rather than on $\sigma(t, x)$ is that $a(t, x)$ actually determines the law of X, while there is some ambiguity with $\sigma(t, x)$. Indeed, for any orthogonal $d \times d$-matrix-valued function D, the function σD yields the same diffusion matrix, $\sigma DD^\top \sigma^\top = \sigma\sigma^\top$, as σ. This insight will be used in the existence and uniqueness discussion of affine diffusions in Sect. 10.5 below.

But for most practical purposes, and for simplicity, the reader may actually assume that $\sigma(t, x)$ itself is continuous in (t, x).

Note that the $\mathbb{R}^n$-valued time-inhomogeneous diffusion X in (4.4) can be regarded as $\mathbb{R}_+ \times \mathbb{R}^n$-valued homogeneous diffusion $(X'_0, \ldots, X'_n)(t) = (t_0 + t, X(t))$. That is, $X'_i = X_i, i = 1, \ldots, n$, and we identify the first component X'_0 with calendar time. Calendar time at inception ($t = 0$) is then $X'_0(0) = t_0$. Accordingly, t measures relative time with respect to t_0. See also Remark 9.1 below.

4.1.5 Stochastic Exponential

We define the stochastic exponential $\mathcal{E}(X)$ of an Itô process X by

$$\mathcal{E}_t(X) = e^{X(t) - \frac{1}{2}\langle X, X\rangle_t}.$$

---

[7]Recall that for every $\mathbb{R}^n$-valued random variable Z and sub-σ-algebra $\mathcal{G} \subset \mathcal{F}$, there exists a regular conditional distribution $\mu(\omega, dz)$ of Z given $\mathcal{G}$. That is, $\mu(\omega, \cdot)$ is a probability measure on $\mathbb{R}^n$ for every $\omega \in \Omega$, $\omega \mapsto \mu(\omega, E)$ is $\mathcal{G}$-measurable for every $E \in \mathcal{B}(\mathbb{R}^n)$, and $\mathbb{E}[f(Z) \mid \mathcal{G}](\omega) = \int_{\mathbb{R}^n} f(z)\mu(\omega, dz)$ for all bounded measurable functions f, for a.e. ω. See e.g. [7, Sect. 44].

The proof of the following fundamental properties follows by elementary stochastic calculus.

Lemma 4.2 *Let X and Y be Itô processes.*

(a) $U = \mathcal{E}(X)$ *is a positive Itô process and the unique solution of the stochastic differential equation*

$$dU = U \, dX, \quad U(0) = e^{X(0)}. \tag{4.5}$$

(b) $\mathcal{E}(X)$ *is a continuous local martingale if X is a local martingale.*
(c) $\mathcal{E}(0) = 1$.
(d) $\mathcal{E}(X)\mathcal{E}(Y) = \mathcal{E}(X + Y) e^{\langle X,Y \rangle}$.
(e) $\mathcal{E}(X)^{-1} = \mathcal{E}(-X) e^{\langle X,X \rangle}$.

Proof → Exercise 4.4. □

4.2 Financial Market

We consider a financial market $S = (S_0, \ldots, S_n)^\top$ with a risk-free asset, or money-market account from Sect. 2.3, given by

$$dS_0 = S_0 \, r \, dt, \quad S_0(0) = 1,$$

and n risky assets, whose price processes satisfy the stochastic differential equations

$$dS_i = S_i \left(\mu_i \, dt + \sigma_i \, dW \right), \quad S_i(0) > 0, \quad i = 1, \ldots, n.$$

The short rates r, the appreciation rates μ_i and volatility row vectors $\sigma_i = (\sigma_{i1}, \ldots, \sigma_{id})$ are assumed to form progressive processes such that

$$X_0(t) = \int_0^t r(s) \, ds \quad \text{and} \quad X_i(t) = \int_0^t \mu_i(s) \, ds + \int_0^t \sigma_i(s) \, dW(s)$$

are well-defined Itô processes, for all $i = 1, \ldots, n$. It then follows from Lemma 4.2 that

$$S_i(t) = S_i(0)\mathcal{E}_t(X_i)$$

are positive Itô processes, for all i.

4.2.1 Self-Financing Portfolios

A *portfolio*, or trading *strategy*, is any $\mathbb{R}^{n+1}$-valued progressive process

$$\phi = (\phi_0, \ldots, \phi_n).$$

Its corresponding *value process* is

$$V = \phi\, S = \sum_{i=0}^{n} \phi_i\, S_i.$$

The portfolio ϕ is called *self-financing* for S if $\phi \in \mathcal{L}(S)$ and there is no in- or outflow of capital during the trading in the $n+1$ financial instruments $S_0, \ldots, S_n$. Formally, this means that trading gains or losses over any period of time are solely due to value changes of the underlying instruments:

$$dV = \phi\, dS = \sum_{i=0}^{n} \phi_i\, dS_i.$$

4.2.2 Numeraires

All prices are interpreted as being given in terms of a *numeraire*, which typically is a local currency such as US dollars. But we may and will express from time to time the prices in terms of other numeraires, such as S_p for some $p \le n$. Often, but not always, we choose S_0 as the numeraire. We write calligraphic letters

$$\mathcal{S} = \frac{S}{S_p} \quad \text{and} \quad \mathcal{V} = \frac{V}{S_p} = \sum_{i=0}^{n} \phi_i\, \mathcal{S}_i$$

for the discounted price vector and value process, respectively. It turns out that, up to integrability, the self-financing property does not depend on the choice of the numeraire:

Lemma 4.3 *Let $\phi \in \mathcal{L}(S) \cap \mathcal{L}(\mathcal{S})$. Then ϕ is self-financing for S if and only if it is self-financing for $\mathcal{S}$, in particular*

$$dV = \phi\, d\mathcal{S} = \sum_{\substack{i=0 \\ i \neq p}}^{n} \phi_i\, d\mathcal{S}_i. \tag{4.6}$$

Proof $\rightarrow$ Exercise 4.5. $\square$

Since $d\mathcal{S}_p \equiv 0$, the number of summands in (4.6) reduces to n. This fact allows us to construct self-financing strategies as follows. Let $V(0)$ denote some given initial wealth, and let $(\phi_0, \ldots, \phi_{p-1}, \phi_{p+1}, \ldots, \phi_n)$ be any $\mathbb{R}^n$-valued progressive process in $\mathcal{L}(\mathcal{S}_0, \ldots, \mathcal{S}_{p-1}, \mathcal{S}_{p+1}, \ldots, \mathcal{S}_n)$. We now construct ϕ_p such that the resulting $\mathbb{R}^{n+1}$-valued process $\phi = (\phi_0, \ldots, \phi_n)$ is self-financing. From Lemma 4.3

we already know that the discounted value process is given by

$$\mathcal{V}(t) = V(0) + \sum_{\substack{i=0 \\ i \neq p}}^{n} \int_0^t \phi_i(s) \, d\mathcal{S}_i(s).$$

It thus remains to define

$$\phi_p(t) = \mathcal{V}(t) - \sum_{\substack{i=0 \\ i \neq p}}^{n} \phi_i(t)\mathcal{S}_i(t).$$

Since $\mathcal{S}_p \equiv 1$, we conclude $\phi = (\phi_0, \ldots, \phi_n) \in \mathcal{L}(\mathcal{S})$ and thus ϕ is self-financing for S. It remains to be checked from case to case whether also $\phi \in \mathcal{L}(S)$.

4.3 Arbitrage and Martingale Measures

An *arbitrage portfolio* is a self-financing portfolio ϕ with value process satisfying

$$V(0) = 0 \quad \text{and} \quad V(T) \geq 0 \quad \text{and} \quad \mathbb{P}[V(T) > 0] > 0$$

for some $T > 0$. If no arbitrage portfolios exist for any $T > 0$ we say the model is *arbitrage-free*.

An example of arbitrage is the following.

Lemma 4.4 *Suppose there exists a self-financing portfolio with value process*

$$dU = U k \, dt,$$

for some progressive process k. If the market is arbitrage-free then necessarily

$$r = k, \qquad d\mathbb{P} \otimes dt\text{-a.s.}$$

Proof Indeed, after discounting with S_0 we obtain

$$\mathcal{U}(t) = \frac{U(t)}{S_0(t)} = U(0)e^{\int_0^t (k(s) - r(s)) \, ds}.$$

Then

$$\psi(t) = 1_{\{k(t) > r(t)\}}$$

yields a self-financing strategy with discounted value process

$$\mathcal{V}(t) = \int_0^t \psi(s) \, d\mathcal{U}(s) = \int_0^t \left(1_{\{k(s) > r(s)\}}(k(s) - r(s))\mathcal{U}(s) \right) ds \geq 0.$$

Hence absence of arbitrage requires

$$0 = \mathbb{E}[\mathcal{V}(T)] = \int_{\mathcal{N}} \underbrace{\left(1_{\{k(\omega,t)>r(\omega,t)\}}(k(\omega,t) - r(\omega,t))\mathcal{U}(\omega,t)\right)}_{>0 \text{ on } \mathcal{N}} d\mathbb{P} \otimes dt,$$

where

$$\mathcal{N} = \{(\omega,t) \mid k(\omega,t) > r(\omega,t)\}$$

is a measurable subset of $\Omega \times [0,T]$. But this can only hold if $\mathcal{N}$ is a $d\mathbb{P} \otimes dt$-nullset. Using the same arguments with changed signs proves the lemma ($\rightarrow$ Exercise 4.6). $\qquad\square$

4.3.1 Martingale Measures

We now investigate when a given model is arbitrage-free. To simplify notation in the sequel we fix S_0 as a numeraire. But it is important to note that the following can be made valid for any choice of numeraire.

Definition 4.1 An *equivalent (local) martingale measure* $(E(L)MM)$ $\mathbb{Q} \sim \mathbb{P}$ has the property that the discounted price processes S_i are $\mathbb{Q}$-(local) martingales for all i.

We need to understand how the Brownian motion W transforms under an equivalent change of measure. This is the content of the following result by Girsanov:

Theorem 4.6 (Girsanov's Change of Measure Theorem) *Let $\gamma \in \mathcal{L}$ be such that the stochastic exponential*

$$\mathcal{E}(\gamma \bullet W) \text{ is a uniformly integrable martingale with } \mathcal{E}_\infty(\gamma \bullet W) > 0. \qquad (4.7)$$

Then

$$\frac{d\mathbb{Q}}{d\mathbb{P}} = \mathcal{E}_\infty(\gamma \bullet W) \qquad (4.8)$$

defines an equivalent probability measure $\mathbb{Q} \sim \mathbb{P}$, and the process

$$W^*(t) = W(t) - \int_0^t \gamma(s)^\top ds \qquad (4.9)$$

is a $\mathbb{Q}$-Brownian motion.

Proof See Theorem 1.12 in [135, Chap. VIII]. $\qquad\square$

Note that the $\mathcal{F}_t$-conditional counterpart of (4.8) reads as

$$\left.\frac{d\mathbb{Q}}{d\mathbb{P}}\right|_{\mathcal{F}_t} = \mathcal{E}_t(\gamma \bullet W) \quad \text{for all } t \geq 0.$$

Sufficient, but not necessary, for (4.7) to hold is the following useful, since explicit, condition by Novikov:

Theorem 4.7 (Novikov's Condition) *If*

$$\mathbb{E}\left[e^{\frac{1}{2}\int_0^\infty \|\gamma(s)\|^2\,ds}\right] < \infty \tag{4.10}$$

then (4.7) holds.

Proof See Proposition 1.15 in [135, Chap. VIII] for uniform integrability of $\mathcal{E}(\gamma \bullet W)$, and Proposition 1.26 in [135, Chap. IV] for finiteness of $(\gamma \bullet W)_\infty$ which is equivalent to $\mathcal{E}_\infty(\gamma \bullet W) > 0$. □

We remark that Novikov's condition is only sufficient but not necessary for (4.7) to hold. It can be too strong for some applications ($\to$ Exercise 10.3).

4.3.2 Market Price of Risk

Let $\mathbb{Q}$ be an ELMM of the form (4.8) and the Girsanov transformed Brownian motion W^* given by (4.9). Integration by parts yields the S-dynamics

$$\begin{aligned} dS_i &= S_i(\mu_i - r)dt + S_i\,\sigma_i\,dW \\ &= S_i\left(\mu_i - r + \sigma_i\gamma^\top\right)dt + S_i\sigma_i\,dW^*, \quad i = 1, \ldots, n. \end{aligned}$$

Since S is a $\mathbb{Q}$-local martingale, Lemma 4.1 implies that its drift term has to vanish. Hence γ satisfies, $d\mathbb{Q} \otimes dt$-a.s.,

$$-\sigma_i\gamma^\top = \mu_i - r \quad \text{for all } i = 1, \ldots, n. \tag{4.11}$$

The economic interpretation of this equation is as follows. On the right-hand side we have the excess of return over the risk free rate r for asset i. On the left-hand side we have a linear combination of the volatilities σ_{ij} of asset i with respect to the individual risk factors W_j with factor loadings $-\gamma_j$. This is why $-\gamma$ is called the *market price of risk* vector. The main point to notice is that $-\gamma$ is the same for all risky assets $i = 1, \ldots, n$.

Conversely, it is clear that if (4.11) has a solution $\gamma \in \mathcal{L}$ such that (4.7) holds (Novikov's condition (4.10) is sufficient) then (4.8) defines an ELMM $\mathbb{Q}$.

Finally note that, in view of Lemma 4.2, S_i can be written as the stochastic exponential

$$S_i = S_i(0)\mathcal{E}(\sigma_i \bullet W^*).$$

Hence if σ_i satisfies the Novikov condition (4.10) for all $i = 1, \ldots, n$ then the ELMM $\mathbb{Q}$ is in fact an EMM.

4.3.3 Admissible Strategies

In the presence of local martingales one has to be alert to pitfalls. For example it is possible to construct a local martingale M with $M(0) = 0$ and $M(1) = 1$. Even worse, M can be chosen to be of the form[8]

$$M(t) = \int_0^t \phi(s) \, dW(s)$$

for some $\phi \in \mathcal{L}$, which looks like the discounted value process of a self-financing strategy in the particular market model[9] with $S = W$. This would certainly be a money-making machine, that is, arbitrage. However, it turns out that M is unbounded from below. In reality, no lender would provide us with an infinite credit line. It would therefore be reasonable to require that discounted value processes be bounded from below. The following fundamental theorem of asset pricing would apply under this assumption, see e.g. Delbaen and Schachermayer [53]. However, to avoid too many mathematical subtleties, we use an alternative admissibility concept instead:

Definition 4.2 A self-financing strategy ϕ is *admissible* if its discounted value process V is a $\mathbb{Q}$-martingale for some ELMM $\mathbb{Q}$.

Be aware that admissibility is sensitive with respect to the choice of numeraire.[10] On the other hand, we have the following useful local martingale property result, which generalizes Theorem 4.1(a):

Lemma 4.5 *The discounted value process V of an admissible strategy is a $\mathbb{Q}$-local martingale under every ELMM $\mathbb{Q}$.*

Proof By assumption, $dV = \phi \, dS$ is the stochastic integral with respect to the continuous $\mathbb{Q}$-local martingale S. The statement now follows from Proposition 2.7 in [135, Chap. IV] and Proposition 1.5 in [135, Chap. VIII].   □

4.3.4 The First Fundamental Theorem of Asset Pricing

The existence of an ELMM rules out arbitrage. This is one direction of what is known as the (first) fundamental theorem of asset pricing.

---

[8] See Dudley's Representation Theorem 12.1 in [154].

[9] For the sake of illustration we omit here the positivity of S. In fact, this is the Bachelier model [6]. See also [127, Sect. 3.3].

[10] See Delbaen and Schachermayer [54] for a more thorough discussion on this.

Lemma 4.6 *Suppose there exists an ELMM* $\mathbb{Q}$*. Then the model is arbitrage-free, in the sense that there exists no admissible arbitrage strategy.*

Proof Indeed, let $\mathcal{V}$ be the discounted value process of an admissible strategy, with $\mathcal{V}(0) = 0$ and $\mathcal{V}(T) \geq 0$. Since $\mathcal{V}$ is a $\mathbb{Q}$-martingale for some ELMM $\mathbb{Q}$, we have

$$0 \leq \mathbb{E}_{\mathbb{Q}}[\mathcal{V}(T)] = \mathcal{V}(0) = 0,$$

whence $\mathcal{V}(T) = 0$. $\qquad\qquad\qquad\qquad\qquad\qquad\qquad\qquad\qquad\qquad\qquad\qquad\square$

As for the converse statement, it turns out that the absence of arbitrage among admissible strategies is not sufficient for the existence of an ELMM. A series of attempts to extend the fundamental theorem of asset pricing beyond the discrete-time case cumulated in the seminal article by Delbaen and Schachermayer [53], which states that "no free lunch with vanishing risk" (some form of asymptotic arbitrage) is equivalent to the existence of an ELMM for our financial market model. The technical details of this theorem are far from trivial and beyond the scope of this book. We content ourselves with the insight that the elementary Lemma 4.6 above gives sufficient conditions for the absence of arbitrage, and this is exactly what one needs for applications. In the sequel, we will thus follow what has become a custom in financial engineering, namely to consider the existence of an ELMM as "essentially equivalent" to the absence of arbitrage.

4.4 Hedging and Pricing

Related to any option, such as a cap, floor or swaption, is a payoff X at some future date T. We thus call a *contingent claim* due at T, or T-*claim* for short, any $\mathcal{F}_T$-measurable random variable X. The two main problems now are:

- How can one hedge against the financial risk involved in trading contingent claims?
- What is a fair price for a contingent claim X?

A contingent claim X due at T is *attainable* if there exists an admissible strategy ϕ which replicates, or hedges, X. That is, its value process V satisfies

$$V(T) = X.$$

Here is a simple example. Suppose S_1 is the price process of the T-bond. Then the contingent claim $X \equiv 1$ due at T is attainable by an obvious buy-and-hold strategy, $\phi_0 \equiv 0$, $\phi_1 = 1_{[0,T]}$, with value process $V = S_1$.

4.4.1 Complete Markets

We now determine which claims are attainable. This is most conveniently carried out in terms of discounted prices.

We first provide another pillar from stochastic analysis, the martingale representation theorem. We know that the stochastic integral with respect to W is a local martingale. The converse holds true if the filtration is not too large:

Theorem 4.8 (Representation Theorem) *Assume that the filtration*

$$(\mathcal{F}_t) \text{ is generated by the Brownian motion } W. \qquad (4.12)$$

Then every $\mathbb{P}$-local martingale M has a continuous modification and there exists $\psi \in \mathcal{L}$ such that

$$M(t) = M(0) + \int_0^t \psi(s)\, dW(s).$$

Consequently, every equivalent probability measure $\mathbb{Q} \sim \mathbb{P}$ can be represented in the form (4.8) *for some $\gamma \in \mathcal{L}$.*

Proof See Theorem 3.5 in [135, Chap. V]. The last statement follows since $M(t) = \mathbb{E}[\frac{d\mathbb{Q}}{d\mathbb{P}} \mid \mathcal{F}_t]$ is a positive martingale. $\qquad\qquad\qquad\qquad\qquad\qquad\square$

Next, we give a formal definition of market completeness.

Definition 4.3 The market model is complete if, on any finite time horizon $T > 0$, every T-claim X with bounded discounted payoff $X/S_0(T)$ is attainable.

Note that completeness does not require the absence of arbitrage ($\rightarrow$ Exercise 4.8), nor does absence of arbitrage imply completeness. However, if an ELMM exists and the martingale representation property from Theorem 4.8 applies, then completeness is equivalent to uniqueness of the ELMM $\mathbb{Q}$. This is also called the second fundamental theorem of asset pricing.

Theorem 4.9 (Second Fundamental Theorem of Asset Pricing) *Assume* (4.12) *holds and there exists an ELMM $\mathbb{Q}$. Then the following are equivalent:*

(a) *The model is complete.*
(b) *The ELMM $\mathbb{Q}$ is unique.*
(c) *The $n \times d$-volatility matrix $\sigma = (\sigma_{ij})$ is $d\mathbb{P} \otimes dt$-a.s. injective.*
(d) *The market price of risk $-\gamma$ is $d\mathbb{P} \otimes dt$-a.s. unique.*

Under any of these conditions, every T-claim X with

$$\mathbb{E}_{\mathbb{Q}} \left[\frac{|X|}{S_0(T)} \right] < \infty \qquad (4.13)$$

is attainable.

Proof (a) $\Rightarrow$ (b): Let $A \in \mathcal{F}_T$. By definition there exists an admissible strategy ϕ with discounted value process $\mathcal{V}$ satisfying $\mathcal{V}(t) = \mathbb{E}_{\mathbb{Q}}[1_A \mid \mathcal{F}_t]$ for some ELMM $\mathbb{Q}$.

This implies that $|\mathcal{V}| \leq 1$. In view of Lemma 4.5, $\mathcal{V}$ is thus a martingale under any ELMM. Now let $\mathbb{Q}'$ be any ELMM. Then $\mathbb{Q}'[A] = \mathcal{V}(0) = \mathbb{Q}[A]$, and hence $\mathbb{Q} = \mathbb{Q}'$.

(b) $\Rightarrow$ (c): See Proposition 8.2.1 in [127].

(c) $\Rightarrow$ (d) $\Rightarrow$ (b): This follows from the linear market price of risk equation (4.11) and the last statement of Theorem 4.8.

(c) $\Rightarrow$ (a): Let X be a claim due at some $T > 0$ satisfying (4.13) for some ELMM $\mathbb{Q}$ (this holds in particular if $X/S_0(T)$ is bounded). We define the $\mathbb{Q}$-martingale

$$Y(t) = \mathbb{E}_{\mathbb{Q}}\left[\frac{X}{S_0(T)}\,\middle|\,\mathcal{F}_t\right], \quad t \leq T.$$

By Bayes' rule ($\to$ Exercise 4.9) we obtain

$$Y(t)D(t) = D(t)\mathbb{E}_{\mathbb{Q}}[Y(T) \mid \mathcal{F}_t] = \mathbb{E}[Y(T)D(T) \mid \mathcal{F}_t],$$

with the density process $D(t) = d\mathbb{Q}/d\mathbb{P}|_{\mathcal{F}_t} = \mathcal{E}_t(\gamma \bullet W)$. Hence YD is a $\mathbb{P}$-martingale and by the representation theorem 4.8 there exists some $\psi \in \mathcal{L}$ such that

$$Y(t)D(t) = Y(0) + \int_0^t \psi(s)\,dW(s).$$

Applying Itô's formula yields

$$d\left(\frac{1}{D}\right) = -\frac{1}{D}\gamma\,dW + \frac{1}{D}\|\gamma\|^2\,dt,$$

and

$$\begin{aligned}
dY &= d\left((YD)\frac{1}{D}\right) = YD\,d\left(\frac{1}{D}\right) + \frac{1}{D}d(YD) + d\left\langle YD, \frac{1}{D}\right\rangle \\
&= \left(\frac{1}{D}\psi - Y\gamma\right)dW - \left(\frac{1}{D}\psi - Y\gamma\right)\gamma^{\mathsf{T}}dt \\
&= \left(\frac{1}{D}\psi - Y\gamma\right)dW^*,
\end{aligned}$$

where $dW^* = dW - \gamma\,dt$ denotes the Girsanov transformed $\mathbb{Q}$-Brownian motion. Note that we just have shown that the martingale representation property also holds for W^* under $\mathbb{Q}$.

Since σ is injective, there exists some $d \times n$-matrix-valued progressive process σ^{-1} such that $\sigma^{-1}\sigma$ equals the $d \times d$-identity matrix. If we define $\phi = (\phi_1, \ldots, \phi_n)$ via

$$\phi_i = \frac{\left(\left(\frac{1}{D}\psi - Y\gamma\right)\sigma^{-1}\right)_i}{S_i},$$

it follows that

$$dY = \left(\frac{1}{D}\psi - Y\gamma\right)\sigma^{-1}\sigma\,dW^* = \sum_{i=1}^{n}\phi_i\mathcal{S}_i\sigma_i\,dW^* = \sum_{i=1}^{n}\phi_i\,d\mathcal{S}_i. \qquad (4.14)$$

Hence ϕ yields an admissible strategy with discounted value process satisfying

$$\mathcal{V}(t) = Y(t) = \mathbb{E}_{\mathbb{Q}}\left[\frac{X}{S_0(T)}\right] + \sum_{i=1}^{n}\int_0^t \phi_i(s)\,d\mathcal{S}_i(s), \qquad (4.15)$$

and in particular $\mathcal{V}(T) = Y(T) = X/S_0(T)$. Notice that ϕ is admissible since $\mathcal{V}$ is by construction a true $\mathbb{Q}$-martingale. This also proves the last statement of the theorem.   □

Remark 4.3 Property (c) implies that completeness requires the number of risk factors d be less than or equal the number of risky assets n. Intuitively speaking, the randomness generated by the d noise factors dW can be fully absorbed by the n discounted price increments $d\mathcal{S}_1, \ldots, d\mathcal{S}_n$, as can be seen from (4.14).

4.4.2 Arbitrage Pricing

In the above complete model the unique arbitrage price $\Pi(t)$ prevailing at $t \leq T$ of a T-claim X which satisfies (4.13) is given by (4.15). That is,

$$\Pi(t) = S_0(t)\mathcal{V}(t) = S_0(t)\mathbb{E}_{\mathbb{Q}}\left[\frac{X}{S_0(T)}\,\Big|\,\mathcal{F}_t\right]. \qquad (4.16)$$

Indeed, any other price would yield arbitrage. We illustrate this only for $t = 0$, but the following argument can be made by conditioning for any $t \leq T$.[11] Suppose the initial market price p of the T-claim X satisfies $p > \Pi(0)$. Then, at $t = 0$, we sell short the claim and receive p. We invest $p - \Pi(0)$ in the money-market account and replicate the claim with the remaining initial capital $\Pi(0)$. At maturity T, we clear our short position in the claim and are left with $p - \Pi(0) > 0$ units of $S_0(T)$, an arbitrage. By changing the sign in the above strategy we show that $p < \Pi(0)$ also leads to arbitrage. Hence $p = \Pi(0)$ is the unique price of the claim at $t = 0$ which is consistent with the absence of arbitrage.

We shall often encounter complete models in the sequel, since our bond market consists of infinitely many traded assets, which makes property (c) in Theorem 4.9 likely to be satisfied, see also Remark 4.3.

However, real markets are generically incomplete, which is mainly because asset price trajectories are not continuous in reality. The above Brownian-motion-driven

---

[11] See e.g. Proposition 2.6.1 in [127].

model, and the completeness coming along with it, is merely an approximation of the reality. Pricing and hedging in incomplete markets becomes a difficult (and interesting!) issue. There is a vast literature on this topic, which is beyond the scope of this book.

In incomplete market situations, such as for the short-rate models below, it is a custom to exogenously specify a particular ELMM $\mathbb{Q}$ (or equivalently, the market price of risk $-\gamma$) and then price a T-claim X satisfying (4.13) by $\mathbb{Q}$-expectation as in (4.16). Such prices are not unique in general, since any other ELMM could yield different prices. However, this is at least a consistent arbitrage pricing rule in the sense that the enlarged market

$$S_0, \ldots, S_n, \Pi \tag{4.17}$$

is arbitrage-free ($\rightarrow$ Exercise 4.10). Now define

$$\pi(t) = \frac{1}{S_0(t)} \frac{d\mathbb{Q}}{d\mathbb{P}}\bigg|_{\mathcal{F}_t}.$$

By Bayes' rule we then have

$$\Pi(t) = S_0(t)\mathbb{E}_{\mathbb{Q}}\left[\frac{X}{S_0(T)}\bigg|\mathcal{F}_t\right] = S_0(t)\frac{\mathbb{E}[\frac{X}{S_0(T)}\frac{d\mathbb{Q}}{d\mathbb{P}}|_{\mathcal{F}_T}\,|\,\mathcal{F}_t]}{\frac{d\mathbb{Q}}{d\mathbb{P}}|_{\mathcal{F}_t}}$$

$$= \frac{\mathbb{E}[X\pi(T)\,|\,\mathcal{F}_t]}{\pi(t)},$$

and, in particular, for the price at $t = 0$

$$\Pi(0) = \mathbb{E}[X\pi(T)].$$

This is why π is called the *state-price density* process.

For example, if $\mathbb{E}_{\mathbb{Q}}[1/S_0(T)] < \infty$, the price of a T-bond is ($\rightarrow$ Exercise 4.10)

$$P(t, T) = \mathbb{E}\left[\frac{\pi(T)}{\pi(t)}\bigg|\mathcal{F}_t\right] = \mathbb{E}_{\mathbb{Q}}\left[\frac{S_0(t)}{S_0(T)}\bigg|\mathcal{F}_t\right].$$

Also one can check ($\rightarrow$ Exercise 4.10) that if $\mathbb{Q}$ is an EMM then $S_i\pi$ are $\mathbb{P}$-martingales.

4.5 Exercises

Exercise 4.1 Show that for any $h \in \mathcal{L}$ there exists a nondecreasing sequence of stopping times $\tau_1 \le \tau_2 \le \cdots$ with $\tau_n \to \infty$ such that the stopped processes $h(t \wedge \tau_n)$ belong to $\mathcal{L}^2$.

Exercise 4.2 Use the properties stated in Theorem 4.1 to prove the following:

(a) The stochastic integral of an elementary process $h = \sum_{i=0}^{m} H^i 1_{(t_i, t_{i+1}]}$, for $\mathbb{R}^d$-valued $\mathcal{F}_{t_i}$-measurable random variables $H^i = (H_1^i, \ldots, H_d^i)$, is

$$(h \bullet W)_t = \sum_{i=0}^{m} H^i (W_{t_{i+1} \wedge t} - W_{t_i \wedge t}).$$

(b) If $h \in \mathcal{L}$ is continuous then the "Riemann sums" approximation holds

$$(h \bullet W)_t = \lim \sum_{i=0}^{m} h(t_i)(W_{t_{i+1} \wedge t} - W_{t_i \wedge t}) \quad \text{in probability},$$

where the limit is taken over any sequence of partitions $0 = t_0 < t_1 < \cdots < t_m = t$ with $\max_i |t_{i+1} - t_i| \to 0$. (Hint: prove this first for h uniformly bounded, and then localize the integral with the stopping times $\tau_n = \inf\{t \mid \|h(t)\| \geq n\}$, $n \geq 1$.)

Exercise 4.3 Prove the integration by parts formula (4.2), using Itô's formula.

Exercise 4.4 The aim of this exercise is to prove Lemma 4.2.

(a) Show that $U = \mathcal{E}(X)$ solves (4.5).
(b) Calculate $d \frac{1}{\mathcal{E}(X)}$.
(c) Let V be another solution of (4.5). Using integration by parts, show that $d \frac{V}{\mathcal{E}(X)} = 0$. Conclude that $V = \mathcal{E}(X)$, and hence uniqueness holds for (4.5).
(d) Now prove the remaining properties in Lemma 4.2.

Exercise 4.5 Prove Lemma 4.3.

Exercise 4.6 Complete the proof of Lemma 4.4 (we have only shown that $k \leq r$ $d\mathbb{P} \otimes dt$-a.s.) and find a self-financing strategy which shows that also $k \geq r$ $d\mathbb{P} \otimes dt$-a.s.

Exercise 4.7 (Black–Scholes model) Fix real constants $r, \mu, \sigma > 0$ and consider the market model with a money-market account B and $n = 1$ risky asset S

$$dB = Br\,dt, \quad B(0) = 1,$$
$$dS = S(\mu\,dt + \sigma\,dW), \quad S(0) > 0,$$

where W is a one-dimensional Brownian motion ($d = 1$). Fix a finite time horizon $T > 0$. Let $\gamma \in \mathcal{L}$ be such that $\mathcal{E}_t(\gamma \bullet W)$ is a martingale for $t \leq T$, and define $\mathbb{Q} \sim \mathbb{P}$ on $\mathcal{F}_T$ by $d\mathbb{Q}/d\mathbb{P} = \mathcal{E}_T(\gamma \bullet W)$.

(a) Find the Girsanov transformed Brownian motion W^* and the Itô decomposition of $\mathcal{S} = S/B$ with respect to W^* for $t \leq T$.

(b) Find $\gamma \in \mathcal{L}$ such that $\mathcal{S}$ is a $\mathbb{Q}$-local martingale for $t \leq T$.

(c) Show that the market price of risk, $-\gamma$, in (b) is unique and that $\mathcal{S}$ is in fact a true $\mathbb{Q}$-martingale for $t \leq T$.

Exercise 4.8 This exercise shows that completeness does not imply the absence of arbitrage. Let $\mathcal{F} = \{\Omega, \emptyset\}$ be the trivial σ-algebra, and consider the deterministic financial market model with zero interest rates, $S_0 \equiv 1$, and $n = 1$ additional asset $S_1(t) = 100 + t$. Show that this model is complete but not free of arbitrage.

Exercise 4.9 (Bayes' rule) Let $\mathbb{Q} \sim \mathbb{P}$ be an equivalent probability measure and denote its density process by $D(t) = d\mathbb{Q}/d\mathbb{P}|_{\mathcal{F}_t}$. Let X be an $\mathcal{F}_T$-measurable random variable satisfying $\mathbb{E}_{\mathbb{Q}}[|X|] < \infty$.

(a) Show that we have the Bayes' rule

$$\mathbb{E}_{\mathbb{Q}}[X \mid \mathcal{F}_t] = \frac{\mathbb{E}[XD(T) \mid \mathcal{F}_t]}{D(t)}, \quad t \leq T.$$

(b) As an application show that an adapted process M is a $\mathbb{Q}$-martingale if and only if DM is a $\mathbb{P}$-martingale.

Exercise 4.10 In the framework of Sect. 4.4.2, show that:

(a) the enlarged market (4.17) is arbitrage-free;
(b) if $\mathbb{E}_{\mathbb{Q}}[1/S_0(T)] < \infty$, the price of a T-bond is $P(t, T) = \mathbb{E}[\frac{\pi(T)}{\pi(t)} \mid \mathcal{F}_t]$;
(c) if $\mathbb{Q}$ is an EMM then $S_i \pi$ are $\mathbb{P}$-martingales.

4.6 Notes

For unexplained terminology and more background on stochastic calculus, the reader is referred to Revuz and Yor [135] and Steele [154]. A good reference on stochastic differential equations is the book of Karatzas and Shreve [106].

The martingale approach to the fundamental theorem of asset pricing was developed in Harrison and Kreps [89], and Harrison and Pliska [88]. It was then extended by, among others, Duffie and Huang [57], Delbaen [52], Schachermayer [140], and Delbaen and Schachermayer [53] for locally bounded (which includes Itô) price processes.

Chapter 5
Short-Rate Models

The earliest stochastic interest rate models were models of the short rates. This chapter gives an introduction to diffusion short-rate models in general, and provides a survey of some standard models. Particular focus is on affine term-structures.

5.1 Generalities

The stochastic setup is as in Sect. 4.1. We consider $\mathbb{P}$ as objective probability measure, and let W be a d-dimensional Brownian motion. We assume that:

(a) the short rates follow an Itô process

$$dr(t) = b(t)\,dt + \sigma(t)\,dW(t)$$

   determining the money-market account $B(t) = e^{\int_0^t r(s)\,ds}$;
(b) no arbitrage: there exists an EMM $\mathbb{Q}$ of the form $d\mathbb{Q}/d\mathbb{P} = \mathcal{E}_\infty(\gamma \bullet W)$, see (4.8), such that the discounted bond price process, $P(t,T)/B(t)$, $t \leq T$, is a $\mathbb{Q}$-martingale and $P(T,T) = 1$ for all $T > 0$.

According to Sect. 4.3, the existence of an ELMM for *all* T-bonds excludes arbitrage among every finite selection of zero-coupon bonds, say $P(t,T_1), \ldots, P(t,T_n)$. To be more general one would have to consider strategies involving a continuum of bonds. This can be done using the appropriate functional analytic methods, but is beyond the scope of this book. The interested reader is referred to [16] or [35].

It is important to note that in (b) we require $\mathbb{Q}$ to be an EMM, and not merely an ELMM, because then we have

$$P(t,T) = \mathbb{E}_{\mathbb{Q}}\left[e^{-\int_t^T r(s)\,ds} \mid \mathcal{F}_t\right] \tag{5.1}$$

(compare this with the results at the end of Sect. 4.4.2).

Let $W^*(t) = W(t) - \int_0^t \gamma(s)^\top ds$ denote the Girsanov transformed $\mathbb{Q}$-Brownian motion, as provided by Theorem 4.6. The following proposition, the proof of which is left as an exercise, is a consequence of the Representation Theorem 4.8.

Proposition 5.1 *Under the above assumptions, the process r satisfies under $\mathbb{Q}$*

$$dr(t) = \left(b(t) + \sigma(t)\gamma(t)^\top\right) dt + \sigma(t)\,dW^*(t). \tag{5.2}$$

D. Filipović, *Term-Structure Models*,
Springer Finance,
DOI 10.1007/978-3-540-68015-4_5, © Springer-Verlag Berlin Heidelberg 2009

Moreover, if the filtration $(\mathcal{F}_t)$ is generated by the Brownian motion W, for any $T > 0$ there exists a progressive $\mathbb{R}^d$-valued process $v(t, T)$, $t \leq T$, such that

$$\frac{dP(t, T)}{P(t, T)} = r(t)\,dt + v(t, T)\,dW^*(t) \tag{5.3}$$

and hence

$$\frac{P(t, T)}{B(t)} = P(0, T)\mathcal{E}_t\left(v(\cdot, T) \bullet W^*\right).$$

Proof → Exercise 5.1.   □

It follows from (5.3) that the T-bond price satisfies under the objective probability measure $\mathbb{P}$

$$\frac{dP(t, T)}{P(t, T)} = \left(r(t) - v(t, T)\,\gamma(t)^\top\right) dt + v(t, T)\,dW(t).$$

This illustrates again the role of the market price of risk, $-\gamma$, as the excess of instantaneous return over $r(t)$ in units of volatility $v(t, T)$.

In a general equilibrium framework, the market price of risk is given endogenously, as it is carried out in the seminal paper by Cox, Ingersoll and Ross [47]. Since our arguments refer only to the absence of arbitrage between primary securities (bonds) and derivatives, we are unable to identify the market price of risk. In other words, we started by specifying the $\mathbb{P}$-dynamics of the short rates, and hence the money-market account $B(t)$. However, the money-market account alone cannot be used to replicate bond payoffs: the model is incomplete. According to the second fundamental theorem of asset pricing (Theorem 4.9), this is also reflected by the non-uniqueness of the EMM or the market price of risk. A priori, $\mathbb{Q}$ can be any equivalent probability measure $\mathbb{Q} \sim \mathbb{P}$.

Summarized: a short-rate model is not fully determined without the exogenous specification of the market price of risk.

It is custom, and we follow this tradition in the next section, to postulate the $\mathbb{Q}$-dynamics of r which implies the $\mathbb{Q}$-dynamics of all bond prices by (5.1). All contingent claims can be priced by taking $\mathbb{Q}$-expectations of their discounted payoffs. The market price of risk, and hence the objective measure $\mathbb{P}$, can in turn be inferred by statistical methods from historical observations of price movements.

5.2 Diffusion Short-Rate Models

We now fix a stochastic basis $(\Omega, \mathcal{F}, (\mathcal{F}_t)_{t \geq 0}, \mathbb{Q})$, where $\mathbb{Q}$ is considered as a martingale measure. In the following, we set $d = 1$ and let W^* denote a one-dimensional $\mathbb{Q}$-Brownian motion.

Let $\mathcal{Z} \subset \mathbb{R}$ be a closed interval with non-empty interior, and b and σ continuous functions on $\mathbb{R}_+ \times \mathcal{Z}$. We assume that for any $(t_0, r_0) \in \mathcal{Z}$ the stochastic differential equation

$$dr(t) = b(t_0 + t, r(t)) \, dt + \sigma(t_0 + t, r(t)) \, dW^*(t), \quad r(0) = r_0 \qquad (5.4)$$

admits a unique $\mathcal{Z}$-valued solution $r = r^{(t_0, r_0)}$. Sufficient conditions for the existence and uniqueness are given in Theorem 4.4. We recall the Markov property of r stated in Theorem 4.5.

We define the T-bond prices $P(t, T)$ as in (5.1). It turns out that $P(t, T)$ can be written as a function of $r(t)$, t and T. This is a general property of certain functionals of a Markov process, which extends the Markov property stated in Theorem 4.5, usually referred to as the Feynman–Kac formula.

Lemma 5.1 *Let $T > 0$ and Φ be a continuous function on $\mathcal{Z}$, and assume that $F = F(t, r) \in C^{1,2}([0, T] \times \mathcal{Z})$ is a solution to the boundary value problem on $[0, T] \times \mathcal{Z}$*

$$\partial_t F(t, r) + b(t, r) \partial_r F(t, r) + \frac{1}{2} \sigma^2(t, r) \partial_r^2 F(t, r) - r F(t, r) = 0,$$
$$F(T, r) = \Phi(r). \qquad (5.5)$$

Then

$$M(t) = F(t, r(t)) e^{-\int_0^t r(u) \, du}, \quad t \leq T,$$

is a local martingale. If in addition either:

(a) $\mathbb{E}_{\mathbb{Q}}[\int_0^T |\partial_r F(t, r(t)) e^{-\int_0^t r(u) \, du} \sigma(t, r(t))|^2 dt] < \infty$, *or*
(b) M *is uniformly bounded,*

then M is a true martingale, and

$$F(t, r(t)) = \mathbb{E}_{\mathbb{Q}}\left[e^{-\int_t^T r(u) \, du} \Phi(r(T)) \mid \mathcal{F}_t \right], \quad t \leq T. \qquad (5.6)$$

Proof We can apply Itô's formula to M and obtain

$$dM(t) = \Big(\partial_t F(t, r(t)) + b(t, r(t)) \partial_r F(t, r(t))$$

$$+ \frac{1}{2} \sigma^2(t, r) \partial_r^2 F(t, r(t)) - r(t) F(t, r(t)) \Big) e^{-\int_0^t r(u) \, du} dt$$

$$+ \partial_r F(t, r(t)) e^{-\int_0^t r(u) \, du} \sigma(t, r(t)) \, dW^*(t)$$

$$= \partial_r F(t, r(t)) e^{-\int_0^t r(u) \, du} \sigma(t, r(t)) \, dW^*(t).$$

Hence M is a local martingale.

It is now clear that either Condition (a) or (b) implies that M is a true martingale. Since

$$M(T) = \Phi(r(T))e^{-\int_0^T r(u)\,du}$$

we obtain

$$F(t, r(t))e^{-\int_0^t r(u)\,du} = M(t) = \mathbb{E}_{\mathbb{Q}}\left[e^{-\int_0^T r(u)\,du}\Phi(r(T)) \mid \mathcal{F}_t\right].$$

Multiplying both sides by $e^{\int_0^t r(u)\,du}$ yields the claim. $\qquad\qquad\qquad\square$

We call (5.5) the *term-structure equation* for Φ. Its solution F gives the price of the T-claim $\Phi(r(T))$. In particular, for $\Phi \equiv 1$ we get the T-bond price $P(t, T)$ as a function of t, $r(t)$ and T

$$P(t, T) = F(t, r(t); T).$$

Remark 5.1 Strictly speaking, we have only shown that *if* a smooth solution F of (5.5) exists and satisfies some additional properties (Condition (a) or (b)) then the time t price of the claim $\Phi(r(T))$ (which is the right-hand side of (5.6)) equals $F(t, r(t))$. Conversely, one can also show that the conditional expectation on the right-hand side of (5.6) *given $r(t) = r$* can be written as $F(t, r)$ where F solves the term-structure equation (5.5) but usually only in a *weak sense*, which in particular means that F may not be in $C^{1,2}([0, T] \times \mathcal{Z})$. This is general Markov semigroup theory, beyond Theorem 4.5, and we will not prove this here.

In any case, we have found a pricing algorithm. But is it computationally efficient? Solving partial differential equations in less than three space dimensions is numerically feasible, and the dimension of $\mathcal{Z}$ is one. However, the nuisance is that we have to solve a partial differential equation for every single zero-coupon bond price function $F(\cdot, \cdot; T)$, $T > 0$. From that we might want to derive the yield or even forward curve. If we do not impose further structural assumptions we may run into regularity problems. Hence short-rate models that admit closed-form solutions to the term-structure equation (5.5), at least for $\Phi \equiv 1$, are favorable.

5.2.1 Examples

This is a, far from complete, list of some of the most popular short-rate models, which will be further discussed in Sect. 5.4 below. If not otherwise stated, the parameters are real-valued.

(a) Vasiček [160]: $\mathcal{Z} = \mathbb{R}$,

$$dr(t) = (b + \beta r(t))\,dt + \sigma\,dW^*(t),$$

(b) Cox–Ingersoll–Ross (henceforth CIR) [47]: $\mathcal{Z} = \mathbb{R}_+$, $b \geq 0$,

$$dr(t) = (b + \beta r(t)) \, dt + \sigma \sqrt{r(t)} \, dW^*(t),$$

(c) Dothan [56]: $\mathcal{Z} = \mathbb{R}_+$,

$$dr(t) = \beta r(t) \, dt + \sigma r(t) \, dW^*(t),$$

(d) Black–Derman–Toy [19]: $\mathcal{Z} = \mathbb{R}_+$,

$$dr(t) = \beta(t) r(t) \, dt + \sigma(t) r(t) \, dW^*(t),$$

(e) Black–Karasinski [17]: $\mathcal{Z} = \mathbb{R}_+$, $\ell(t) = \log r(t)$,

$$d\ell(t) = (b(t) + \beta(t)\ell(t)) \, dt + \sigma(t) \, dW^*(t),$$

(f) Ho–Lee [92]: $\mathcal{Z} = \mathbb{R}$,

$$dr(t) = b(t) \, dt + \sigma \, dW^*(t),$$

(g) Hull–White extended Vasiček [96]: $\mathcal{Z} = \mathbb{R}$,

$$dr(t) = (b(t) + \beta(t) r(t)) \, dt + \sigma(t) \, dW^*(t),$$

(h) Hull–White extended CIR [96]: $\mathcal{Z} = \mathbb{R}_+$, $b(t) \geq 0$,

$$dr(t) = (b(t) + \beta(t) r(t)) \, dt + \sigma(t) \sqrt{r(t)} \, dW^*(t).$$

5.2.2 Inverting the Forward Curve

Once the short-rate model is chosen, the initial term-structure

$$T \mapsto P(0, T) = F(0, r(0); T)$$

and hence the initial forward curve is fully specified by the term structure equation (5.5).

Conversely, one may want to invert the term-structure equation (5.5) to match a given initial forward curve. Say we have chosen the Vasiček model. Then the implied T-bond price is a function of the current short-rate level $r(0)$ and the three model parameters b, β and σ:

$$P(0, T) = F(0, r(0); T, b, \beta, \sigma).$$

But $F(0, r(0); T, b, \beta, \sigma)$ is just a parameterized curve family with three degrees of freedom. It turns out that it is often too restrictive and will provide a poor fit of the current data in terms of accuracy, e.g. a least-squares criterion.

Therefore the class of time-inhomogeneous short-rate models, such as the Hull–White extensions, was introduced. By letting the parameters depend on time one gains infinite degree of freedom and hence a perfect fit of any given curve. Usually, the functions $b(t)$ etc. are fully determined by the initial term-structure. This will now be made more explicit in the following sections.

5.3 Affine Term-Structures

As we have argued after Remark 5.1, short-rate models that admit closed-form expressions for the implied bond prices $F(t, r; T)$ are favorable.

Among the most tractable models are those where bond prices are of the form

$$F(t, r; T) = \exp(-A(t, T) - B(t, T)r),$$

for some smooth functions A and B. Such models are said to provide an *affine term-structure (ATS)*. Notice that $F(T, r; T) = 1$ implies $A(T, T) = B(T, T) = 0$.

One nice thing about short-rate ATS models is that they can be completely characterized.

Proposition 5.2 *The short-rate model (5.4) provides an ATS if and only if its diffusion and drift terms are of the form*

$$\sigma^2(t, r) = a(t) + \alpha(t)r \quad \text{and} \quad b(t, r) = b(t) + \beta(t)r, \tag{5.7}$$

for some continuous functions a, α, b, β, and the functions A and B satisfy the system of ordinary differential equations, for all $t \le T$,

$$\partial_t A(t, T) = \frac{1}{2}a(t)B^2(t, T) - b(t)B(t, T), \quad A(T, T) = 0, \tag{5.8}$$

$$\partial_t B(t, T) = \frac{1}{2}\alpha(t)B^2(t, T) - \beta(t)B(t, T) - 1, \quad B(T, T) = 0. \tag{5.9}$$

Proof We insert $F(t, r; T) = \exp(-A(t, T) - B(t, T)r)$ in the term-structure equation (5.5) to infer that the short-rate model (5.4) provides an ATS if and only if

$$\frac{1}{2}\sigma^2(t, r)B^2(t, T) - b(t, r)B(t, T) = \partial_t A(t, T) + (\partial_t B(t, T) + 1)r \tag{5.10}$$

holds for all $t \le T$ and $r \in \mathcal{Z}$.

It follows by inspection that the specification (5.7)–(5.9) satisfies (5.10). This proves one direction of the proposition.

We now show the necessity of (5.7)–(5.9). Fix $t \ge 0$, and suppose first that the functions $B(t, \cdot)$ and $B^2(t, \cdot)$ are linearly independent. Then we can find $T_1 > T_2 > t$ such that the matrix

$$M = \begin{pmatrix} B^2(t, T_1) & -B(t, T_1) \\ B^2(t, T_2) & -B(t, T_2) \end{pmatrix}$$

is invertible. From (5.10) we derive

$$\begin{pmatrix} \sigma^2(t,r)/2 \\ b(t,r) \end{pmatrix} = M^{-1} \left(\begin{pmatrix} \partial_t A(t,T_1) \\ \partial_t A(t,T_2) \end{pmatrix} + \begin{pmatrix} \partial_t B(t,T_1)+1 \\ \partial_t B(t,T_2)+1 \end{pmatrix} r \right).$$

Thus $\sigma^2(t,r)$ and $b(t,r)$ are affine functions of r, which shows (5.7). Plugging this in, the left-hand side of (5.10) reads

$$\frac{1}{2}a(t)B^2(t,T) - b(t)B(t,T) + \left(\frac{1}{2}\alpha(t)B^2(t,T) - \beta(t)B(t,T) \right) r.$$

Terms containing r must match. This implies (5.8)–(5.9).

It remains to consider the case where $B(t,\cdot) = c(t)B^2(t,\cdot)$ for some constant $c(t)$. But this implies $B(t,\cdot) \equiv B(t,t) = 0$, and in view of (5.10) thus $\partial_t B(t,T) = -1$. Hence the set of elements t for which the functions $B(t,\cdot)$ and $B^2(t,\cdot)$ are linearly independent is open and dense in $\mathbb{R}_+$. By continuity of $\sigma(t,r)$ and $b(t,r)$ we conclude that (5.7), and consequently (5.8)–(5.9), holds for all t. □

The functions a, α, b, β in (5.7) can be further specified. They have to be such that $a(t) + \alpha(t)r \geq 0$ and $r(t)$ does not leave the state space $\mathcal{Z}$. In fact, it can be shown that every non-degenerate (that is, $\sigma(t,r) \not\equiv 0$) short-rate ATS model can be transformed via affine transformation into one of the two cases (see also Theorem 10.2 below):

(a) $\mathcal{Z} = \mathbb{R}$: necessarily $\alpha(t) = 0$ and $a(t) \geq 0$, and b, β are arbitrary. This is the Hull–White extension of the Vasiček model.
(b) $\mathcal{Z} = \mathbb{R}_+$: necessarily $a(t) = 0$, $\alpha(t) \geq 0$ and $b(t) \geq 0$ (otherwise the process would cross zero), and β is arbitrary. This is the Hull–White extension of the CIR model.

Looking at the list in Sect. 5.2.1 we see that all short-rate models except the Dothan, Black–Derman–Toy and Black–Karasinski models have an ATS.

5.4 Some Standard Models

In the following, we discuss some of the standard short-rate models listed in Sect. 5.2.1 above in more detail.

5.4.1 Vasiček Model

The solution to

$$dr = (b + \beta r)\,dt + \sigma\,dW^*$$

is explicitly given by ($\rightarrow$ Exercise 5.3)

$$r(t) = r(0)e^{\beta t} + \frac{b}{\beta}\left(e^{\beta t} - 1\right) + \sigma e^{\beta t}\int_0^t e^{-\beta s}\,dW^*(s).$$

It follows that $r(t)$ is a Gaussian process with mean

$$\mathbb{E}_{\mathbb{Q}}\left[r(t)\right] = r(0)e^{\beta t} + \frac{b}{\beta}\left(e^{\beta t} - 1\right)$$

and variance

$$\mathrm{Var}_{\mathbb{Q}}[r(t)] = \sigma^2 e^{2\beta t}\int_0^t e^{-2\beta s}\,ds = \frac{\sigma^2}{2\beta}\left(e^{2\beta t} - 1\right). \tag{5.11}$$

Hence

$$\mathbb{Q}[r(t) < 0] > 0,$$

which is not satisfactory, although this probability is usually very small.

Vasiček assumed the market price of risk to be constant, on a finite time horizon, so that also the objective $\mathbb{P}$-dynamics of $r(t)$ is of the above form ($\rightarrow$ Exercise 5.3).

If $\beta < 0$ then $r(t)$ is mean-reverting with mean reversion level $b/|\beta|$, see Fig. 5.1, and $r(t)$ converges to a Gaussian random variable with mean $b/|\beta|$ and variance $\sigma^2/(2|\beta|)$, for $t \to \infty$.

Equations (5.8)–(5.9) become

$$\partial_t A(t, T) = \frac{\sigma^2}{2}B^2(t, T) - bB(t, T), \quad A(T, T) = 0,$$

$$\partial_t B(t, T) = -\beta B(t, T) - 1, \quad B(T, T) = 0.$$

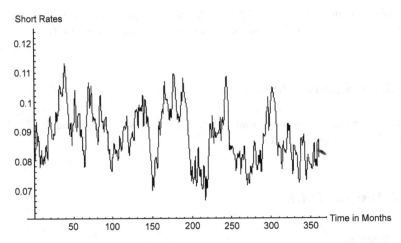

Fig. 5.1 Vasiček short-rate process for $\beta = -0.86$, $b/|\beta| = 0.09$ (mean reversion level), $\sigma = 0.0148$ and $r(0) = 0.08$

The explicit solution is

$$B(t, T) = \frac{1}{\beta} \left(e^{\beta(T-t)} - 1 \right)$$

and A is given as ordinary integral

$$A(t, T) = A(T, T) - \int_t^T \partial_s A(s, T)\, ds$$

$$= -\frac{\sigma^2}{2} \int_t^T B^2(s, T)\, ds + b \int_t^T B(s, T)\, ds$$

$$= \frac{\sigma^2 (4e^{\beta(T-t)} - e^{2\beta(T-t)} - 2\beta(T-t) - 3)}{4\beta^3} + b \frac{e^{\beta(T-t)} - 1 - \beta(T-t)}{\beta^2}.$$

We recall that zero-coupon bond prices are given in closed-form by

$$P(t, T) = \exp(-A(t, T) - B(t, T) r(t)).$$

We will see below that it is possible to derive a closed-form expression also for bond options.

5.4.2 CIR Model

It is worth mentioning that the CIR stochastic differential equation

$$dr(t) = (b + \beta r(t))\, dt + \sigma \sqrt{r(t)}\, dW^*(t), \quad r(0) \geq 0,$$

has a unique nonnegative solution, see Lemma 10.6 below. Even more, if $b \geq \sigma^2/2$ then $r > 0$ whenever $r(0) > 0$, see [110, Proposition 6.2.4] or Exercise 10.12 below.

The ATS equation (5.9) now becomes non linear:

$$\partial_t B(t, T) = \frac{\sigma^2}{2} B^2(t, T) - \beta B(t, T) - 1, \quad B(T, T) = 0.$$

This is called a *Riccati equation*. It is good news that the explicit solution is known:

$$B(t, T) = \frac{2(e^{\gamma(T-t)} - 1)}{(\gamma - \beta)(e^{\gamma(T-t)} - 1) + 2\gamma}$$

where $\gamma = \sqrt{\beta^2 + 2\sigma^2}$. Integration yields

$$A(t, T) = -\frac{2b}{\sigma^2} \log \left(\frac{2\gamma e^{(\gamma-\beta)(T-t)/2}}{(\gamma - \beta)(e^{\gamma(T-t)} - 1) + 2\gamma} \right).$$

Hence also in the CIR model we have closed-form expressions for the bond prices. Moreover, we will see below that also bond option prices are explicit. Together with the fact that it yields nonnegative interest rates, this is the main reason why the CIR model is so popular.

5.4.3 Dothan Model

Dothan [56] starts from a drift-less geometric Brownian motion under the objective probability measure $\mathbb{P}$

$$dr(t) = \sigma r(t)\,dW(t)$$

with $\mathbb{P}$-Brownian motion W. The market price of risk is chosen to be constant, on a finite time horizon, which yields

$$dr(t) = \beta r(t)\,dt + \sigma r(t)\,dW^*(t)$$

as $\mathbb{Q}$-dynamics. This is easily integrated:

$$r(t) = r(s)\exp\left(\left(\beta - \sigma^2/2\right)(t-s) + \sigma(W^*(t) - W^*(s))\right), \quad s \le t.$$

Thus the $\mathcal{F}_s$-conditional distribution of $r(t)$ is lognormal with mean and variance ($\to$ Exercise 5.5)

$$\mathbb{E}_{\mathbb{Q}}[r(t) \mid \mathcal{F}_s] = r(s)e^{\beta(t-s)},$$

$$\mathrm{Var}_{\mathbb{Q}}[r(t) \mid \mathcal{F}_s] = r^2(s)e^{2\beta(t-s)}\left(e^{\sigma^2(t-s)} - 1\right).$$

The Dothan and all lognormal short-rate models (Black–Derman–Toy and Black–Karasinski) yield positive interest rates. But no closed-form expressions for bond prices or options are available, with one exception: Dothan admits a "semi-explicit" expression for the bond prices, see [27, Sect. 3.2.2].

A major drawback of lognormal models is the explosion of the money-market account. Let Δt be small, then

$$\mathbb{E}_{\mathbb{Q}}[B(\Delta t)] = \mathbb{E}_{\mathbb{Q}}\left[e^{\int_0^{\Delta t} r(s)\,ds}\right] \approx \mathbb{E}_{\mathbb{Q}}\left[e^{\frac{r(0)+r(\Delta t)}{2}\Delta t}\right].$$

We face an expectation of the type

$$\mathbb{E}_{\mathbb{Q}}\left[e^{e^Y}\right]$$

where Y is Gaussian distributed. But such an expectation is infinite. This means that in arbitrarily small time the money-market account grows to infinity on average. Similarly, one shows that the price of a Eurodollar future[1] is infinite for all lognormal models.

---

[1] Eurodollar futures are defined in Sect. 8.2.1 below.

The idea of lognormal rates was taken up in the mid-nineties by Sandmann and Sondermann [139] and many others, which finally led to the so-called market models with lognormal LIBOR or swap rates, which will be studied in Chap. 11 below.

5.4.4 Ho–Lee Model

For the Ho–Lee model

$$dr(t) = b(t)\, dt + \sigma\, dW^*(t)$$

the ATS equations (5.8)–(5.9) become

$$\partial_t A(t, T) = \frac{\sigma^2}{2} B^2(t, T) - b(t) B(t, T), \quad A(T, T) = 0,$$

$$\partial_t B(t, T) = -1, \quad B(T, T) = 0.$$

Hence

$$B(t, T) = T - t,$$

$$A(t, T) = -\frac{\sigma^2}{6}(T - t)^3 + \int_t^T b(s)(T - s)\, ds.$$

The forward curve is thus

$$f(t, T) = \partial_T A(t, T) + \partial_T B(t, T) r(t) = -\frac{\sigma^2}{2}(T - t)^2 + \int_t^T b(s)\, ds + r(t).$$

Let $f_0(T)$ be the observed (estimated) initial forward curve. Then

$$b(s) = \partial_s f_0(s) + \sigma^2 s$$

gives a perfect fit of $f_0(T)$. Plugging this back into the ATS yields

$$f(t, T) = f_0(T) - f_0(t) + \sigma^2 t(T - t) + r(t).$$

We can also integrate this expression to get

$$P(t, T) = e^{-\int_t^T f_0(s)\, ds + f_0(t)(T-t) - \frac{\sigma^2}{2} t(T-t)^2 - (T-t)r(t)}.$$

It is interesting to see that

$$r(t) = r(0) + \int_0^t b(s)\, ds + \sigma W^*(t) = f_0(t) + \frac{\sigma^2 t^2}{2} + \sigma W^*(t).$$

That is, $r(t)$ fluctuates along the modified initial forward curve, and we have

$$f_0(t) = \mathbb{E}_{\mathbb{Q}}[r(t)] - \frac{\sigma^2 t^2}{2}.$$

5.4.5 Hull–White Model

The Hull–White extensions of Vasiček and CIR can be fitted to the initial yield and
volatility curve. However, this flexibility has its price: the model cannot be handled
analytically in general. We therefore restrict ourself to the following extension of
the Vasiček model that was analyzed by Hull and White [96]:

$$dr(t) = (b(t) + \beta r(t)) \, dt + \sigma \, dW^*(t).$$

In this model we choose the constants β and σ to obtain a nice volatility structure,
whereas $b(t)$ is chosen in order to match the initial forward curve.

Equation (5.9) for $B(t, T)$ is just as in the Vasiček model

$$\partial_t B(t, T) = -\beta B(t, T) - 1, \quad B(T, T) = 0$$

with explicit solution

$$B(t, T) = \frac{1}{\beta} \left(e^{\beta(T-t)} - 1 \right).$$

Equation (5.8) for $A(t, T)$ now reads

$$A(t, T) = -\frac{\sigma^2}{2} \int_t^T B^2(s, T) \, ds + \int_t^T b(s) B(s, T) \, ds.$$

We consider the initial forward curve (notice that $\partial_T B(s, T) = -\partial_s B(s, T)$)

$$f_0(T) = \partial_T A(0, T) + \partial_T B(0, T) r(0)$$

$$= \frac{\sigma^2}{2} \int_0^T \partial_s B^2(s, T) \, ds + \int_0^T b(s) \partial_T B(s, T) \, ds + \partial_T B(0, T) r(0)$$

$$= \underbrace{-\frac{\sigma^2}{2\beta^2} \left(e^{\beta T} - 1 \right)^2}_{=:g(T)} + \underbrace{\int_0^T b(s) e^{\beta(T-s)} \, ds + e^{\beta T} r(0)}_{=:\phi(T)}.$$

The function ϕ satisfies

$$\partial_T \phi(T) = \beta \phi(T) + b(T), \quad \phi(0) = r(0).$$

It follows that

$$b(T) = \partial_T \phi(T) - \beta \phi(T)$$

$$= \partial_T (f_0(T) + g(T)) - \beta (f_0(T) + g(T)).$$

Plugging in and performing some calculations ($\to$ Exercise 5.6) eventually yields

$$f(t, T) = f_0(T) - e^{\beta(T-t)} f_0(t) - \frac{\sigma^2}{2\beta^2} \left(e^{\beta(T-t)} - 1 \right) \left(e^{\beta(T-t)} - e^{\beta(T+t)} \right)$$

$$+ e^{\beta(T-t)} r(t). \tag{5.12}$$

5.5 Exercises

Exercise 5.1 Proceed as in the proof of (c) $\Rightarrow$ (a) in Theorem 4.9 and prove Proposition 5.1.

Exercise 5.2 We take as given a diffusion short-rate model with $\mathbb{Q}$-dynamics as specified in (5.4). Consider a T-claim $Z = \Phi(r(T)) > 0$ and suppose the assumptions of Lemma 5.1 are satisfied.

(a) Show that the price process $\Pi(t) = F(t, r(t))$ of Z has a local rate of return equal to the short rate. In other words, show that Π is of the form

$$d\Pi(t) = \Pi(t)r(t)\,dt + \Pi(t)\sigma_\Pi(t)\,dW^*(t).$$

(b) Calculate σ_Π in terms of F.

Exercise 5.3 Show that $r(t) = r(0)e^{\beta t} + \frac{b}{\beta}(e^{\beta t} - 1) + \sigma e^{\beta t}\int_0^t e^{-\beta s}\,dW^*(s)$ is the unique solution to the Vasiček short-rate equation $dr = (b + \beta r)\,dt + \sigma\,dW^*$, for some given initial short rate $r(0)$.

Exercise 5.4 The aim of this exercise is to construct a solution to a particular type of square-root stochastic differential equation. Let $W = (W_1, \ldots, W_d)$ be a d-dimensional Brownian motion. There exists (why?) a unique solution $X = (X_1, \ldots, X_d)$ of the system of stochastic differential equations

$$dX_i(t) = cX_i(t)\,dt + \rho\,dW_i(t), \quad X_i(0) = x_i, \quad i = 1, \ldots, d,$$

for some constant real coefficients c and ρ. Show that there exists a Brownian motion B such that the *nonnegative* process

$$Y = X_1^2 + \cdots + X_d^2$$

satisfies the stochastic differential equation

$$dY(t) = (b + \beta Y(t))dt + \sigma\sqrt{Y(t)}\,dB(t)$$

where $b = d\rho^2$, $\beta = 2c$ and $\sigma = 2\rho$ (hint: use Lévy's characterization theorem 4.2 to show that $dB = \sum_{i=1}^n \frac{X_i}{\sqrt{Y}}\,dW_i$ defines a Brownian motion). Note that this approach only works for b being an integer multiple of ρ^2.

Exercise 5.5 Show that in the Dothan short-rate model, for $t > s$, the $\mathcal{F}_s$-conditional distribution of $r(t)$ is lognormal with mean and variance

$$\mathbb{E}_{\mathbb{Q}}[r(t) \mid \mathcal{F}_s] = r(s)e^{\beta(t-s)},$$

$$\mathrm{Var}_{\mathbb{Q}}[r(t) \mid \mathcal{F}_s] = r^2(s)e^{2\beta(t-s)}\left(e^{\sigma^2(t-s)} - 1\right).$$

Exercise 5.6 Derive (5.12).

Exercise 5.7 *Swap a fixed rate vs. a short rate.* Consider the following version of an interest rate swap contract between two parties A and B. The payments are as follows:

- A hypothetically invests the principal amount K at time $t = 0$ and lets it grow at fixed continuously compounded rate of interest R (to be determined below) over the time interval $[0, T]$.
- At T the principal will have grown to K_A. A will then pay the surplus $K_A - K$ to B.
- B hypothetically invests the principal amount K at the stochastic short rate of interest $r(t)$ over the interval $[0, T]$.
- At T the principal will have grown to K_B. B will then pay the surplus $K_B - K$ to A.

(a) Draft a figure to illustrate the cash flows.
(b) Calculate the prices at $t = 0$ of these cash flows.
(c) The *swap rate* for this contract is defined as the value R of the fixed rate which gives this contract the value zero at $t = 0$. Compute the swap rate.

5.6 Notes

One of the earliest models of the term-structure was analyzed in a seminal paper by Vasiček [160]. This chapter follows closely the outline of Björk [13, Chaps. 21 and 22]. Further references are Brigo and Mercurio [27, Chap. 3] and Musiela and Rutkowski [127, Sect. 10.1]. Exercise 5.7 is from [13, Chap. 21].

Chapter 6
Heath–Jarrow–Morton (HJM) Methodology

As we have seen in Chap. 5, short-rate models are not always flexible enough to calibrating them to the observed initial term-structure. In the late eighties, Heath, Jarrow and Morton (henceforth HJM) [90] proposed a new framework for modeling the entire forward curve directly. This chapter provides the essentials of the HJM framework.

6.1 Forward Curve Movements

The stochastic setup is as in Sect. 4.1. We consider $\mathbb{P}$ as objective probability measure, and let W be a d-dimensional Brownian motion.

We assume that we are given an $\mathbb{R}$-valued and $\mathbb{R}^d$-valued stochastic process $\alpha = \alpha(\omega, t, T)$ and $\sigma = (\sigma_1(\omega, t, T), \ldots, \sigma_d(\omega, t, T))$, respectively, with two indices, t, T, such that

(HJM.1) α and σ are $\text{Prog} \otimes \mathcal{B}$-measurable;
(HJM.2) $\int_0^T \int_0^T |\alpha(s, t)| \, ds \, dt < \infty$ for all T;
(HJM.3) $\sup_{s, t \leq T} \|\sigma(s, t)\| < \infty$ for all T.[1]

For a given integrable initial forward curve $T \mapsto f(0, T)$ it is then assumed that, for every T, the forward rate process $f(\cdot, T)$ follows the Itô dynamics

$$f(t, T) = f(0, T) + \int_0^t \alpha(s, T) \, ds + \int_0^t \sigma(s, T) \, dW(s), \quad t \leq T. \qquad (6.1)$$

This is a very general setup. The only substantive economic restrictions are the continuous sample paths assumption for the forward rate process, and the finite number, d, of random drivers $W_1, \ldots, W_d$.

The integrals in (6.1) are well defined by **(HJM.1)**–**(HJM.3)**. Note that $\alpha(t, T)$ and $\sigma(t, T)$ enter the dynamic equation (6.1) and the sequel only for $t \leq T$; we can and will set them equal to zero for all $t > T$ without loss of generality. Moreover, it follows from Corollary 6.3 below that the short-rate process

$$r(t) = f(t, t) = f(0, t) + \int_0^t \alpha(s, t) \, ds + \int_0^t \sigma(s, t) \, dW(s)$$

---

[1] Note that this is a ω-wise boundedness assumption.

D. Filipović, *Term-Structure Models*,
Springer Finance,
DOI 10.1007/978-3-540-68015-4_6, © Springer-Verlag Berlin Heidelberg 2009

has a progressive modification—again denoted by $r(t)$—satisfying $\int_0^t |r(s)|\,ds <$ ∞ a.s. for all t. Hence the money-market account $B(t) = e^{\int_0^t r(s)\,ds}$ is well defined. More can be said about the zero-coupon bond prices $P(t,T) = e^{-\int_t^T f(t,u)\,du}$:

Lemma 6.1 *For every maturity T, the zero-coupon bond price follows an Itô process of the form*

$$P(t,T) = P(0,T) + \int_0^t P(s,T)\,(r(s)+b(s,T))\,ds + \int_0^t P(s,T)v(s,T)\,dW(s),$$
$$(6.2)$$

for $t \leq T$, where

$$v(s,T) = -\int_s^T \sigma(s,u)\,du, \qquad\qquad (6.3)$$

is the T-bond volatility and

$$b(s,T) = -\int_s^T \alpha(s,u)\,du + \frac{1}{2}\|v(s,T)\|^2.$$

Proof Using the classical Fubini Theorem and Theorem 6.2 below for stochastic integrals twice, we calculate

$$\log P(t,T)$$

$$= -\int_t^T f(t,u)\,du$$

$$= -\int_t^T f(0,u)\,du - \int_t^T \int_0^t \alpha(s,u)\,ds\,du - \int_t^T \int_0^t \sigma(s,u)\,dW(s)\,du$$

$$= -\int_t^T f(0,u)\,du - \int_0^t \int_t^T \alpha(s,u)\,du\,ds - \int_0^t \int_t^T \sigma(s,u)\,du\,dW(s)$$

$$= -\int_0^T f(0,u)\,du - \int_0^t \int_s^T \alpha(s,u)\,du\,ds - \int_0^t \int_s^T \sigma(s,u)\,du\,dW(s)$$

$$\quad + \int_0^t f(0,u)\,du + \int_0^t \int_s^t \alpha(s,u)\,du\,ds + \int_0^t \int_s^t \sigma(s,u)\,du\,dW(s)$$

$$= -\int_0^T f(0,u)\,du + \int_0^t \left(b(s,T) - \frac{1}{2}\|v(s,T)\|^2\right) ds + \int_0^t v(s,T)\,dW(s)$$

$$\quad + \underbrace{\int_0^t \left(f(0,u) + \int_0^u \alpha(s,u)\,ds + \int_0^u \sigma(s,u)\,dW(s)\right) du}_{=r(u)}$$

$$= \log P(0,T) + \int_0^t \left(r(s)+b(s,T) - \frac{1}{2}\|v(s,T)\|^2\right) ds + \int_0^t v(s,T)\,dW(s).$$

Itô's formula now implies (6.2) ($\to$ Exercise 6.2). $\square$

As a corollary, we derive the dynamic equation of the discounted bond price process as follows:

Corollary 6.1 *We have, for $t \leq T$,*

$$\frac{P(t,T)}{B(t)} = P(0,T) + \int_0^t \frac{P(s,T)}{B(s)} b(s,T) \, ds + \int_0^t \frac{P(s,T)}{B(s)} v(s,T) \, dW(s).$$

Proof Itô's formula ($\rightarrow$ Exercise 6.2). $\qquad\qquad\qquad\qquad\qquad\qquad\qquad\qquad$ □

6.2 Absence of Arbitrage

In this section we investigate the restrictions on the dynamics (6.1) under the assumption of no arbitrage. In what follows we let $\mathbb{Q} \sim \mathbb{P}$ be an equivalent probability measure of the form (4.8) for some $\gamma \in \mathcal{L}$. With $dW^* = dW - \gamma^\top dt$ we denote the Girsanov transformed $\mathbb{Q}$-Brownian motion, see Theorem 4.6. According to Definition 4.1, we call $\mathbb{Q}$ an ELMM for the bond market if the discounted bond price process $\frac{P(t,T)}{B(t)}$ is a $\mathbb{Q}$-local martingales for $t \leq T$, for all T.

Theorem 6.1 (HJM Drift Condition) $\mathbb{Q}$ *is an ELMM if and only if*

$$b(t,T) = -v(t,T)\gamma(t)^\top \quad \text{for all } T, \, d\mathbb{P} \otimes dt\text{-a.s.} \tag{6.4}$$

In this case, the $\mathbb{Q}$-dynamics of the forward rates $f(t,T)$ are of the form

$$f(t,T) = f(0,T) + \int_0^t \underbrace{\left(\sigma(s,T) \int_s^T \sigma(s,u)^\top du \right)}_{\text{HJM drift}} ds + \int_0^t \sigma(s,T) \, dW^*(s),$$

$$\tag{6.5}$$

and the discounted T-bond price satisfies

$$\frac{P(t,T)}{B(t)} = P(0,T)\mathcal{E}_t(v(\cdot,T) \bullet W^*) \tag{6.6}$$

for $t \leq T$.

Proof In view of Corollary 6.1 we find that

$$d\frac{P(t,T)}{B(t)} = \frac{P(t,T)}{B(t)} \left(b(t,T) + v(t,T)\gamma(t)^\top \right) dt + \frac{P(t,T)}{B(t)} v(t,T) \, dW^*(t).$$

Hence $\frac{P(t,T)}{B(t)}$, $t \leq T$, is a $\mathbb{Q}$-local martingale if and only if $b(t,T) = -v(t,T)\gamma(t)^\top$ $d\mathbb{P} \otimes dt$-a.s. Since $v(t,T)$ and $b(t,T)$ are both continuous in T, we deduce that $\mathbb{Q}$ is an ELMM if and only if (6.4) holds.

Differentiating both sides of (6.4) in T yields

$$-\alpha(t, T) + \sigma(t, T) \int_t^T \sigma(t, u)^\top du = \sigma(t, T) \gamma(t)^\top \quad \text{for all } T, \, d\mathbb{P} \otimes dt\text{-a.s.}$$

Inserting this in (6.1) gives (6.5). Equation (6.6) now follows from Lemma 4.2. □

Remark 6.1 It follows from (6.2) and (6.4) that

$$dP(t, T) = P(t, T)\left(r(t) - v(t, T)\gamma(t)^\top\right)dt + P(t, T)v(t, T)\,dW(t).$$

Whence the interpretation of $-\gamma$ as the market price of risk for the bond market.

The striking feature of the HJM framework is that the distribution of $f(t, T)$ and $P(t, T)$ under $\mathbb{Q}$ only depends on the volatility process $\sigma(t, T)$, and not on the $\mathbb{P}$-drift $\alpha(t, T)$. Hence option pricing only depends on σ. This situation is similar to the Black–Scholes stock price model ($\to$ Exercise 4.7).

We can give sufficient conditions for $\frac{P(t,T)}{B(t)}$ to be a true $\mathbb{Q}$-martingale.

Corollary 6.2 *Suppose that (6.4) holds. Then $\mathbb{Q}$ is an EMM if either*

(a) *the Novikov condition*

$$\mathbb{E}_\mathbb{Q}\left[e^{\frac{1}{2}\int_0^T \|v(t,T)\|^2 \, dt}\right] < \infty \quad \text{for all } T \tag{6.7}$$

holds; or

(b) *the forward rates are nonnegative: $f(t, T) \geq 0$ for all $t \leq T$.*

Proof By Theorem 4.7, the Novikov condition (6.7) is sufficient for $\frac{P(t,T)}{B(t)}$ in (6.6) to be a $\mathbb{Q}$-martingale.

If $f(t, T) \geq 0$, then $0 \leq P(t, T) \leq 1$ and $B(t) \geq 1$. Hence $0 \leq \frac{P(t,T)}{B(t)} \leq 1$. Since a uniformly bounded local martingale is a true martingale, the corollary is proved. □

6.3 Short-Rate Dynamics

What is the interplay between the short-rate models in Chap. 5 and the present HJM framework? Let us consider the simplest HJM model: a constant $\sigma(t, T) \equiv \sigma > 0$. Suppose that $\mathbb{Q}$ is an ELMM. Then (6.5) implies

$$f(t, T) = f(0, T) + \sigma^2 t\left(T - \frac{t}{2}\right) + \sigma W^*(t).$$

Hence for the short rates we obtain

$$r(t) = f(t, t) = f(0, t) + \frac{\sigma^2 t^2}{2} + \sigma W^*(t).$$

This is just the Ho–Lee model of Sect. 5.4.4.

In general, we have the following:

Proposition 6.1 *Suppose that $f(0, T)$, $\alpha(t, T)$ and $\sigma(t, T)$ are differentiable in T with $\int_0^T |\partial_u f(0, u)| \, du < \infty$ and such that* **(HJM.1)**–**(HJM.3)** *are satisfied when $\alpha(t, T)$ and $\sigma(t, T)$ are replaced by $\partial_T \alpha(t, T)$ and $\partial_T \sigma(t, T)$, respectively.*

Then the short-rate process is an Itô process of the form

$$r(t) = r(0) + \int_0^t \zeta(u) \, du + \int_0^t \sigma(u, u) \, dW(u), \qquad (6.8)$$

where

$$\zeta(u) = \alpha(u, u) + \partial_u f(0, u) + \int_0^u \partial_u \alpha(s, u) \, ds + \int_0^u \partial_u \sigma(s, u) \, dW(s).$$

Proof Recall first that

$$r(t) = f(t, t) = f(0, t) + \int_0^t \alpha(s, t) \, ds + \int_0^t \sigma(s, t) \, dW(s).$$

Applying the Fubini Theorem 6.2 below to the stochastic integral gives

$$\int_0^t \sigma(s, t) \, dW(s) = \int_0^t \sigma(s, s) \, dW(s) + \int_0^t (\sigma(s, t) - \sigma(s, s)) \, dW(s)$$

$$= \int_0^t \sigma(s, s) \, dW(s) + \int_0^t \int_s^t \partial_u \sigma(s, u) \, du \, dW(s)$$

$$= \int_0^t \sigma(s, s) \, dW(s) + \int_0^t \int_0^u \partial_u \sigma(s, u) \, dW(s) \, du.$$

Moreover, from the classical Fubini Theorem we deduce in a similar way that

$$\int_0^t \alpha(s, t) \, ds = \int_0^t \alpha(s, s) \, ds + \int_0^t \int_0^u \partial_u \alpha(s, u) \, ds \, du,$$

and finally

$$f(0, t) = r(0) + \int_0^t \partial_u f(0, u) \, du.$$

Combining these formulas, we obtain (6.8). $\qquad \square$

6.4 HJM Models

In the preceding sections we have studied the stochastic behavior of the forward rate process $f(t, T)$ for some generic drift and volatility processes $\alpha(\omega, t, T)$

and $\sigma(\omega, t, T)$. For modeling purposes we would prefer a forward rate dependent volatility coefficient

$$\sigma(\omega, t, T) = \sigma(t, T, f(\omega, t, T))$$

for some appropriate function σ. The simplest choice is a deterministic function $\sigma(t, T)$ which does not depend on ω. This results in Gaussian distributed forward rates $f(t, T)$ and leads to simple bond option price formulas, as we will see in Sect. 7.2 below. A particular case is the constant $\sigma(t, T) \equiv \sigma$, which corresponds to the Ho–Lee model as we have seen in Sect. 6.3 above.

It is shown in [90] and [125] that, for any continuous initial forward curve $f(0, T)$, there exists a unique jointly continuous solution $f(t, T)$ of

$$df(t, T) = \left(\sigma(t, T, f(t, T)) \int_t^T \sigma(t, u, f(t, u)) \, du \right) dt + \sigma(t, T, f(t, T)) \, dW(t) \tag{6.9}$$

if $\sigma(t, T, f)$ is uniformly bounded, jointly continuous, and Lipschitz continuous in the last argument. It is remarkable that the boundedness condition on σ cannot be substantially weakened as the following example shows.

6.4.1 Proportional Volatility

We consider the special case of a single Brownian motion ($d = 1$) and where $\sigma(t, T, f(t, T)) = \sigma f(t, T)$ for some constant $\sigma > 0$. This volatility function is positive and Lipschitz continuous but not bounded. The solution of (6.9), if it existed, must satisfy ($\rightarrow$ Exercise 6.3)

$$f(t, T) = f(0, T) e^{\sigma^2 \int_0^t \int_s^T f(s,u) \, du \, ds} e^{\sigma W(t) - \frac{\sigma^2}{2} t}. \tag{6.10}$$

Following the arguments in Avellaneda and Laurence [5, Sect. 13.6], we now sketch that there is no finite-valued solution to expression (6.10).

Indeed, assume for simplicity that the initial forward curve is flat, i.e. $f(0, T) \equiv 1$, and $\sigma = 1$. Differentiating both sides of (6.10) with respect to T, we obtain

$$\partial_T f(t, T) = f(t, T) \int_0^t f(s, T) \, ds = \frac{1}{2} \partial_t \left(\int_0^t f(s, T) \, ds \right)^2.$$

Integrating this equation with respect to t from $t = 0$ to 1, and interchanging the order of differentiation and integration,[2] yields

$$\partial_T \int_0^1 f(s, T) \, ds = \frac{1}{2} \left(\int_0^1 f(s, T) \, ds \right)^2.$$

---

[2]This argumentation is somehow sketchy. A full rigorous proof that (6.10) is not finite valued can be found in Morton [125].

Solving this differential equation path-wise for $X(T) = \int_0^1 f(s, T) ds$, $T \geq 1$, we obtain as unique solution

$$X(T) = \frac{X(1)}{1 - \frac{X(1)}{2}(T - 1)}.$$

In view of (6.10), we have $X(1) > 0$. Hence $X(T) \uparrow \infty$ for $T \uparrow \tau$ where $\tau = 1 + \frac{2}{X(1)}$ is a finite random time. We conclude that $f(\omega, t, \tau(\omega))$ must become $+\infty$ for some $t \leq 1$, for almost all ω.

The nonexistence of HJM models with proportional volatility encouraged the development of the so-called LIBOR market models, which will be further discussed in Chap. 11 below.

6.5 Fubini's Theorem

In this section we prove Fubini's theorem for stochastic integrals. For the classical version of Fubini's theorem, we refer to the standard textbooks in integration theory.

Theorem 6.2 (Fubini's theorem for Stochastic Integrals) *Consider the $\mathbb{R}^d$-valued stochastic process $\phi = \phi(\omega, t, s)$ with two indices, $0 \leq t, s \leq T$, satisfying the following properties:*[3]

(a) *ϕ is $\mathrm{Prog}_T \otimes \mathcal{B}[0, T]$-measurable;*
(b) *$\sup_{t,s} \|\phi(t, s)\| < \infty$.*[4]

Then $\lambda(t) = \int_0^T \phi(t, s) ds \in \mathcal{L}$, and there exists a $\mathcal{F}_T \otimes \mathcal{B}[0, T]$-measurable modification $\psi(s)$ of $\int_0^T \phi(t, s) dW(t)$ with $\int_0^T \psi^2(s) ds < \infty$ a.s.
Moreover, $\int_0^T \psi(s) ds = \int_0^T \lambda(t) dW(t)$, that is,

$$\int_0^T \left(\int_0^T \phi(t, s) dW(t) \right) ds = \int_0^T \left(\int_0^T \phi(t, s) ds \right) dW(t). \tag{6.11}$$

Proof Without loss of generality, we can put $d = 1$, as we just have to prove (6.11) componentwise.

We assume first that (b) is replaced by

(b') *$|\phi| \leq C$ for some finite constant C.*

Then clearly $\lambda \in \mathcal{L}$. Denote by $\mathcal{H}$ the set of all ϕ satisfying (a) and (b') and for which the theorem holds. We will show that $\mathcal{H}$ contains all ϕ satisfying (a) and (b').

---

[3] Prog_T denotes the progressive σ-algebra Prog restricted to $\Omega \times [0, T]$.
[4] Note that this is a ω-wise boundedness assumption.

Let K be some bounded progressive process and f some bounded $\mathcal{B}[0, T]$-measurable function. Then $\phi(\omega, t, s) = K(\omega, t) f(s)$ satisfies

$$\int_0^T \phi(t, s)\, ds = K(t) \int_0^T f(s)\, ds, \qquad \int_0^T \phi(t, s)\, dW(t) = f(s) \int_0^T K(t)\, dW(t)$$

and thus $\phi \in \mathcal{H}$. It follows from elementary measure theory that processes of the form Kf generate the σ-algebra $\mathrm{Prog}_T \otimes \mathcal{B}[0, T]$.

Next, we let $\phi_n \in \mathcal{H}$ and suppose that $\phi_n \uparrow \phi$ for some bounded $\mathrm{Prog}_T \otimes \mathcal{B}[0, T]$-measurable process ϕ. We can assume that $\sup_{t,s} |\phi_n| \leq N$, for some finite constant N that does not depend on n. Define

$$\psi_n(s) = \int_0^T \phi_n(t, s)\, dW(t).$$

From the Itô isometry and dominated convergence it follows that

$$\mathbb{E}\left[\left(\psi_n(s) - \int_0^T \phi(t, s)\, dW(t)\right)^2\right] = \mathbb{E}\left[\int_0^T |\phi_n(t, s) - \phi(t, s)|^2\, dt\right] \to 0$$

$$(6.12)$$

for $n \to \infty$, for all $s \leq T$. Define $A = \{(\omega, s) \mid \lim_n \psi_n(\omega, s) \text{ exists}\}$. Then A is $\mathcal{F}_T \otimes \mathcal{B}[0, T]$-measurable and so is the process

$$\psi(\omega, s) = \begin{cases} \lim_n \psi_n(\omega, s), & \text{if } (\omega, s) \in A, \\ 0, & \text{otherwise.} \end{cases} \qquad (6.13)$$

In view of (6.12) we have $\psi(s) = \int_0^T \phi(t, s)\, dW(t)$ a.s. for all $s \leq T$. Thus, $\psi(s)$ has the desired properties. From Jensen's integral inequality, the Itô isometry and dominated convergence we then have, on one hand,

$$\mathbb{E}\left[\left(\int_0^T \psi_n(s)\, ds - \int_0^T \psi(s)\, ds\right)^2\right]$$

$$\leq T \int_0^T \mathbb{E}\left[(\psi_n(s) - \psi(s))^2\right] ds$$

$$= T \int_0^T \mathbb{E}\left[\int_0^T |\phi_n(t, s) - \phi(t, s)|^2\, dt\right] ds \to 0 \quad \text{for } n \to \infty. \quad (6.14)$$

On the other hand,

$$\mathbb{E}\left[\left(\int_0^T \left(\int_0^T \phi_n(t, s)\, ds\right) dW(t) - \int_0^T \lambda(t)\, dW(t)\right)^2\right]$$

$$= \mathbb{E}\left[\int_0^T \left|\int_0^T \phi_n(t, s)\, ds - \int_0^T \phi(t, s)\, ds\right|^2 dt\right]$$

$$\leq T\mathbb{E}\left[\int_0^T \int_0^T |\phi_n(t, s) - \phi(t, s)|^2\, ds\, dt\right] \to 0 \quad \text{for } n \to \infty. \quad (6.15)$$

Combining (6.14) and (6.15) shows that (6.11) also holds for ϕ, and thus $\phi \in \mathcal{H}$.

Since $\mathcal{H}$ is also a vector space, it follows from the monotone class theorem 6.3 below that $\mathcal{H}$ contains all bounded $\text{Prog}_T \otimes \mathcal{B}[0, T]$-measurable processes, which proves the theorem under the assumption (b').

For the general case, we define the nondecreasing sequence of stopping times

$$\tau_n = \inf\left\{t \mid \sup_s |\phi(t, s)| > n\right\} \wedge T.$$

Then $\phi_n(t, s) = \phi(t, s) 1_{\{t \leq \tau_n\}}$ satisfies (b'). From the above step, we thus obtain $\lambda_n \in \mathcal{L}$ and some $\mathcal{F}_T \otimes \mathcal{B}[0, T]$-measurable $\psi_n(s)$ with $\psi_n(s) = \int_0^{T \wedge \tau_n} \phi(t, s) \, dW(t)$ a.s. for all $s \leq T$. Since $\tau_n \uparrow T$, the process ψ is well defined by setting $\psi(s) = \psi_n(s)$ for $s \leq \tau_n$ and has the desired properties. Moreover, $\lambda_n(t) = \lambda(t) 1_{\{t \leq \tau_n\}}$, and we infer that $\lambda \in \mathcal{L}$ and (6.11) holds on $\{\tau_n = T\}$ for all $n \geq 1$. Since $\mathbb{P}[\tau_n < T] \to 0$, letting $n \to \infty$, the theorem is proved. $\qquad\square$

Corollary 6.3 *Let ϕ be as in Theorem 6.2. Then the process*

$$\int_0^s \phi(t, s) \, dW(t), \quad s \in [0, T],$$

has a progressive modification $\pi(s)$ with $\int_0^T \pi^2(s) \, ds < \infty$ a.s.

Proof For $\phi(\omega, t, s) = K(\omega, t) f(s)$, with bounded progressive process K and bounded measurable function f, the process

$$\int_0^s \phi(t, s) \, dW(t) = f(s) \int_0^s K(t) \, dW(t)$$

is clearly progressive and path-wise square integrable. Now use a similar monotone class and localization argument as in the proof of Theorem 6.2 ($\to$ Exercise 6.4). $\square$

Here we recall the monotone class theorem, which is proved in e.g. [154, Sect. 12.6].

Theorem 6.3 (Monotone Class Theorem) *Suppose the set $\mathcal{H}$ consists of real-valued bounded functions defined on a set Ω with the following properties:*

(a) *$\mathcal{H}$ is a vector space;*
(b) *$\mathcal{H}$ contains the constant function 1_Ω;*
(c) *if $f_n \in \mathcal{H}$ and $f_n \uparrow f$ monotone, for some bounded function f on Ω, then $f \in \mathcal{H}$.*

If $\mathcal{H}$ contains a collection $\mathcal{M}$ of real-valued functions, which is closed under multiplication (that is, $f, g \in \mathcal{M}$ implies $fg \in \mathcal{M}$). Then $\mathcal{H}$ contains all real-valued bounded functions that are measurable with respect to the σ-algebra which is generated by $\mathcal{M}$ (that is, $\sigma\{f^{-1}(A) \mid A \in \mathcal{B}, f \in \mathcal{M}\}$).

6.6 Exercises

Exercise 6.1 Using the monotone class theorem 6.3, show that a process X is progressive if and only if X is Prog-measurable.

Exercise 6.2 Complete the proofs of Lemma 6.1 and Corollary 6.1.

Exercise 6.3 Show that the solution to the proportional volatility HJM model would equal (6.10) if it existed.

Exercise 6.4 Complete the proof of Corollary 6.3.

Exercise 6.5 The goal of this exercise is to show that parallel shifts of the forward curve creates arbitrage. Consider first the one-period model for the forward curve

$$f(0, t) = 0.04, \quad t \geq 0,$$

$$f(\omega, 1, t) = \begin{cases} 0.06, & t \geq 1, \omega = \omega_1, \\ 0.02, & t \geq 1, \omega = \omega_2, \end{cases}$$

where $\Omega = \{\omega_1, \omega_2\}$ with $\mathbb{P}[\omega_i] > 0$, $i = 1, 2$.

(a) Show that the matrix

$$\begin{pmatrix} P(0, 1) & P(0, 2) & P(0, 3) \\ P(\omega_1, 1, 1) & P(\omega_1, 1, 2) & P(\omega_1, 1, 3) \\ P(\omega_2, 1, 1) & P(\omega_2, 1, 2) & P(\omega_2, 1, 3) \end{pmatrix}$$

is invertible.

(b) Use (a) to find an arbitrage strategy with value process $V(0) = 0$ and $V(\omega_i, 1) = 1$ for both ω_i.

Next, we extend the one-period finding to the continuous time HJM framework with a one-dimensional driving Brownian motion W. An HJM forward curve evolution by parallel shifts is then of the form

$$f(t, T) = h(T - t) + Z(t)$$

for some deterministic initial curve $f(0, T) = h(T)$ and some Itô process $dZ(t) = b(t) dt + \rho(t) dW(t)$ with $Z(0) = 0$.

(c) Show that the HJM drift condition implies $b(t) \equiv b$, $\rho^2(t) \equiv a$, and

$$h(x) = -\frac{a}{2}x^2 + bx + c$$

for some constants $a \geq 0$, and $b, c \in \mathbb{R}$.

(d) How is this model related to the Ho–Lee model from Sect. 5.4.4?

(e) Argue that, for generic initial curves $f(0, T)$, non-trivial forward curve evolutions by parallel shifts are excluded by the HJM drift condition.

Exercise 6.6 Consider the Hull–White extended Vasiček short-rate dynamics under the EMM $\mathbb{Q} \sim \mathbb{P}$

$$dr(t) = (b(t) + \beta r(t)) \, dt + \sigma \, dW^*(t),$$

where W^* is a standard real-valued $\mathbb{Q}$-Brownian motion, β and $\sigma > 0$ are constants, and $b(t)$ is a deterministic continuous function. Using the results from Sect. 5.4.5, find the corresponding HJM forward rate dynamics

$$f(t, T) = f(0, T) + \int_0^t \alpha(s, T) \, ds + \int_0^t \sigma(s, T) \, dW^*(s).$$

(a) What are $f(0, T), \alpha(s, T), \sigma(s, T)$?
(b) Verify your findings in (a), by checking whether $\alpha(s, T)$ satisfies the HJM drift condition.
(c) Discuss the role of $b(s)$. Do $\alpha(s, T)$ and $\sigma(s, T)$ depend on $b(s)$?
(d) What does this imply for the Vasiček model $(b(s) \equiv b)$?
(e) Verify Proposition 6.1 by showing that $dr(t) = \zeta(t) \, dt + \sigma(t, t) \, dW^*(t)$, where $\zeta(t)$ is given by f, α, σ as in Proposition 6.1.

6.7 Notes

The approach in Sect. 6.4 has been carried out by Heath, Jarrow and Morton [90], and in more depth and generality by Morton [125], and also in [68] and [35]. The proof of Fubini's Theorem 6.2 for stochastic integrals follows along the line of arguments in Protter [132, Sect. IV.6], however cannot be immediately deduced from [132, Theorem 64], as it requires a localization step carried out above.

Chapter 7
Forward Measures

In this chapter we replace the risk-free numeraire by another traded asset, such as the T-bond. This change of numeraire technique proves most useful for option pricing and provides the basis for the market models studied below. We derive explicit option price formulas for Gaussian HJM models. This includes the Vasiček short-rate model and some extension of the Black–Scholes model with stochastic interest rates.

7.1 T-Bond as Numeraire

We consider the HJM setup from Chap. 6 and assume there exists an EMM $\mathbb{Q}$ for the bond market such that all discounted T-bond price processes follow $\mathbb{Q}$-martingales. As usual, we denote by W^* the respective $\mathbb{Q}$-Brownian motion.

Fix $T > 0$. Since $P(0, T)B(T) > 0$ and

$$\mathbb{E}_{\mathbb{Q}}\left[\frac{1}{P(0, T)B(T)}\right] = \mathbb{E}_{\mathbb{Q}}\left[\frac{P(T, T)}{P(0, T)B(T)}\right] = 1$$

we can define an equivalent probability measure $\mathbb{Q}^T \sim \mathbb{Q}$ on $\mathcal{F}_T$ by

$$\frac{d\mathbb{Q}^T}{d\mathbb{Q}} = \frac{1}{P(0, T)B(T)}.$$

For $t \leq T$ we have

$$\frac{d\mathbb{Q}^T}{d\mathbb{Q}}\bigg|_{\mathcal{F}_t} = \mathbb{E}_{\mathbb{Q}}\left[\frac{d\mathbb{Q}^T}{d\mathbb{Q}}\,\bigg|\,\mathcal{F}_t\right] = \frac{P(t, T)}{P(0, T)B(t)}.$$

$\mathbb{Q}^T$ is called the T-forward measure. From (6.6) we infer

$$\frac{d\mathbb{Q}^T}{d\mathbb{Q}}\bigg|_{\mathcal{F}_t} = \mathcal{E}_t\left(v(\cdot, T) \bullet W^*\right), \quad t \leq T. \tag{7.1}$$

Hence Girsanov's Theorem 4.6 implies that

$$W^T(t) = W^*(t) - \int_0^t v(s, T)^\top ds, \quad t \leq T,$$

is a $\mathbb{Q}^T$-Brownian motion.

D. Filipović, *Term-Structure Models,*
Springer Finance,
DOI 10.1007/978-3-540-68015-4_7, © Springer-Verlag Berlin Heidelberg 2009

Here is a fundamental property of the T-forward measure for the financial modeling:

Lemma 7.1 *For any $S > 0$, the T-bond discounted S-bond price process*

$$\frac{P(t, S)}{P(t, T)} = \frac{P(0, S)}{P(0, T)} \mathcal{E}_t \left(\sigma_{S,T} \bullet W^T \right), \quad t \leq S \wedge T$$

is a $\mathbb{Q}^T$-martingale, where we define

$$\sigma_{S,T}(t) = -\sigma_{T,S}(t) = v(t, S) - v(t, T) = \int_S^T \sigma(t, u)\, du. \qquad (7.2)$$

Moreover, the T- and S-forward measures are related by

$$\left.\frac{d\mathbb{Q}^S}{d\mathbb{Q}^T}\right|_{\mathcal{F}_t} = \frac{P(t, S)\, P(0, T)}{P(t, T)\, P(0, S)} = \mathcal{E}_t \left(\sigma_{S,T} \bullet W^T \right), \quad t \leq S \wedge T.$$

Proof Let $u \leq t \leq S \wedge T$. Bayes' rule gives

$$\mathbb{E}_{\mathbb{Q}^T}\left[\frac{P(t, S)}{P(t, T)} \,\Big|\, \mathcal{F}_u \right] = \frac{\mathbb{E}_{\mathbb{Q}}\left[\frac{P(t,T)}{P(0,T)B(t)} \frac{P(t,S)}{P(t,T)} \,|\, \mathcal{F}_u \right]}{\frac{P(u,T)}{P(0,T)B(u)}} = \frac{\frac{P(u,S)}{B(u)}}{\frac{P(u,T)}{B(u)}} = \frac{P(u, S)}{P(u, T)},$$

which proves that $P(t, S)/P(t, T)$ is a martingale. The stochastic exponential representation follows from Lemma 4.2 and (6.6) ($\to$ Exercise 7.1). The second claim follows from the identity

$$\left.\frac{d\mathbb{Q}^S}{d\mathbb{Q}^T}\right|_{\mathcal{F}_t} = \left.\frac{d\mathbb{Q}^S}{d\mathbb{Q}}\right|_{\mathcal{F}_t} \left.\frac{d\mathbb{Q}}{d\mathbb{Q}^T}\right|_{\mathcal{F}_t}. \qquad \square$$

We thus have received an entire collection of EMMs. Each $\mathbb{Q}^T$ corresponds to a different numeraire, namely the T-bond. Since $\mathbb{Q}$ is related to the risk-free asset, one often calls $\mathbb{Q}$ the *risk-neutral (or spot) measure*.

T-forward measures give simpler pricing formulas. Indeed, let X be a T-claim such that

$$\mathbb{E}_{\mathbb{Q}}\left[\frac{|X|}{B(T)} \right] < \infty. \qquad (7.3)$$

Its arbitrage price at time $t \leq T$ is then given by

$$\pi(t) = B(t)\mathbb{E}_{\mathbb{Q}}\left[\frac{X}{B(T)} \,\Big|\, \mathcal{F}_t \right].$$

To compute $\pi(t)$ we have to know the joint distribution of $1/B(T)$ and X, and integrate with respect to that distribution. Thus we have to compute a double integral, which in most cases turns out to be rather hard work. If $1/B(T)$ and X were independent under $\mathbb{Q}$ conditional on $\mathcal{F}_t$ we would have

$$\pi(t) = P(t, T)\mathbb{E}_{\mathbb{Q}}[X \mid \mathcal{F}_t],$$

a much nicer formula, since:

- we only have to compute the single integral $\mathbb{E}_\mathbb{Q}[X \mid \mathcal{F}_t]$;
- the bond price $P(t, T)$ can be observed at time t and does not have to be computed within the model.

However, as we are mainly interested in pricing interest rate sensitive claims, independence of $1/B(T)$ and X would be a very stringent and unrealistic assumption. The good news is that the above formula holds—not under $\mathbb{Q}$ though, but under $\mathbb{Q}^T$:

Proposition 7.1 *Let X be a T-claim such that* (7.3) *holds. Then*

$$\mathbb{E}_{\mathbb{Q}^T}[|X|] < \infty \qquad (7.4)$$

and

$$\pi(t) = P(t, T)\mathbb{E}_{\mathbb{Q}^T}[X \mid \mathcal{F}_t]. \qquad (7.5)$$

Proof Bayes' rule yields

$$\mathbb{E}_{\mathbb{Q}^T}[|X|] = \mathbb{E}_\mathbb{Q}\left[\frac{|X|}{P(0, T)B(T)}\right] < \infty \quad \text{(by (7.3))},$$

whence (7.4). Moreover, again by Bayes' rule,

$$\pi(t) = P(0, T)B(t)\mathbb{E}_\mathbb{Q}\left[\frac{X}{P(0, T)B(T)} \;\middle|\; \mathcal{F}_t\right]$$

$$= P(0, T)B(t)\frac{P(t, T)}{P(0, T)B(t)}\mathbb{E}_{\mathbb{Q}^T}[X \mid \mathcal{F}_t]$$

$$= P(t, T)\mathbb{E}_{\mathbb{Q}^T}[X \mid \mathcal{F}_t],$$

which proves (7.5). $\qquad\square$

As a first application, we now show that the expectation hypothesis holds under the forward measure, as announced in Sect. 2.2.3. That is, the forward rate $f(t, T)$ is given as conditional expectation of the future short rate $r(T)$ under the T-forward measure. Indeed, equation (6.5) now reads

$$f(t, T) = f(0, T) + \int_0^t \sigma(s, T)\,dW^T(s). \qquad (7.6)$$

Hence, if $\sigma(\cdot, T) \in \mathcal{L}^2$ then $f(t, T)$, $t \leq T$, is a $\mathbb{Q}^T$-martingale. Summarizing we have thus proved:

Lemma 7.2 (Expectation Hypothesis) *If* $\sigma(\cdot, T) \in \mathcal{L}^2$, *the expectation hypothesis holds under the T-forward measure:*

$$f(t, T) = \mathbb{E}_{\mathbb{Q}^T}[r(T) \mid \mathcal{F}_t] \quad \text{for } t \leq T.$$

Remark 7.1 A word of warning: in view of equation (7.6) it is tempting to "specify" a forward rate model by postulating the dynamics of $f(\cdot, T)$ under $\mathbb{Q}^T$ for each maturity T separately without reference to some underlying $\mathbb{Q}$. However, it is far from clear whether a common risk-neutral measure $\mathbb{Q}$, tying all $\mathbb{Q}^T$'s, exists in this case. On the other hand, we note that this is exactly the approach in the LIBOR market model in Chap. 11 below. The important difference being that there one considers finitely many maturities only.

As a second application, we give a proof of the Dybvig–Ingersoll–Ross theorem [62], which states that long rates can never fall. Recall the zero-coupon yield $R(t, T) = \frac{1}{T-t} \int_t^T f(t, s)\, ds$. We define the asymptotic long rate

$$R_\infty(t) = \lim_{T \to \infty} R(t, T)$$

if it exists.

Lemma 7.3 (Dybvig–Ingersoll–Ross Theorem) *For all $s < t$ the long rates satisfy $R_\infty(s) \le R_\infty(t)$ if they exist.*

Proof Let $s < t$ be such that $R_\infty(s)$ and $R_\infty(t)$ exist. Then $p(u) = \lim_{T \to \infty} P(t, T)^{\frac{1}{T}} = e^{-R_\infty(u)}$ exist for $u \in \{s, t\}$, and it remains to prove that $p(s) \ge p(t)$.

Under the t-forward measure $\mathbb{Q}^t$, we have

$$\frac{P(s, T)}{P(s, t)} = \mathbb{E}_{\mathbb{Q}^t}[P(t, T) \mid \mathcal{F}_s],$$

and thus

$$p(s) = \lim_{T \to \infty} \mathbb{E}_{\mathbb{Q}^t}[P(t, T) \mid \mathcal{F}_s]^{\frac{1}{T}}.$$

Now let $X \ge 0$ be any bounded random variable with $\mathbb{E}_{\mathbb{Q}^t}[X] = 1$. Using the $\mathcal{F}_s$-conditional versions of Fatou's lemma, Hölder's inequality and dominated convergence, we obtain

$$\mathbb{E}_{\mathbb{Q}^t}\left[X\, p(t)\right] = \mathbb{E}_{\mathbb{Q}^t}\left[\liminf_{T \to \infty} X\, P(t, T)^{\frac{1}{T}}\right]$$

$$\le \mathbb{E}_{\mathbb{Q}^t}\left[\liminf_{T \to \infty} \mathbb{E}_{\mathbb{Q}^t}\left[X\, P(t, T)^{\frac{1}{T}} \mid \mathcal{F}_s\right]\right]$$

$$\le \mathbb{E}_{\mathbb{Q}^t}\left[\liminf_{T \to \infty} \mathbb{E}_{\mathbb{Q}^t}\left[X^{\frac{T}{T-1}} \mid \mathcal{F}_s\right]^{\frac{T-1}{T}} \mathbb{E}_{\mathbb{Q}^t}[P(t, T) \mid \mathcal{F}_s]^{\frac{1}{T}}\right]$$

$$= \mathbb{E}_{\mathbb{Q}^t}\left[X\, p(s)\right].$$

Since X was arbitrary with the stated properties, we conclude that $p(t) \le p(s)$, and the lemma is proved. $\qquad\qquad\square$

7.2 Bond Option Pricing

We now consider a European call option on an S-bond with expiry date $T < S$ and strike price K. Its arbitrage price at time $t = 0$[1] is

$$\pi = \mathbb{E}_{\mathbb{Q}}\left[e^{-\int_0^T r(s)\,ds}\,(P(T, S) - K)^+\right].$$

We decompose

$$\pi = \mathbb{E}_{\mathbb{Q}}\left[B(T)^{-1}P(T, S)1_{\{P(T,S)\geq K\}}\right] - K\mathbb{E}_{\mathbb{Q}}\left[B(T)^{-1}1_{\{(P(T,S)\geq K\}}\right]$$

$$= P(0, S)\mathbb{Q}^S[P(T, S) \geq K] - KP(0, T)\mathbb{Q}^T[P(T, S) \geq K]. \qquad (7.7)$$

Now observe that

$$\mathbb{Q}^S[P(T, S) \geq K] = \mathbb{Q}^S\left[\frac{P(T, T)}{P(T, S)} \leq \frac{1}{K}\right],$$

$$\mathbb{Q}^T[P(T, S) \geq K] = \mathbb{Q}^T\left[\frac{P(T, S)}{P(T, T)} \geq K\right].$$

In view of Lemma 7.1, this suggests that we look at those models for which $\sigma_{T,S}$ is deterministic, and hence $\frac{P(T,T)}{P(T,S)}$ and $\frac{P(T,S)}{P(T,T)}$ are lognormally distributed under the respective forward measures. We thus assume that

$$\sigma(t, T) = (\sigma_1(t, T), \ldots, \sigma_d(t, T)) \text{ are deterministic functions of } (t, T), \qquad (7.8)$$

and hence forward rates $f(t, T)$ are Gaussian distributed.

We thus obtain the following closed-form option price formula.

Proposition 7.2 *Under the above Gaussian assumption* (7.8), *the bond option price is*

$$\pi = P(0, S)\Phi[d_1] - KP(0, T)\Phi[d_2],$$

where Φ is the standard Gaussian cumulative distribution function,

$$d_{1,2} = \frac{\log[\frac{P(0,S)}{KP(0,T)}] \pm \frac{1}{2}\int_0^T \|\sigma_{T,S}(s)\|^2\,ds}{\sqrt{\int_0^T \|\sigma_{T,S}(s)\|^2\,ds}},$$

and $\sigma_{T,S}(s)$ is given in (7.2).

---

[1]For simplicity, we consider $t = 0$ here. The following results carry over to $t \geq 0$.

Proof It is enough to observe that

$$\frac{\log \frac{P(T,T)}{P(T,S)} - \log \frac{P(0,T)}{P(0,S)} + \frac{1}{2} \int_0^T \|\sigma_{T,S}(s)\|^2 \, ds}{\sqrt{\int_0^T \|\sigma_{T,S}(s)\|^2 \, ds}}$$

and

$$\frac{\log \frac{P(T,S)}{P(T,T)} - \log \frac{P(0,S)}{P(0,T)} + \frac{1}{2} \int_0^T \|\sigma_{T,S}(s)\|^2 \, ds}{\sqrt{\int_0^T \|\sigma_{T,S}(s)\|^2 \, ds}}$$

are standard Gaussian distributed under $\mathbb{Q}^S$ and $\mathbb{Q}^T$, respectively. $\square$

7.2.1 Example: Vasiček Short-Rate Model

For the Vasiček short-rate model ($d = 1$)

$$dr = (b + \beta r) \, dt + \sigma \, dW^*$$

we obtain from the results in Sect. 5.4.1 ($\to$ Exercise 6.6) that

$$df(t, T) = \alpha(t, T) \, dt + \sigma e^{\beta(T-t)} \, dW^*, \tag{7.9}$$

for the corresponding drift term $\alpha(t, T)$. Hence the corresponding forward rate volatility $\sigma(t, T) = \sigma e^{\beta(T-t)}$ is deterministic, and the above option price formula in Proposition 7.2 applies. A similar closed-form expression is available for the price of a put option ($\to$ Exercise 7.4), and hence an explicit price formula for caps.

For $\beta = -0.86$, $b/|\beta| = 0.09$ (mean reversion level), $\sigma = 0.0148$ and $r(0) = 0.08$, as in Fig. 5.1, one gets the ATM cap prices and Black volatilities shown in Table 7.1 and Fig. 7.1 ($\to$ Exercise 7.5). The tenor is as follows: $t_0 = 0$ (today), $T_0 = 1/4$ (first reset date), and $T_i - T_{i-1} \equiv 1/4$, $i = 1, \ldots, 119$ (maturity of the last cap is $T_{119} = 30$).

In contrast to Fig. 2.1, it seems that the Vasiček model cannot produce humped volatility curves.

7.3 Black–Scholes Model with Gaussian Interest Rates

The aim of this section is to develop a European call option price formula for the generalized Black–Scholes model with stochastic short rates within a Gaussian HJM model. The purpose of this section is twofold. First, it illustrates the general change of numeraire technique for option pricing as developed in Geman et al. [77]. And second, it provides the stage for the Black–Scholes model (see Exercise 4.7) in the main text of this book.

Table 7.1 Vasiček ATM cap prices and Black volatilities

Maturity	ATM prices	ATM vols
1	0.00215686	0.129734
2	0.00567477	0.106348
3	0.00907115	0.0915455
4	0.0121906	0.0815358
5	0.01503	0.0743607
6	0.017613	0.0689651
7	0.0199647	0.0647515
8	0.0221081	0.0613624
10	0.025847	0.0562337
12	0.028963	0.0525296
15	0.0326962	0.0485755
20	0.0370565	0.0443967
30	0.0416089	0.0402203

Fig. 7.1 Vasiček ATM cap Black volatilities

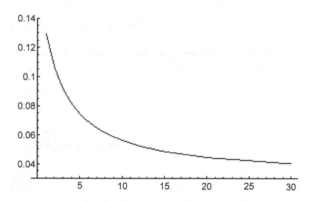

In the classical Black–Scholes model [18], there is one risky asset (stock) S and the money-market account B following the $\mathbb{Q}$-dynamics

$$dB = Br\,dt, \quad B(0) = 1,$$
$$dS = Sr\,dt + S\Sigma\,dW^*, \quad S(0) > 0,$$

with constant volatility[2] $\Sigma = (\Sigma_1, \ldots, \Sigma_d) \in \mathbb{R}^d$ and constant short rate r.

We now generalize this, and let r be stochastic within the Gaussian HJM setup (7.8). Next, we consider a European call option on S with maturity T and

---

[2]In fact, the standard Black–Scholes model assumes $d = 1$.

strike price K. Its arbitrage price at time $t = 0$ is[3]

$$\pi = \mathbb{E}_{\mathbb{Q}}\left[\frac{1}{B(T)}(S(T) - K)^+\right]$$

$$= \mathbb{E}_{\mathbb{Q}}\left[\frac{S(T)}{B(T)}1_{\{S(T)\geq K\}}\right] - K\mathbb{E}_{\mathbb{Q}}\left[\frac{1}{B(T)}1_{\{S(T)\geq K\}}\right].$$

As introduced in Sect. 7.1, we let $\mathbb{Q}^T$ denote the T-forward measure, which corresponds to the T-bond as numeraire. Similarly, we may choose S as numeraire and introduce the EMM $\mathbb{Q}^{(S)} \sim \mathbb{Q}$ on $\mathcal{F}_T$ via

$$\frac{d\mathbb{Q}^{(S)}}{d\mathbb{Q}} = \frac{S(T)}{S(0)\,B(T)} = \mathcal{E}_T(\Sigma \bullet W^*).$$

We denote the respective Girsanov transformed $\mathbb{Q}^{(S)}$-Brownian motion by

$$W^{(S)}(t) = W^*(t) - \Sigma^\top t, \quad t \leq T.$$

Proposition 7.1 carries over to $\mathbb{Q}^{(S)}$, and we obtain

$$\pi = S(0)\mathbb{Q}^{(S)}[S(T) \geq K] - K P(0, T)\mathbb{Q}^T[S(T) \geq K].$$

It remains to compute the probabilities of the event $\{S(T) \geq K\}$ under the measures $\mathbb{Q}^{(S)}$ and $\mathbb{Q}^T$. As in Sect. 7.2, we start by writing

$$\mathbb{Q}^{(S)}[S(T) \geq K] = \mathbb{Q}^{(S)}\left[\frac{P(T, T)}{S(T)} \leq \frac{1}{K}\right],$$

$$\mathbb{Q}^T[S(T) \geq K] = \mathbb{Q}^T\left[\frac{S(T)}{P(T, T)} \geq K\right].$$

Next, we observe that $\frac{P(t,T)}{S(t)}$ is a $\mathbb{Q}^{(S)}$-martingale and $\frac{S(t)}{P(t,T)}$ is a $\mathbb{Q}^T$ martingale for $t \leq T$. We find their respective representations by an application of Itô's formula.[4] We start with the second term, for which we obtain from (6.6) that

$$d\frac{S(t)}{P(t)} = \frac{1}{P(t)}\,dS(t) - \frac{S(t)}{P(t)^2}\,dP(t) - \frac{1}{P(t)^2}\,d\langle S, P\rangle_t + \frac{S(t)}{P(t)^3}\,d\langle P, P\rangle_t$$

$$= (\cdots)\,dt + \frac{S(t)}{P(t)}\,(\Sigma - v(t, T))\,dW^*(t),$$

where the T-bond volatility $v(t, T)$ is defined in (6.3), and we omitted the parameter T in $P(t, T)$ for notational simplicity. Note that we do not have to explicitly compute the drift term. Indeed, since the volatility process is unaffected by the change

---

[3]The following can easily be extended to arbitrary $t \geq 0$.

[4]For didactic reasons (to practice stochastic calculus) this approach slightly deviates from the one in Sect. 7.2.

of measure from $\mathbb{Q}$ to $\mathbb{Q}^T$, we conclude that the $\mathbb{Q}^T$-dynamics are given by

$$d\frac{S(t)}{P(t,T)} = \frac{S(t)}{P(t,T)}\,(\Sigma - v(t,T))\,dW^T(t).$$

Hence, in view of Lemma 4.2,

$$\frac{S(T)}{P(T,T)} = \frac{S(0)}{P(0,T)}\mathcal{E}_T\Big((\Sigma - v(\cdot,T)) \bullet W^T\Big)$$

is lognormally distributed under $\mathbb{Q}^T$.

Along similar calculations, we obtain that

$$\frac{P(T,T)}{S(T)} = \frac{P(0,T)}{S(0)}\mathcal{E}_T\Big(-(\Sigma - v(\cdot,T)) \bullet W^{(S)}\Big) \tag{7.10}$$

is lognormally distributed under $\mathbb{Q}^{(S)}$.

We thus obtain the following generalized Black–Scholes option price formula.

Proposition 7.3 *In the above generalized Black–Scholes model, the option price is*

$$\pi = S(0)\Phi[d_1] - K P(0,T)\Phi[d_2],$$

where Φ is the standard Gaussian cumulative distribution function and

$$d_{1,2} = \frac{\log[\frac{S(0)}{K P(0,T)}] \pm \frac{1}{2}\int_0^T \|\Sigma - v(t,T)\|^2\,dt}{\sqrt{\int_0^T \|\Sigma - v(t,T)\|^2\,dt}}. \tag{7.11}$$

Note that $v(t,T) = 0$ yields the classical Black–Scholes option price formula for constant short rate.

Proof Follows as in the proof of Proposition 7.2 ($\to$ Exercise 7.8). □

7.3.1 Example: Black–Scholes–Vasiček Model

We now make the generalized Black–Scholes option price formula in Proposition 7.3 more explicit for the Vasiček short-rate model. Let $d = 2$ be the dimension of the driving Brownian motion W^*. We represent the Vasiček short-rate dynamics this time by

$$dr = (b + \beta r)\,dt + \sigma\,dW^*,$$

where $\sigma = (\sigma_1, \sigma_2)$ is in $\mathbb{R}^2$. Note that this corresponds to the standard representation $dr = (b + \beta r)\,dt + \|\sigma\|\,d\mathcal{W}^*$ for the one-dimensional $\mathbb{Q}$-Brownian motion

$$\mathcal{W}^* = \frac{\sigma_1\,W_1^* + \sigma_2\,W_2^*}{\|\sigma\|}.$$

Hence, from (7.9) we obtain the $\mathbb{R}^2$-valued T-bond volatility

$$v(t, T) = -\sigma \int_t^T e^{\beta(T-s)} \, ds = \frac{\sigma}{\beta} \left(1 - e^{\beta(T-t)} \right).$$

A tedious, but elementary, computation yields

$$\int_0^T \| \Sigma - v(t, T) \|^2 \, dt$$

$$= \| \Sigma \|^2 T + 2 \Sigma \sigma^\top \frac{e^{\beta T} - 1 - \beta T}{\beta^2} + \| \sigma \|^2 \frac{e^{2\beta T} - 4 e^{\beta T} + 2\beta T + 3}{2\beta^3} \quad (7.12)$$

for the aggregate volatility in (7.11).

Let us finally discuss the relation between the option price π and the instantaneous covariation $d\langle S, r \rangle / dt = \Sigma \sigma^\top$ between the stock price S and the short rates r. In fact, π is monotone increasing in $\int_0^T \| \Sigma - v(t, T) \|^2 \, dt$, which again is increasing in $\Sigma \sigma^\top$, as seen from (7.12). Indeed, the function $e^x - 1 - x$ is positive for all real $x \neq 0$. We conclude that the option price π increases with increasing covariation between S and r. For negative covariation, π may even be smaller than the classical Black–Scholes option price with constant short rates ($\sigma = 0$).

7.4 Exercises

Exercise 7.1 The aim of this exercise is to complete the proof of Lemma 7.1.

(a) Prove the representation formula for $P(t, S) / P(t, T)$ by applying Lemma 4.2 to (6.6).
(b) Let $\mathbb{P} \sim \mathbb{Q} \sim \mathbb{R}$ be equivalent probability measures on some measurable space $(\Omega, \mathcal{F})$, and let $\mathcal{G} \subset \mathcal{F}$ be a sub-σ-algebra. Show that

$$\left. \frac{d\mathbb{R}}{d\mathbb{P}} \right|_{\mathcal{G}} = \left. \frac{d\mathbb{R}}{d\mathbb{Q}} \right|_{\mathcal{G}} \left. \frac{d\mathbb{Q}}{d\mathbb{P}} \right|_{\mathcal{G}}.$$

Exercise 7.2 The aim of this exercise is to verify the Dybvig–Ingersoll–Ross theorem (Lemma 7.3) for specific models.

(a) Show that the Vasiček short-rate model admits a long rate $R_\infty(t)$ if $\beta \leq 0$. Verify that it is nondecreasing.
(b) Same for the CIR model, without restrictions on β.
(c) Show that the HJM model with $d = 1$ and $\sigma(t, T) = (1 + T - t)^{-1/2}$ admits a strictly increasing long rate $R_\infty(t)$.

Exercise 7.3 Let $F(t; T, S)$ be the simple forward rate for $[T, S]$ prevailing at t. Show that $F(t; T, S)$, $t \leq T$, is a martingale with respect to some forward measure $\mathbb{Q}^u$; that is,

$$F(t; T, S) = \mathbb{E}_{\mathbb{Q}^u}[F(T; T, S) \mid \mathcal{F}_t].$$

What is u?

Exercise 7.4 Show that the price of a put option

$$p = \mathbb{E}_{\mathbb{Q}}\left[e^{-\int_0^T r(s)\,ds}(K - P(T, S))^+\right]$$

in the Gaussian setup of Proposition 7.2 is given by

$$p = KP(0, T)\Phi[-d_2] - P(0, S)\Phi[-d_1]$$

where $d_{1,2}$ are defined as in Proposition 7.2.

Exercise 7.5 Using Exercise 7.4, derive the ATM cap prices and Black volatilities in Table 7.1.

Exercise 7.6 Derive call and put bond option prices in the Ho–Lee short-rate model.

Exercise 7.7 Consider the Gaussian HJM model with a $d = 2$-dimensional driving Brownian motion $W = (W_1, W_2)^\top$, and forward rate dynamics

$$df(t, T) = \alpha(t, T)\,dt + \sigma_1(T - t)\,dW_1(t) + \sigma_2 e^{-a(T-t)}\,dW_2(t),$$

where σ_1, σ_2 and a are positive constants and $\alpha(t, T)$ the corresponding HJM drift.

(a) Derive the bond price dynamics.
(b) Compute the price of a European call option on an underlying bond.

Exercise 7.8 Complete the proof of Proposition 7.3. This includes the derivation of (7.10) by applying Itô's formula to $\frac{P(t,T)}{S(t)}$.

Exercise 7.9 (Jamshidian Decomposition) Consider an affine diffusion short-rate model $r(t)$ with zero-coupon bond prices

$$P(t, T) = e^{-A(t,T)-B(t,T)r(t)}$$

for some deterministic functions $A, B > 0$.

(a) Show that in this model, the price of a put option on a coupon bond is identical to the price of a portfolio of put options on zero-coupon bonds with appropriate strike prices. Hint: the function

$$r \mapsto p(r) = \sum_{i=1}^{n} c_i e^{-A(T_0,T_i)-B(T_0,T_i)r}$$

is strictly monotone in r. Show that $p(r) > p(r^*)$ if and only if $e^{-A(T_0,T_i)-B(T_0,T_i)r} > e^{-A(T_0,T_i)-B(T_0,T_i)r^*}$, for all i, for $r, r^* \in \mathbb{R}$.

(b) Now consider the Vasiček short-rate model with parameters as in Fig. 7.1. Use (a) to price a put option on the coupon bond with: coupon dates $T_i = (1 + i)/4$, $i = 0, \ldots, 3$, maturity of the bond $T_3 = 1$, coupons $c_i = 4$, nominal $N = 100$, the maturity of the option is $T_0 = 1/4$ (the option payoff does not include the coupon payment c_0), and the strike price of the option is $K =$ price of the underlying coupon bond at $t = 0$ divided by $P(0, T_0)$ ("at-the-money").

7.5 Notes

This chapter follows along the lines of Chap. 24 in [13]. The proof of Lemma 7.3 is adapted from Hubalek et al. [93]. The corresponding Exercise 7.2 is taken from Carmona and Tehranchi [35, Sect. 6.3.2]. Generalizations have recently been proposed by Goldammer and Schmock [82] and Kardaras and Platen [107]. Section 7.3.1 is based on [65], where a full sensitivity analysis is given with respect to all model parameters. The Black–Scholes call option formula with Vasiček short rates seems to have appeared in the finance literature for the first time in [133], albeit without probabilistic derivation. Exercises 7.6 and 7.7 are from Chap. 24 in [13]. Exercise 7.9 is based on the paper by Jamshidian [101].

Chapter 8
Forwards and Futures

In this chapter, we discuss two common types of term contracts: forwards, which are mainly traded over the counter (OTC), and futures, which are actively traded on many exchanges. The underlying is in both cases a T-claim $\mathcal{Y}$, for some fixed future date T. This can be an exchange rate, an interest rate, a commodity such as copper, any traded or non-traded asset, an index, etc. We discuss interest rate futures and futures rates in a separate section and relate them to forward rates in the Gaussian HJM model.

8.1 Forward Contracts

We consider the HJM setup from Chap. 6, and let $\mathcal{Y}$ denote a T-claim.

A *forward contract* on $\mathcal{Y}$, contracted at t, with time of delivery $T > t$, and with the *forward price* $f(t; T, \mathcal{Y})$ is defined by the following payment scheme:

- at T, the holder of the contract (long position) pays $f(t; T, \mathcal{Y})$ and receives $\mathcal{Y}$ from the underwriter (short position);
- at t, the forward price is chosen such that the present value of the forward contract is zero, thus

$$\mathbb{E}_{\mathbb{Q}}\left[e^{-\int_t^T r(s)\,ds}\left(\mathcal{Y} - f(t; T, \mathcal{Y})\right) \mid \mathcal{F}_t\right] = 0.$$

This is equivalent to

$$f(t; T, \mathcal{Y}) = \frac{1}{P(t,T)}\mathbb{E}_{\mathbb{Q}}\left[e^{-\int_t^T r(s)\,ds}\mathcal{Y} \mid \mathcal{F}_t\right] = \mathbb{E}_{\mathbb{Q}^T}\left[\mathcal{Y} \mid \mathcal{F}_t\right].$$

For example, the forward price at t of:

(a) a dollar delivered at T is 1;
(b) an S-bond delivered at $T \le S$ is $\frac{P(t,S)}{P(t,T)}$;
(c) any traded asset S delivered at T is $\frac{S(t)}{P(t,T)}$.

The forward price $f(s; T, \mathcal{Y})$ has to be distinguished from the (spot) price at time s of the forward contract entered at time $t \le s$, which is

$$\mathbb{E}_{\mathbb{Q}}\left[e^{-\int_s^T r(u)\,du}\left(\mathcal{Y} - f(t; T, \mathcal{Y})\right) \mid \mathcal{F}_s\right]$$

$$= \mathbb{E}_{\mathbb{Q}}\left[e^{-\int_s^T r(u)\,du}\mathcal{Y} \mid \mathcal{F}_s\right] - P(s,T)f(t; T, \mathcal{Y}).$$

D. Filipović, *Term-Structure Models*,
Springer Finance,
DOI 10.1007/978-3-540-68015-4_8, © Springer-Verlag Berlin Heidelberg 2009

8.2 Futures Contracts

A *futures contract* on $\mathcal{Y}$ with time of delivery T is defined as follows:

- at every $t \leq T$, there is a market quoted *futures price* $F(t; T, \mathcal{Y})$, which makes the futures contract on $\mathcal{Y}$, if entered at t, equal to zero;
- at T, the holder of the contract (long position) pays $F(T; T, \mathcal{Y})$ and receives $\mathcal{Y}$ from the underwriter (short position);
- during any infinitesimal time interval $(t, t + \Delta t]$ the holder of the contract receives (or pays, if negative) the amount $F(t; T, \mathcal{Y}) - F(t + \Delta t; T, \mathcal{Y})$ (this is called *marking to market* or *resettlement*).

Hence there is a continuous cash flow between the two parties of a futures contract. They are required to keep a certain amount of money as a safety margin.

The volumes in which futures are traded are huge. One of the reasons for this is that in many markets it is difficult to trade, or hedge, directly in the underlying object. This might be an index which includes many different illiquid instruments, or a commodity such as copper, gas or electricity. Holding a short position in a futures does not force you to physically deliver the underlying object if you exit the contract before delivery date. Selling short makes it possible to hedge against the underlying (see also Exercise 8.2).

Suppose $\mathbb{E}_{\mathbb{Q}}[|\mathcal{Y}|] < \infty$. Then the futures price process is given by the $\mathbb{Q}$-martingale

$$F(t; T, \mathcal{Y}) = \mathbb{E}_{\mathbb{Q}}[\mathcal{Y} \mid \mathcal{F}_t]. \tag{8.1}$$

In the following we give a heuristic argument for (8.1) based on the above characterization of a futures contract.

First, our model economy is driven by a Brownian motion and changes in a continuous way. Hence there is no reason to believe that futures prices evolve discontinuously, and we may assume that $F(t) = F(t; T, \mathcal{Y})$ is an Itô process.

Now suppose we enter the futures contract at time $t < T$. We face a continuous resettlement cash flow in the interval $(t, T]$. The cumulative discounted cash flow equals $V = \lim_N V_N$ with

$$V_N = \sum_{i=1}^{N} \frac{1}{B(t_i)} (F(t_i) - F(t_{i-1}))$$

where the limit is taken over a sequence of partitions $t = t_0 < \cdots < t_N = T$ with $\max_i |t_i - t_{i-1}| \to 0$ for $N \to \infty$. We can rewrite V_N as

$$V_N = \sum_{i=1}^{N} \frac{1}{B(t_{i-1})} (F(t_i) - F(t_{i-1})) + \sum_{i=1}^{N} \left(\frac{1}{B(t_i)} - \frac{1}{B(t_{i-1})} \right) (F(t_i) - F(t_{i-1})).$$

Note that $1/B \in \mathcal{L}(F)$, by continuity of B. It then follows from elementary stochastic calculus (see Sect. 4.1, and Exercise 4.2(b)) that V_N converges in probability towards

$$V = \int_t^T \frac{1}{B(s)} \, dF(s) + \int_t^T d\left\langle \frac{1}{B}, F \right\rangle_s = \int_t^T \frac{1}{B(s)} \, dF(s).$$

The second equality holds, since the co-variation of the absolutely continuous process $1/B$ and the Itô process F is zero.

The futures contract has present value zero, that is,

$$\mathbb{E}_{\mathbb{Q}} \left[\int_t^T \frac{1}{B(s)} \, dF(s) \mid \mathcal{F}_t \right] = 0.$$

We conclude that the Itô process

$$M(t) = \int_0^t \frac{1}{B(s)} \, dF(s) = \mathbb{E}_{\mathbb{Q}} \left[\int_0^T \frac{1}{B(s)} \, dF(s) \mid \mathcal{F}_t \right]$$

is a $\mathbb{Q}$-martingale for $t \leq T$. If, moreover,

$$\mathbb{E}_{\mathbb{Q}} \left[\int_0^T B(s)^2 \, d\langle M, M \rangle_s \right] = \mathbb{E}_{\mathbb{Q}} \left[\langle F, F \rangle_T \right] < \infty$$

then $B \in \mathcal{L}^2(M)$ and

$$F(t) = \int_0^t B(s) \, dM(s)$$

is a $\mathbb{Q}$-martingale for $t \leq T$, which implies (8.1).

8.2.1 Interest Rate Futures

Interest rate futures contracts may be divided into futures on short-term instruments and futures on coupon bonds. We only consider an example from the first group.

Eurodollars are deposits of US dollars in institutions outside of the US. LIBOR is the interbank rate of interest for Eurodollar loans. The *Eurodollar futures* contract is tied to the LIBOR. It was introduced by the International Money Market (IMM) of the Chicago Mercantile Exchange (CME) in 1981, and is designed to protect its owner from fluctuations in the 3-month ($=1/4$ year) LIBOR. The maturity (delivery) months are March, June, September and December.

Fix a maturity date T and let $L(T)$ denote the 3-month LIBOR for the period $[T, T + 1/4]$, prevailing at T. The market *quote* of the Eurodollar futures contract on $L(T)$ at time $t \leq T$ is

$$1 - L_F(t, T) \quad [100 \text{ per cent}]$$

where $L_F(t, T)$ is the corresponding *futures rate* (compare with the example in Sect. 3.2.2). As t tends to T, $L_F(t, T)$ tends to $L(T)$. The *futures price*, used for the marking to market, is defined by

$$F(t; T, L(T)) = 1 - \frac{1}{4} L_F(t, T) \quad \text{[million dollars]}.$$

Consequently, a change of 1 basis point (0.01%) in the futures rate $L_F(t, T)$ leads to a cash flow of

$$10^6 \times 10^{-4} \times \frac{1}{4} = 25 \quad \text{[dollars]}.$$

We also see that the final price $F(T; T, L(T)) = 1 - \frac{1}{4} L(T) = \mathcal{Y}$ is not $P(T, T + 1/4) = 1 - \frac{1}{4} L(T) P(T, T + 1/4)$ as one might suppose. In fact, the underlying $\mathcal{Y}$ is a synthetic value. At maturity there is no physical delivery. Instead, settlement is made in cash.

On the other hand, since

$$1 - \frac{1}{4} L_F(t, T) = F(t; T, L(T))$$

$$= \mathbb{E}_{\mathbb{Q}} [F(T; T, L(T)) \mid \mathcal{F}_t] = 1 - \frac{1}{4} \mathbb{E}_{\mathbb{Q}} [L(T) \mid \mathcal{F}_t],$$

we obtain an explicit formula for the futures rate

$$L_F(t, T) = \mathbb{E}_{\mathbb{Q}} [L(T) \mid \mathcal{F}_t].$$

8.3 Forward vs. Futures in a Gaussian Setup

Let S be the price process of a traded asset. Hence the $\mathbb{Q}$-dynamics of S is of the form

$$\frac{dS(t)}{S(t)} = r(t) \, dt + \rho(t) \, dW^*(t),$$

for some volatility process ρ. Fix a delivery date T. The forward and futures prices of S for delivery at T are

$$f(t; T, S(T)) = \frac{S(t)}{P(t, T)}, \qquad F(t; T, S(T)) = \mathbb{E}_{\mathbb{Q}}[S(T) \mid \mathcal{F}_t].$$

Under Gaussian assumption we can establish the relationship between the two prices.

Proposition 8.1 *Suppose $\rho(t)$ and $v(t, T)$ are deterministic functions in t, where $v(t, T)$ denotes the T-bond volatility (6.3). Then*

$$F(t; T, S(T)) = f(t; T, S(T)) e^{\int_t^T (v(s,T) - \rho(s)) v(s,T)^\top ds}$$

for $t \leq T$.

Hence, if the instantaneous covariation of $S(t)$ and $P(t, T)$ is negative,

$$\frac{d\langle S, P(\cdot, T)\rangle_t}{dt} = S(t) P(t, T) \rho(t) v(t, T)^\top \le 0,$$

then the futures price dominates the forward price.

Proof Write $\mu(s) = v(s, T) - \rho(s)$. It is clear that

$$f(t; T, S(T)) = \frac{S(0)}{P(0, T)} \mathcal{E}_t(\mu \bullet W^*) \exp\left(\int_0^t \mu(s) v(s, T)^\top ds\right).$$

Since $\mathcal{E}(\mu \bullet W^*)$ is a $\mathbb{Q}$-martingale and $\rho(s)$ and $v(s, T)$ are deterministic, we obtain

$$F(t; T, S(T)) = \mathbb{E}_{\mathbb{Q}}[f(T; T, S(T)) \mid \mathcal{F}_t] = f(t; T, S(T)) e^{\int_t^T \mu(s) v(s, T)^\top ds},$$

as desired.   □

Similarly, one can show ($\to$ Exercise 8.3):

Lemma 8.1 *In the Gaussian HJM framework (7.8) we have the following relations (convexity adjustments) between instantaneous forward and futures rates:*

$$f(t, T) = \mathbb{E}_{\mathbb{Q}}[r(T) \mid \mathcal{F}_t] - \int_t^T \left(\sigma(s, T) \int_s^T \sigma(s, u)^\top du\right) ds,$$

and simple forward and futures rates

$$F(t; T, S) = \mathbb{E}_{\mathbb{Q}}[F(T, S) \mid \mathcal{F}_t]$$
$$- \frac{P(t, T)}{(S - T) P(t, S)} \left(e^{\int_t^T (\int_T^S \sigma(s,v) dv \int_s^S \sigma(s,u)^\top du) ds} - 1\right),$$

for $t \le T < S$, respectively.

Hence, if

$$\sigma(s, v) \sigma(s, u)^\top \ge 0 \quad \text{for all } s \le u \wedge v$$

then futures rates are always greater than the corresponding forward rates.

8.4 Exercises

Exercise 8.1 Consider the trivial HJM model where the forward rates $f(t, T)$ are deterministic. Show that all forward measures collapse into the risk-neutral measure, $\mathbb{Q}^T = \mathbb{Q}$, and forward prices equal futures prices.

Exercise 8.2 Consider the Black–Scholes model in Exercise 4.7. Show that the European call option on the stock S with payoff $(S(T) - K)^+$ at maturity T can be replicated by a portfolio based on the money-market account B and the futures contract on $S(T)$.

Exercise 8.3 Prove Lemma 8.1.

8.5 Notes

This chapter follows [13, Chap. 26], see also Hull [95]. Section 8.2.1, in particular, follows [163, Sect. 5.4]. Exercise 8.2 is taken from [13, Chap. 26].

Chapter 9
Consistent Term-Structure Parametrizations

Practitioners and academics alike have a vital interest in parameterized term-structure models. In this chapter, we take up a point left open at the end of Chap. 3, and exploit whether parameterized curve families $\phi(\cdot, z)$, used for estimating the forward curve, go well with arbitrage-free interest rate models. According to Table 3.4, taken from the BIS document [11], there is a rich source of cross-sectional data, that is, daily estimations of the parameter z, for the Nelson–Siegel and Svensson families. This suggests that calibrating a diffusion process Z for the parameter z would lead to an accurate factor model for the forward curve. Conditions for the absence of arbitrage can be formulated in terms of the drift and diffusion of Z and derivatives of ϕ. These conditions turn out to be surprisingly restrictive in some cases.

9.1 Multi-factor Models

We have seen in Sect. 5.2 that every time-homogeneous diffusion short-rate model $r(t)$ induces forward rates of the form

$$f(t, T) = \phi(T - t, r(t)),$$

for some deterministic function ϕ. This simple form has its obvious computational merits (see for instance Exercise 7.9). On the other hand, it has rather unrealistic implications. For example the family of attainable forward curves

$$\{\phi(\cdot, r) \mid r \in \mathbb{R}\}$$

is only one-dimensional. In other words, the movements of the entire term-structure are explained by the single state variable $r(t)$. This is too restrictive with regard to the principal component analysis of the term-structure from Sect. 3.4.3, which implies that (at least) two or three factors are needed for a statistically accurate description of the forward curve movements.[1]

---

[1] Carmona and Tehranchi [35] also discuss the aspect of maturity-specific risk. If the number d of driving Brownian motions is too low then hedging instruments can be chosen independently of the claim to be hedged. For instance if $d = 1$, such as in a short-rate model, then an option on a bond of maturity five years could be perfectly hedged by the money-market account and a bond of maturity 30 years.

D. Filipović, *Term-Structure Models,*
Springer Finance,
DOI 10.1007/978-3-540-68015-4_9, © Springer-Verlag Berlin Heidelberg 2009

To gain more flexibility, we now allow for multiple factors. Fix $m \geq 1$ and a closed state space $\mathcal{Z} \subset \mathbb{R}^m$ with non-empty interior. An m factor model is an interest rate model of the form

$$f(t, T) = \phi(T - t, Z(t))$$

where ϕ is a deterministic function and Z is a $\mathcal{Z}$-valued diffusion state process solving the stochastic differential equation

$$dZ(t) = b(Z(t)) \, dt + \rho(Z(t)) \, dW^*(t),$$

$$Z(0) = z.$$

Here W^* is a d-dimensional Brownian motion defined on a filtered probability space $(\Omega, \mathcal{F}, (\mathcal{F}_t), \mathbb{Q})$, satisfying the usual conditions. We assume that:

(**A1**) $\phi \in C^{1,2}(\mathbb{R}_+ \times \mathcal{Z})$;
(**A2**) the function $b : \mathcal{Z} \to \mathbb{R}^m$ is continuous, and $\rho : \mathcal{Z} \to \mathbb{R}^{m \times d}$ is measurable and such that the diffusion matrix

$$a(z) = \rho(z)\rho(z)^\top$$

is continuous in $z \in \mathcal{Z}$ (see Remark 4.2);
(**A3**) the above stochastic differential equation has a $\mathcal{Z}$-valued solution $Z = Z^z$, for every $z \in \mathcal{Z}$;
(**A4**) $\mathbb{Q}$ is the risk-neutral local martingale measure for the induced bond prices

$$P(t, T) = \exp\left(-\int_0^{T-t} \phi(x, Z^z(t)) \, dx \right),$$

for all $z \in \mathcal{Z}$.

The short rates are now given by $r(t) = \phi(0, Z(t))$. Hence Assumption (**A4**) holds if and only if

$$\frac{\exp(-\int_0^{T-t} \phi(x, Z^z(t)) \, dx)}{\exp(\int_0^t \phi(0, Z^z(s)) \, ds)}, \quad t \leq T, \tag{9.1}$$

is a $\mathbb{Q}$-local martingale, for all $z \in \mathcal{Z}$. Applying Itô's formula to (9.1), the consistency condition in Proposition 9.1 below could now be derived by a direct calculation ($\rightarrow$ Exercise 9.1). Here, instead, we first show that the above factor model can be embedded into the HJM framework of Chap. 6 and then make use of the HJM drift condition, Theorem 6.1.

Remark 9.1 Time-inhomogeneous models are included by identifying one component, say Z_1, with calendar time. We therefore set $dZ_1 = dt$, which is equivalent to $b_1 \equiv 1$ and $\rho_{1j} \equiv 0$ for $j = 1, \ldots, d$. Calendar time at inception is now $Z_1(0) = z_1$, and t, T, etc. accordingly denote relative time with respect to z_1. The requirement (**A4**) for all $z \in \mathcal{Z}$ now means, in particular, that absence of arbitrage holds relative to any initial calendar time z_1. See also Exercise 9.4 below.

9.2 Consistency Condition

Since the function $(x, z) \mapsto \phi(x, z)$ is in $C^{1,2}(\mathbb{R}_+ \times \mathcal{Z})$ we can apply Itô's formula and obtain

$$
df(t, T) = \Bigg(-\partial_x \phi(T - t, Z(t)) + \sum_{i=1}^{m} b_i(Z(t)) \partial_{z_i} \phi(T - t, Z(t))
$$
$$
+ \frac{1}{2} \sum_{i,j=1}^{m} a_{ij}(Z(t)) \partial_{z_i} \partial_{z_j} \phi(T - t, Z(t)) \Bigg) dt
$$
$$
+ \sum_{i=1}^{m} \sum_{j=1}^{d} \partial_{z_i} \phi(T - t, Z(t)) \rho_{ij}(Z(t)) \, dW_j^*(t).
$$

Hence the induced forward rate model is of the HJM type (6.1) with

$$
\alpha(t, T) = -\partial_x \phi(T - t, Z(t)) + \sum_{i=1}^{m} b_i(Z(t)) \partial_{z_i} \phi(T - t, Z(t))
$$
$$
+ \frac{1}{2} \sum_{i,j=1}^{m} a_{ij}(Z(t)) \partial_{z_i} \partial_{z_j} \phi(T - t, Z(t)) \tag{9.2}
$$

and

$$
\sigma_j(t, T) = \sum_{i=1}^{m} \partial_{z_i} \phi(T - t, Z(t)) \rho_{ij}(Z(t)), \quad j = 1, \dots, d, \tag{9.3}
$$

satisfying **(HJM.1)–(HJM.3)** ($\to$ Exercise 9.2).

From Theorem 6.1 we know that **(A4)** is equivalent to the HJM drift condition (6.4), which reads here as

$$
-\phi(T - t, Z(t)) + \phi(0, Z(t)) + \sum_{i=1}^{m} b_i(Z(t)) \partial_{z_i} \Phi(T - t, Z(t))
$$
$$
+ \frac{1}{2} \sum_{i,j=1}^{m} a_{ij}(Z(t)) \partial_{z_i} \partial_{z_j} \Phi(T - t, Z(t))
$$
$$
= \frac{1}{2} \sum_{j=1}^{d} \left(\sum_{i=1}^{m} \partial_{z_i} \Phi(T - t, Z(t)) \rho_{ij}(Z(t)) \right)^2
$$
$$
= \frac{1}{2} \sum_{k,l=1}^{m} a_{kl}(Z(t)) \partial_{z_k} \Phi(T - t, Z(t)) \partial_{z_l} \Phi(T - t, Z(t)),
$$

where we define

$$\Phi(x, z) = \int_0^x \phi(u, z)\, du.$$

This has to hold a.s. for all $t \leq T$ and initial points $z = Z(0)$. Letting $t \to 0$, and replacing T by x, we thus get the following result.

Proposition 9.1 (Consistency Condition) *Under the above assumptions* **(A1)**– **(A3)**, *there is equivalence between* **(A4)** *and*

$$\partial_x \Phi(x, z) = \phi(0, z) + \sum_{i=1}^m b_i(z) \partial_{z_i} \Phi(x, z)$$

$$+ \frac{1}{2} \sum_{i,j=1}^m a_{ij}(z) \left(\partial_{z_i} \partial_{z_j} \Phi(x, z) - \partial_{z_i} \Phi(x, z) \partial_{z_j} \Phi(x, z) \right) \quad (9.4)$$

for all $(x, z) \in \mathbb{R}_+ \times \mathcal{Z}$.

This result motivates the following terminology:

Definition 9.1 The pair of characteristics $\{a, b\}$ and the forward curve parametrization ϕ are *consistent* if **(A4)**, or equivalently the consistency condition (9.4), holds.

There are two ways to approach equation (9.4). First, one takes $\phi(0, z)$, a and b as given and looks for a solution Φ for the partial differential equation (9.4) with initial condition $\Phi(0, z) = 0$. Or, one takes ϕ as given (a parametric estimation method for the term-structure) and tries to find a and b such that the partial differential equation (9.4) is satisfied for all (x, z). This is an *inverse problem*. It turns out that the latter approach is quite restrictive on possible choices of a and b.

Proposition 9.2 *Suppose that the functions*

$$\partial_{z_i} \Phi(\cdot, z) \quad and \quad \frac{1}{2} \left(\partial_{z_i} \partial_{z_j} \Phi(\cdot, z) - \partial_{z_i} \Phi(\cdot, z) \partial_{z_j} \Phi(\cdot, z) \right),$$

for $1 \leq i \leq j \leq m$, *are linearly independent for all z in some dense subset* $\mathcal{D} \subset \mathcal{Z}$. *Then there exists one and only one consistent pair* $\{a, b\}$.

Proof Set $M = m + m(m+1)/2$, the number of unknown functions b_k and $a_{kl} = a_{lk}$. Let $z \in \mathcal{D}$. Then there exists a sequence $0 \leq x_1 < \cdots < x_M$ such that the $M \times M$-matrix with kth row vector built by

$$\partial_{z_i} \Phi(x_k, z) \quad and \quad \frac{1}{2} \left(\partial_{z_i} \partial_{z_j} \Phi(x_k, z) - \partial_{z_i} \Phi(x_k, z) \partial_{z_j} \Phi(x_k, z) \right),$$

for $1 \leq i \leq j \leq m$, is invertible. Thus, $b(z)$ and $a(z)$ are uniquely determined by (9.4). This holds for each $z \in \mathcal{D}$. By continuity of b and a hence for all $z \in \mathcal{Z}$. $\square$

This result has the following important practical implications: suppose that the parameterized curve family

$$\{\phi(\cdot, z) \mid z \in \mathcal{Z}\}$$

is used for daily estimation of the forward curve in terms of the state variable z. Then the above proposition tells us that, under the stated assumption, any consistent $\mathbb{Q}$-diffusion model Z for z is fully determined by ϕ.

Moreover, the corresponding diffusion matrix, $a(z)$, of Z is not affected by any equivalent measure transformation. Consequently, statistical calibration is only possible for the drift of the model (or equivalently, for the market price of risk), since the observations of z are made under the objective measure $\mathbb{P} \sim \mathbb{Q}$, where $d\mathbb{Q}/d\mathbb{P}$ is left unspecified by our consistency considerations.

9.3 Affine Term-Structures

We first look at the simplest case, namely a time-homogeneous *affine term-structure* (*ATS*)

$$\phi(x, z) = g_0(x) + g_1(x)z_1 + \cdots + g_m(x)z_m. \tag{9.5}$$

Here the second-order z-derivatives vanish, and (9.4) reduces to

$$g_0(x) - g_0(0) + \sum_{i=1}^{m} z_i (g_i(x) - g_i(0)) = \sum_{i=1}^{m} b_i(z)G_i(x) - \frac{1}{2} \sum_{i,j=1}^{m} a_{ij}(z)G_i(x)G_j(x), \tag{9.6}$$

where we define

$$G_i(x) = \int_0^x g_i(u)\, du.$$

Now if the $m + m(m+1)/2$ functions

$$G_1, \ldots, G_m, \ G_1 G_1, \ G_1 G_2, \ldots, G_m G_m \tag{9.7}$$

are linearly independent, we can invert and solve the linear equation (9.6) for a and b, as we did in the proof of Proposition 9.2. Since the left-hand side of (9.6) is affine is z, we obtain that also a and b are affine of the form

$$a_{ij}(z) = a_{ij} + \sum_{k=1}^{m} \alpha_{k;ij} z_k,$$

$$b_i(z) = b_i + \sum_{j=1}^{m} \beta_{ij} z_j, \tag{9.8}$$

for some constants a_{ij}, $\alpha_{k;ij}$, b_i and β_{ij}. Plugging this back into (9.6) and matching constant terms and terms containing z_ks we obtain

$$\partial_x G_0(x) = g_0(0) + \sum_{i=1}^{m} b_i G_i(x) - \frac{1}{2} \sum_{i,j=1}^{m} a_{ij} G_i(x) G_j(x), \tag{9.9}$$

$$\partial_x G_k(x) = g_k(0) + \sum_{i=1}^{m} \beta_{ki} G_i(x) - \frac{1}{2} \sum_{i,j=1}^{m} \alpha_{k;ij} G_i(x) G_j(x). \tag{9.10}$$

We have thus proved:

Proposition 9.3 *Suppose the functions in (9.7) are linearly independent. If the pair $\{a, b\}$ is consistent with the ATS (9.5) then a and b are necessarily affine of the form (9.8). Moreover, the functions G_i solve the system of Riccati equations (9.9)–(9.10) with initial conditions $G_i(0) = 0$.*

Conversely, suppose a and b are affine of the form (9.8), and let $g_i(0)$ be some given constants. If the functions G_i solve the system of Riccati equations (9.9)–(9.10) with initial conditions $G_i(0) = 0$, then the ATS (9.5) is consistent with $\{a, b\}$.

This proposition extends to the multi-factor case that we found in Sect. 5.3 for the time-homogeneous one-factor case with

$$A(t, T) = G_0(T - t) \quad \text{and} \quad B(t, T) = G_1(T - t).$$

Note that we did not have to assume linear independence of the respective functions (9.7) in Proposition 5.2. However, this assumption becomes necessary as soon as $m \geq 2$ ($\rightarrow$ Exercise 9.3).

Note also that here we have the freedom to choose the constants $g_i(0)$ which are related to the short rates by

$$r(t) = f(t, t) = g_0(0) + g_1(0) Z_1(t) + \cdots + g_m(0) Z_m(t). \tag{9.11}$$

A typical choice is $g_1(0) = 1$ and all the other $g_i(0) = 0$, whence $Z_1(t)$ is the—in general non-Markovian—short-rate process.

9.4 Polynomial Term-Structures

We extend the ATS setup and consider a *polynomial term-structure (PTS)*

$$\phi(x, z) = \sum_{|\mathbf{i}|=0}^{n} g_{\mathbf{i}}(x) z^{\mathbf{i}}, \tag{9.12}$$

where we use the multi-index notation $\mathbf{i} = (i_1, \ldots, i_m)$, $|\mathbf{i}| = i_1 + \cdots + i_m$ and $z^{\mathbf{i}} = z_1^{i_1} \cdots z_m^{i_m}$. Here n denotes the *degree* of the PTS; that is, there exists an index $\mathbf{i}$ with $|\mathbf{i}| = n$ and $g_{\mathbf{i}} \neq 0$.

Obviously, for $n = 1$ we are back to an ATS. For $n = 2$ we have a *quadratic term-structure (QTS)*, which has also been intensively studied in the literature, see the notes section for some references.

The following question now arises naturally: do we gain something by looking at $n = 3$ and higher-degree PTS models? The answer is, surprisingly, no. In fact, we now shall show the amazing result that there is no consistent PTS for $n > 2$.

To make the results better accessible to the reader, we first state and prove them for the case $m = 1$. This avoids multi-index notation. The general case will then be treated in the following section.

9.4.1 Special Case: $m = 1$

In this case, we simply identify bold $\mathbf{i}$ with $i \equiv |\mathbf{i}| = i_1 \in \{0, \ldots, n\}$. In particular, the PTS (9.12) now reads

$$\phi(x, z) = \sum_{i=0}^{n} g_i(x) z^i.$$

We denote the integral of g_i by

$$G_i(x) = \int_0^x g_i(u)\, du.$$

Theorem 9.1 (Maximal Degree Problem I) *Suppose that G_i and $G_i G_j$ are linearly independent functions, for $1 \le i \le j \le n$, and that $\rho \not\equiv 0$.*

Then consistency implies $n \in \{1, 2\}$. Moreover, $b(z)$ and $a(z)$ are polynomials in z with $\deg b(z) \le 1$ in any case (QTS and ATS), and $\deg a(z) = 0$ if $n = 2$ (QTS) and $\deg a(z) \le 1$ if $n = 1$ (ATS).

Proof Equation (9.4) can be rewritten

$$\sum_{i=0}^{n} (g_i(x) - g_i(0)) z^i = \sum_{i=0}^{n} G_i(x) B_i(z) - \sum_{i,j=0}^{n} G_i(x) G_j(x) A_{ij}(z), \qquad (9.13)$$

where we define

$$B_i(z) = b(z) i z^{i-1} + \frac{1}{2} a(z) i (i-1) z^{i-2},$$

$$A_{ij}(z) = \frac{1}{2} a(z) i j z^{i-1} z^{j-1}.$$

By assumption we can solve the linear equation (9.13) for B and A, and thus $B_i(z)$ and $A_{ij}(z)$ are polynomials in z of order less than or equal n. In particular, this holds for

$$B_1(z) = b(z) \quad \text{and} \quad 2A_{11}(z) = a(z).$$

But then, since $a \neq 0$ by assumption, $2A_{nn}(z) = a(z)n^2 z^{2n-2}$ cannot be a polynomial of order less than or equal n unless $2n - 2 \leq n$, which implies $n \leq 2$. The theorem is thus proved for $n = 1$. For $n = 2$, we obtain $\deg a(z) = 0$ and thus $B_2(z) = 2b(z)z + a(z)$. Hence also in this case $\deg b(z) \leq 1$, and the theorem is proved for $m = 1$. $\qquad\qquad\qquad\qquad\qquad\qquad\qquad\qquad\qquad\qquad\qquad\qquad\qquad$ $\square$

We can relax the linear independence hypothesis on G_i, $G_i G_j$ in Theorem 9.1 as follows.

Theorem 9.2 (Maximal Degree Problem II) *Suppose that:*

(a) $\sup \mathcal{Z} = \infty$;
(b) b *and* ρ *satisfy a linear growth condition*

$$|b(z)| + |\rho(z)| \leq C(1 + |z|), \quad z \in \mathcal{Z},$$

    for some finite constant C;
(c) $a(z)$ *is asymptotically bounded away from zero:*

$$\liminf_{z \to \infty} a(z) > 0.$$

Then consistency implies $n \in \{1, 2\}$.

Note that the linear growth condition (b) is standard for asserting non-explosion of the diffusion Z.

Proof Again, we consider equation (9.4), which reads

$$\sum_{i=0}^{n} (g_i(x) - g_i(0)) z^i$$

$$= b(z) \sum_{i=0}^{n} G_i(x) i z^{i-1}$$

$$+ \frac{1}{2} a(z) \left(\sum_{i,j=0}^{n} G_i(x) i(i-1) z^{i-2} - \left(\sum_{i=0}^{n} G_i(x) i z^{i-1} \right)^2 \right). \quad (9.14)$$

We argue by contradiction and assume that $n > 2$, which implies $2n - 2 > n$. Dividing (9.14) by z^{2n-2}, for $z \neq 0$, yields

$$\frac{1}{2} a(z) \frac{\left(\sum_{i=0}^{n} G_i(x) i z^{i-1} \right)^2}{z^{2n-2}}$$

$$= \frac{b(z) \sum_{i=0}^{n} G_i(x) i z^{i-1}}{z} \frac{1}{z^{2n-3}} + \frac{a(z)}{2z^2} \frac{\sum_{i,j=0}^{n} G_i(x) i(i-1) z^{i-2}}{z^{2n-4}}$$

$$- \frac{\sum_{i=0}^{n} (g_i(x) - g_i(0)) z^i}{z^{2n-2}}.$$

By assumption (a) this holds for all z large enough. The right-hand side converges to zero, for $z \to \infty$, by assumption (b). Taking the lim inf of the left-hand side yields by (c), that

$$\frac{1}{2} \liminf_{z \to \infty} a(z) G_n^2(x) n^2 > 0,$$

a contradiction. Thus $n \leq 2$. $\qquad\square$

9.4.2 General Case: $m \geq 1$

For $\mu \in \{1, \ldots, n\}$ and $k \in \{1, \ldots, m\}$ we write $(\mu)_k$ for the multi-index with μ at the kth position and zeros elsewhere. Let $\mathbf{i}_1, \mathbf{i}_2, \ldots, \mathbf{i}_N$ be a numbering of the set of multi-indices

$$I = \{\mathbf{i} = (i_1, \ldots, i_m) \mid |\mathbf{i}| \leq n\}, \quad \text{where } N = |I| = \sum_{|\mathbf{i}|=0}^{n} 1.$$

As above, we denote the integral of $g_\mathbf{i}$ by

$$G_\mathbf{i}(x) = \int_0^x g_\mathbf{i}(u)\, du.$$

Theorem 9.3 (Maximal Degree Problem I) *Suppose that $G_{\mathbf{i}_\mu}$ and $G_{\mathbf{i}_\mu} G_{\mathbf{i}_\nu}$ are linearly independent functions, $1 \leq \mu \leq \nu \leq N$, and that $\rho \not\equiv 0$.*

Then consistency implies $n \in \{1, 2\}$. Moreover, $b(z)$ and $a(z)$ are polynomials in z with $\deg b(z) \leq 1$ in any case (QTS and ATS), and $\deg a(z) = 0$ if $n = 2$ (QTS) and $\deg a(z) \leq 1$ if $n = 1$ (ATS).

Proof Define the functions

$$B_\mathbf{i}(z) = \sum_{k=1}^{m} b_k(z) \frac{\partial z^\mathbf{i}}{\partial z_k} + \frac{1}{2} \sum_{k,l=1}^{m} a_{kl}(z) \frac{\partial^2 z^\mathbf{i}}{\partial z_k \partial z_l}, \tag{9.15}$$

$$A_{\mathbf{i}\mathbf{j}}(z) = A_{\mathbf{j}\mathbf{i}}(z) = \frac{1}{2} \sum_{k,l=1}^{m} a_{kl}(z) \frac{\partial z^\mathbf{i}}{\partial z_k} \frac{\partial z^\mathbf{j}}{\partial z_l}. \tag{9.16}$$

Equation (9.4) can be rewritten

$$\sum_{\mu=1}^{N} \left(g_{\mathbf{i}_\mu}(x) - g_{\mathbf{i}_\mu}(0)\right) z^{\mathbf{i}_\mu} = \sum_{\mu=1}^{N} G_{\mathbf{i}_\mu}(x) B_{\mathbf{i}_\mu}(z) - \sum_{\mu,\nu=1}^{N} G_{\mathbf{i}_\mu}(x) G_{\mathbf{i}_\nu}(x) A_{\mathbf{i}_\mu \mathbf{i}_\nu}(z). \tag{9.17}$$

By assumption we can solve this linear equation for B and A, and thus $B_\mathbf{i}(z)$ and $A_{\mathbf{i}\mathbf{j}}(z)$ are polynomials in z of order less than or equal n. In particular, we have

$$B_{(1)_k}(z) = b_k(z),$$

$$2 A_{(1)_k (1)_l}(z) = a_{kl}(z), \quad k, l \in \{1, \ldots, m\}, \tag{9.18}$$

hence $b(z)$ and $a(z)$ are polynomials in z with $\deg b(z), \deg a(z) \le n$. An easy calculation shows that

$$2A_{(n)_k(n)_k}(z) = a_{kk}(z)n^2 z_k^{2n-2}, \quad k \in \{1, \ldots, m\}. \tag{9.19}$$

We may assume that $a_{kk} \not\equiv 0$, since $\rho \not\equiv 0$. But then the right-hand side of (9.19) cannot be a polynomial in z of order less than or equal n unless $n \le 2$. This proves the first part of the theorem.

If $n = 1$ there is nothing more to prove. Now let $n = 2$. Notice that by definition

$$\deg_\mu a_{kl}(z) \le (\deg_\mu a_{kk}(z) + \deg_\mu a_{ll}(z))/2,$$

where $\deg_\mu$ denotes the degree of dependence on the single component z_μ. Equation (9.19) yields $\deg_k a_{kk}(z) = 0$. Hence $\deg_l a_{kl}(z) \le 1$. Consider

$$2A_{(1)_k+(1)_l,(1)_k+(1)_l}(z) = a_{kk}(z)z_l^2 + 2a_{kl}(z)z_k z_l + a_{ll}(z)z_k^2, \quad k, l \in \{1, \ldots, m\}.$$

From the preceding arguments it is now clear that also $\deg_l a_{kk}(z) = 0$, and hence $\deg a(z) = 0$. We finally have

$$B_{(1)_k+(1)_l}(z) = b_k(z)z_l + b_l(z)z_k + a_{kl}(z), \quad k, l \in \{1, \ldots, m\},$$

from which we conclude that $\deg b(z) \le 1$. □

As above, we can relax the hypothesis on G in Theorem 9.3 as follows.

Theorem 9.4 (Maximal Degree Problem II) *Suppose that*:

(a) $\mathcal{Z}$ *is a cone*;
(b) b *and* ρ *satisfy a linear growth condition*

$$\|b(z)\| + \|\rho(z)\| \le C(1 + \|z\|), \quad z \in \mathcal{Z}, \tag{9.20}$$

for some finite constant C;
(c) $a(z)$ *becomes uniformly elliptic for* $\|z\|$ *large enough*:

$$\langle a(z)v, v \rangle \ge k(z)\|v\|^2, \quad v \in \mathbb{R}^m, \tag{9.21}$$

for some function $k : \mathcal{Z} \to \mathbb{R}_+$ *with*

$$\liminf_{z \in \mathcal{Z}, \|z\| \to \infty} k(z) > 0. \tag{9.22}$$

Then consistency implies $n \in \{1, 2\}$.

Note that the linear growth condition (9.20) is standard for asserting non-explosion of the diffusion Z.

Proof We shall make use of the basic inequality

$$|z^{\mathbf{i}}| \le \|z\|^{|\mathbf{i}|}, \quad z \in \mathbb{R}^m. \tag{9.23}$$

This is immediate, since

$$\frac{|z^{\mathbf{i}}|}{\|z\|^{|\mathbf{i}|}} = \left(\frac{|z_1|}{\|z\|}\right)^{i_1} \cdots \left(\frac{|z_m|}{\|z\|}\right)^{i_m} \le 1, \quad z \in \mathbb{R}^m \setminus \{0\}.$$

Now define

$$\Gamma_k(x, z) = \sum_{\mu=1}^{N} G_{\mathbf{i}_\mu}(x) \frac{\partial z^{\mathbf{i}_\mu}}{\partial z_k}, \tag{9.24}$$

$$\Lambda_{kl}(x, z) = \Lambda_{lk}(x, z) = \sum_{\mu=1}^{N} G_{\mathbf{i}_\mu}(x) \frac{\partial^2 z^{\mathbf{i}_\mu}}{\partial z_k \partial z_l}. \tag{9.25}$$

Then (9.4) can be rewritten as

$$\sum_{|\mathbf{i}|=0}^{n} (g_{\mathbf{i}}(x) - g_{\mathbf{i}}(0)) z^{\mathbf{i}} = \sum_{k=1}^{m} b_k(z) \Gamma_k(x, z)$$

$$+ \frac{1}{2} \sum_{k,l=1}^{m} a_{kl}(z) \left(\Lambda_{kl}(x, z) - \Gamma_k(x, z)\Gamma_l(x, z)\right). \tag{9.26}$$

We now argue by contradiction and suppose that $n > 2$. We have from (9.24)

$$\Gamma_k(x, z) = \sum_{|\mathbf{i}|=n} G_{\mathbf{i}}(x) i_k z^{\mathbf{i}-(1)_k} + \cdots = P_k(x, z) + \cdots,$$

where $P_k(x, z)$ is a homogeneous polynomial in z of order $n - 1$, and $\cdots$ stands for lower-order terms in z. By assumptions there exist $x \in \mathbb{R}_+$ and $k \in \{1, \dots, m\}$ such that $P_k(x, \cdot) \ne 0$. Choose $z^* \in \mathcal{Z} \setminus \{0\}$ with $P_k(x, z^*) \ne 0$ and set $z_\alpha = \alpha z^*$, for $\alpha > 0$. In view of (a), we have $z_\alpha \in \mathcal{Z}$ and

$$\Gamma_k(x, z_\alpha) = \alpha^{n-1} P_k(x, z^*) + \cdots,$$

where $\cdots$ denotes lower-order terms in α. Consequently,

$$\lim_{\alpha \to \infty} \frac{\Gamma_k(x, z_\alpha)}{\|z_\alpha\|^{n-1}} = \frac{P_k(x, z^*)}{\|z^*\|^{n-1}} \ne 0. \tag{9.27}$$

Combining (9.21) and (9.22) with (9.27) we conclude that

$$L = \liminf_{\alpha \to \infty} \frac{1}{\|z_\alpha\|^{2n-2}} \langle a(z_\alpha)\Gamma(x, z_\alpha), \Gamma(x, z_\alpha) \rangle$$

$$\ge \liminf_{\alpha \to \infty} k(z_\alpha) \frac{\|\Gamma(x, z_\alpha)\|^2}{\|z_\alpha\|^{2n-2}} > 0. \tag{9.28}$$

On the other hand, by (9.26),

$$
L \le \sum_{|\mathbf{i}|=0}^{n} |g_{\mathbf{i}}(x) - g_{\mathbf{i}}(0)| \frac{|z_\alpha^{\mathbf{i}}|}{\|z_\alpha\|^{2n-2}}
$$

$$
+ \frac{\|b(z_\alpha)\|}{\|z_\alpha\|} \frac{\|\Gamma(x, z_\alpha)\|}{\|z_\alpha\|^{2n-3}} + \frac{1}{2} \frac{\|a(z_\alpha)\|}{\|z_\alpha\|^{2}} \frac{\|\Lambda(x, z_\alpha)\|}{\|z_\alpha\|^{2n-4}},
$$

for all $\alpha > 0$. In view of (9.24), (9.25), (9.20) and (9.23), the right-hand side converges to zero for $\alpha \to \infty$. This contradicts (9.28), hence $n \le 2$. $\qquad\square$

9.5 Exponential–Polynomial Families

In this section, we consider the Nelson–Siegel and Svensson families. For a discussion of general exponential–polynomial families see [68]. See also the notes section for more references.

9.5.1 Nelson–Siegel Family

Recall the form of the Nelson–Siegel curves

$$
\phi_{NS}(x, z) = z_1 + (z_2 + z_3 x) e^{-z_4 x}.
$$

The next result is somewhat disillusioning.

Proposition 9.4 *The unique solution to (9.4) for* ϕ_{NS} *is*

$$
a(z) = 0, \qquad b_1(z) = b_4(z) = 0, \qquad b_2(z) = z_3 - z_2 z_4, \qquad b_3(z) = -z_3 z_4.
$$

The corresponding state process is

$$
Z_1(t) \equiv z_1,
$$
$$
Z_2(t) = (z_2 + z_3 t) e^{-z_4 t},
$$
$$
Z_3(t) = z_3 e^{-z_4 t},
$$
$$
Z_4(t) \equiv z_4,
$$

where $Z(0) = (z_1, \ldots, z_4)$ *denotes the initial point. Hence there is no non-trivial consistent diffusion process* Z *for the Nelson–Siegel family.*

Proof $\to$ Exercise 9.6. $\qquad\square$

9.5.2 Svensson Family

The Svensson forward curves are

$$\phi_S(x, z) = z_1 + (z_2 + z_3 x)e^{-z_5 x} + z_4 x e^{-z_6 x}.$$

There are two more factors than in the Nelson–Siegel family. The consistency result is thus not as stringent as above.

Proposition 9.5 *The only non-trivial consistent HJM model for the Svensson family is the Hull–White extended Vasiček short-rate model*

$$dr(t) = \left(z_1 z_5 + z_3 e^{-z_5 t} + z_4 e^{-2z_5 t} - z_5 r(t)\right)dt + \sqrt{z_4 z_5}\, e^{-z_5 t}\, dW^*(t),$$

where $(z_1, \ldots, z_5)$ *are given by the initial forward curve*

$$f(0, x) = z_1 + (z_2 + z_3 x)e^{-z_5 x} + z_4 x e^{-2z_5 x}$$

and W^* *is some* $\mathbb{Q}$-*standard Brownian motion. The form of the corresponding state process* Z *is given in* (9.31)–(9.33) *in the proof below (see also Exercise 9.7).*

This proposition states, in particular, that the Svensson family admits no consistently varying exponents, see (9.31). This suggests that the exponents, z_5 and $z_6 = 2z_5$, be rather considered as model parameters than factors. This hypothesis has been empirically tested on US bond price data in [146]. The findings indicated that constant exponents hypothesis could not be falsified, while the relation $z_6 = 2z_5$ does. However, it also turned out that the statistical properties of the time series for the factor z are very sensitive on the numerical term-structure estimation procedure. It could be well the case, as shown in [8], that an inter-temporal smoothing device, consistent with the structure in Proposition 9.5, substantially improves parameter stability and smoothness. To date, this is an open problem.

Proof The consistency equation (9.4) becomes, after differentiating both sides in x,

$$q_1(x) + q_2(x)e^{-z_5 x} + q_3(x)e^{-z_6 x}$$
$$+ q_4(x)e^{-2z_5 x} + q_5(x)e^{-(z_5 + z_6)x} + q_6(x)e^{-2z_6 x} = 0, \qquad (9.29)$$

for some polynomials $q_1, \ldots, q_6$. Indeed, we assume for the moment that

$$z_5 \neq z_6, \qquad z_5 + z_6 \neq 0 \quad \text{and} \quad z_i \neq 0 \quad \text{for all } i = 1, \ldots, 6. \qquad (9.30)$$

Then the terms involved in (9.29) are

$$\partial_x \phi_S(x, z) = (-z_2 z_5 + z_3 - z_3 z_5 x)e^{-z_5 x} + (z_4 - z_4 z_6 x)e^{-z_6 x},$$

$$\nabla_z \phi_S(x, z) = \begin{pmatrix} 1 \\ e^{-z_5 x} \\ x e^{-z_5 x} \\ x e^{-z_6 x} \\ (-z_2 x - z_3 x^2) e^{-z_5 x} \\ -z_4 x^2 e^{-z_6 x} \end{pmatrix},$$

$$\partial_{z_i} \partial_{z_j} \phi_S(x, z) = 0 \quad \text{for } 1 \leq i, j \leq 4,$$

$$\nabla_z \partial_{z_5} \phi_S(x, z) = \begin{pmatrix} 0 \\ -x e^{-z_5 x} \\ -x^2 e^{-z_5 x} \\ 0 \\ (z_2 x^2 + z_3 x^3) e^{-z_5 x} \\ 0 \end{pmatrix}, \qquad \nabla_z \partial_{z_6} \phi_S(x, z) = \begin{pmatrix} 0 \\ 0 \\ 0 \\ -x^2 e^{-z_6 x} \\ 0 \\ z_4 x^3 e^{-z_6 x} \end{pmatrix},$$

$$\int_0^x \nabla_z \phi_S(u, z)\, du = \begin{pmatrix} x \\ -\frac{1}{z_5} e^{-z_5 x} + \frac{1}{z_5} \\ \left(-\frac{x}{z_5} - \frac{1}{z_5^2}\right) e^{-z_5 x} + \frac{1}{z_5^2} \\ \left(-\frac{x}{z_6} - \frac{1}{z_6^2}\right) e^{-z_6 x} + \frac{1}{z_6^2} \\ \left(\frac{z_3}{z_5} x^2 + \left(\frac{z_2}{z_5} + \frac{2 z_3}{z_5^2}\right) x + \frac{z_2}{z_5^2} + \frac{2 z_3}{z_5^3}\right) e^{-z_5 x} - \frac{z_2}{z_5^2} - \frac{z_3}{z_5^3} \\ \left(\frac{z_4}{z_6} x^2 + \frac{2 z_4}{z_6^2} x + \frac{2 z_4}{z_6^3}\right) e^{-z_6 x} - \frac{z_4}{z_6^3} \end{pmatrix}.$$

Straightforward calculations lead to

$$q_1(x) = -a_{11}(z) x + \cdots,$$

$$q_4(x) = a_{55}(z) \frac{z_3^2}{z_5} x^4 + \cdots,$$

$$q_6(x) = a_{66}(z) \frac{z_4^2}{z_6} x^4 + \cdots,$$

$$\deg q_2, \ \deg q_3, \ \deg q_5 \leq 3,$$

where $\cdots$ stands for lower-order terms in x. Because of (9.30) we conclude that

$$a_{11}(z) = a_{55}(z) = a_{66}(z) = 0.$$

But a is a positive semi-definite symmetric matrix. Hence

$$a_{1j}(z) = a_{j1}(z) = a_{5j}(z) = a_{j5}(z) = a_{6j}(z) = a_{j6}(z) = 0, \quad j = 1, \ldots, 6.$$

Taking this into account, expression (9.29) simplifies considerably. We are left with

$$q_1(x) = b_1(z),$$

$$\deg q_2, \ \deg q_3 \leq 1,$$

$$q_4(x) = a_{33}(z)\frac{1}{z_5}x^2 + \cdots,$$

$$q_5(x) = a_{34}(z)\left(\frac{1}{z_5} + \frac{1}{z_6}\right)x^2 + \cdots,$$

$$q_6(x) = a_{44}(z)\frac{1}{z_6}x^2 + \cdots.$$

Because of (9.30) we know that the exponents $-2z_5$, $-(z_5 + z_6)$ and $-2z_6$ are mutually different. Hence

$$b_1(z) = a_{3j}(z) = a_{j3}(z) = a_{4j}(z) = a_{j4}(z) = 0, \quad j = 1, \ldots, 6.$$

Only $a_{22}(z)$ is left as positive candidate among the components of $a(z)$. The remaining terms are

$$q_2(x) = (b_3(z) + z_3 z_5)x + b_2(z) - z_3 - \frac{a_{22}(z)}{z_5} + z_2 z_5,$$

$$q_3(x) = (b_4(z) + z_4 z_6)x - z_4,$$

$$q_4(x) = a_{22}(z)\frac{1}{z_5},$$

while $q_1 = q_5 = q_6 = 0$.

If $2z_5 \neq z_6$ then also $a_{22}(z) = 0$. If $2z_5 = z_6$ then the condition $q_3 + q_4 = q_2 = 0$ leads to

$$a_{22}(z) = z_4 z_5,$$

$$b_2(z) = z_3 + z_4 - 2z_5 z_2,$$

$$b_3(z) = -z_5 z_3,$$

$$b_4(z) = -2z_5 z_4.$$

We derived the above results under the assumption (9.30). But the set of z where (9.30) holds is dense $\mathcal{Z}$. By continuity of $a(z)$ and $b(z)$ in z, the above results thus extend for all $z \in \mathcal{Z}$. In particular, all Z_i's but Z_2 are deterministic; Z_1, Z_5 and Z_6 are even constant.

Thus, since

$$a(z) = 0 \quad \text{if } 2z_5 \neq z_6,$$

we only have a non-trivial process Z if

$$Z_6(t) \equiv 2Z_5(t) \equiv 2Z_5(0). \tag{9.31}$$

In that case we have, writing for short $z_i = Z_i(0)$,

$$Z_1(t) \equiv z_1,$$

$$Z_3(t) = z_3 e^{-z_5 t},$$

$$(9.32)$$

$$Z_4(t) = z_4 e^{-2z_5 t}$$

and

$$dZ_2(t) = \left(z_3 e^{-z_5 t} + z_4 e^{-2z_5 t} - z_5 Z_2(t)\right) dt + \sum_{j=1}^{d} \rho_{2j}(t) \, dW_j^*(t), \qquad (9.33)$$

where $\rho_{2j}(t)$ (not necessarily deterministic) are such that

$$\sum_{j=1}^{d} \rho_{2j}^2(t) = a_{22}(Z(t)) = z_4 z_5 e^{-2z_5 t}.$$

By Lévy's characterization theorem we have that

$$\mathcal{W}^*(t) = \sum_{j=1}^{d} \int_0^t \frac{\rho_{2j}(s)}{\sqrt{z_4 z_5} e^{-z_5 s}} \, dW_j^*(s)$$

is a real-valued standard Brownian motion. Hence the corresponding short-rate process

$$r(t) = \phi_S(0, Z(t)) = z_1 + Z_2(t)$$

satisfies

$$dr(t) = \left(z_1 z_5 + z_3 e^{-z_5 t} + z_4 e^{-2z_5 t} - z_5 r(t)\right) dt + \sqrt{z_4 z_5} e^{-z_5 t} \, d\mathcal{W}^*(t).$$

Hence the proposition is proved.   □

9.6 Exercises

Exercise 9.1 Derive the consistency equation (9.4) directly by applying Itô's formula to (9.1).

Exercise 9.2 Show that α and σ in (9.2) and (9.3) satisfy **(HJM.1)–(HJM.3)**. Hint: show that continuity of $\sigma \sigma^{\top}$ implies that σ is bounded on compacts.

Exercise 9.3 The aim of this exercise is to show that the independence assumption of the functions in (9.7) cannot be omitted in Proposition 9.3.

(a) Show first that the time-homogeneous version of Proposition 5.2 can be derived as a corollary of Proposition 9.3 for $m = 1$.

(b) Consider the two-factor state process

$$dZ_1 = Z_2 \, dt + \sqrt{\min\{1, Z_1\}} \, dW^*,$$

$$dZ_2 = \min\{1, Z_1\} \, dt$$

with state space $\mathcal{Z} = \mathbb{R}_+^2$. You can assume, without proving it, that there exists an $\mathbb{R}_+^2$-valued solution $Z = Z^z$ with $Z(0) = z$, for every $z \in \mathbb{R}_+^2$. Show that this state process is not affine but is nevertheless consistent with the ATS $\phi(x, z) = z_1 + z_2 x$.

Exercise 9.4 Extend Proposition 9.3 to the case of a time-inhomogeneous ATS:

$$\phi(x, z) = g_0(x, z_0) + g_1(x, z_0)z_1 + \cdots + g_m(x, z_0)z_m,$$

where the state vector $z = (z_0, z_1, \ldots, z_m) \in \mathbb{R}_+ \times \mathcal{Z}$ is extended accordingly by calendar time z_0.

Exercise 9.5 Find the general form of a one-factor ($m = 1$) QTS model $f(t, T) = g_0(T - t) + g_1(T - t) Z(t) + g_2(T - t) Z(t)^2$ for some real-valued ($\mathcal{Z} = \mathbb{R}$) diffusion state process Z. Show there are specifications which yield an ATS.

Exercise 9.6 Consider the Nelson–Siegel family $\phi_{NS}(x, z) = z_1 + (z_2 + z_3 x)e^{-z_4 x}$.

(a) Check whether the linear independence assumption of Proposition 9.2 is satisfied.

(b) Give a full proof of Proposition 9.4.

Exercise 9.7 Consider the Hull–White extended Vasiček short-rate model

$$dr(t) = \left(z_1 z_5 + z_3 e^{-z_5 t} + z_4 e^{-2z_5 t} - z_5 r(t)\right) dt + \sqrt{z_4 z_5} \, e^{-z_5 t} \, dW^*(t),$$

which is consistent with the Svensson family given in Proposition 9.5. Show that the zero-coupon bond price equals $P(t, T) = e^{-A(t,T) - B(t,T)r(t)}$ where $r(t) = z_1 + Z_2(t)$ with

$$Z_2(t) = e^{-z_5 t}\left(z_2 + z_3 t + \frac{z_4}{z_5}\left(1 - e^{-z_5 t}\right)\right) + \sqrt{z_4 z_5} \, e^{-z_5 t} \, W^*(t)$$

and

$$A(t, T) = \frac{z_1}{z_5}\left(e^{-z_5(T-t)} - 1 + z_5(T - t)\right) + \frac{z_3 e^{-z_5 T}}{z_5^2}\left(e^{z_5(T-t)} - 1 - z_5(T - t)\right)$$

$$+ \frac{z_4 e^{-2z_5 T}}{4z_5^2}\left(e^{2z_5(T-t)} - 1 - 2z_5(T - t)\right),$$

$$B(t, T) = \frac{1}{z_5}\left(1 - e^{-z_5(T-t)}\right).$$

Exercise 9.8 Theorem 3.1 suggests that we look at the Lorimier family of forward curves

$$\phi_L(x, z) = \zeta_0^\top z + \sum_{i=1}^N h_i(x) \zeta_i^\top z,$$

which is parameterized by z in some state space $\mathcal{Z} \subset \mathbb{R}^N$. Here $h_i(x)$ are the second-order splines given by (3.6), and $\zeta_0^\top, \ldots, \zeta_N^\top \in \mathbb{R}^N$ denote the respective last N components of the row vectors of A^{-1} in (3.11). Further denote by $T_1, \ldots, T_N$ the tenor structure from Theorem 3.1. The aim of this exercise is to show that there exists no non-trivial consistent HJM model.

(a) Verify that $\sum_{i=1}^N T_i \zeta_i = 0$. Hence $\zeta_1, \ldots, \zeta_N$ are linearly dependent, and $\phi_L(0, z) = \zeta_0^\top z$.

(b) Show that both sides of the consistency equation (9.4) are locally polynomial functions of maximal order 6 on the intervals $x \in (T_i, T_{i+1})$.

(c) Now argue by backward induction: show first that the 6th-order term is $\zeta_N^\top a(z) \zeta_N$ for $x \in (T_{N-1}, T_N)$. Infer by induction that $\zeta_k^\top a(z) \zeta_k = 0$ for all $1 \le k \le N$, and finally also $\zeta_0^\top a(z) \zeta_0 = 0$.

(d) Conclude that $\zeta_k^\top b(z) = 0$ for all $0 \le k \le N$, and that $\zeta_k^\top z = 0$ for all $1 \le k \le N$.

(e) Finally, argue that only the constant HJM model $f(t, T) \equiv f(0, T) \equiv \zeta_0^\top z$ is consistent with the Lorimier family.

9.7 Notes

The consistency problem for term-structure models was introduced into the mathematical finance literature by Björk and Christensen [14]. Björk and Svensson [15] translated the problem into a geometric framework, see also Björk [12] and [68]. This initiated a series of papers by Björk and his coauthors. Björk and Svensson [15] and Filipović and Teichmann [73, 74] found that the only generically consistent term-structure parametrizations are the time-inhomogeneous affine ones, such as the Hull–White extended Vasiček and CIR short-rate models (see also Exercise 9.4). Generic here means that any initial forward curve be admitted. Time-homogeneous and inhomogeneous affine processes have subsequently been studied in detail in Duffie et al. [61] and Filipović [71], respectively. See also Chap. 10 following below.

Proposition 9.3 on the characterization of affine term structure models is due to Duffie and Kan [58]. The corresponding Exercise 9.3 is from [69, Sect. 8]. Quadratic term-structure models have been studied in the context of both theoretical analysis and empirical testing in e.g. Ahn, Dittmar and Gallant [1], Boyle and Tian [22], Chen et al. [39, 40], Cheng and Scaillet [41], Gombani and Runggaldier [83], Leippold and Wu [115, 116], and Gourieroux and Sufana [85], to mention a few. The results on the polynomial term-structures in Sect. 9.4 are from [70]. The consistency results for the Nelson–Siegel and Svensson family are from Björk and Christensen [14] and [66, 67], see also the book [68]. Sharef and Filipović [147] provide consistent exponential–polynomial forward rate models with two exponential terms to have more than one non-trivial factor. The consistency problem for the Lorimier family in Exercise 9.8 was first solved by Mykhaylo Shkolnikov [148].

Chapter 10
Affine Processes

We have seen in Sects. 5.3 and 9.3 above that an affine diffusion induces an affine term-structure. In this chapter, we discuss the class of affine processes in more detail. Their nice analytical properties make them favorite for a broad range of financial applications, including term-structure modeling, option pricing and credit risk modeling.

10.1 Definition and Characterization of Affine Processes

As in Sect. 9.1, we fix a dimension $d \geq 1$ and a closed state space $\mathcal{X} \subset \mathbb{R}^d$ with non-empty interior. We let $b : \mathcal{X} \to \mathbb{R}^d$ be continuous, and $\rho : \mathcal{X} \to \mathbb{R}^{d \times d}$ be measurable and such that the diffusion matrix

$$a(x) = \rho(x)\rho(x)^\top$$

is continuous in $x \in \mathcal{X}$ (see Remark 4.2). Let W denote a d-dimensional Brownian motion defined on a filtered probability space $(\Omega, \mathcal{F}, (\mathcal{F}_t), \mathbb{P})$. Throughout, we assume that for every $x \in \mathcal{X}$ there exists a unique solution $X = X^x$ of the stochastic differential equation

$$\begin{aligned} dX(t) &= b(X(t))\,dt + \rho(X(t))\,dW(t), \\ X(0) &= x. \end{aligned} \tag{10.1}$$

Definition 10.1 We call X *affine* if the $\mathcal{F}_t$-conditional characteristic function of $X(T)$ is exponential affine in $X(t)$, for all $t \leq T$. That is, there exist $\mathbb{C}$- and $\mathbb{C}^d$-valued functions $\phi(t, u)$ and $\psi(t, u)$, respectively, with jointly continuous t-derivatives such that $X = X^x$ satisfies

$$\mathbb{E}\left[e^{u^\top X(T)} \mid \mathcal{F}_t \right] = e^{\phi(T-t,u) + \psi(T-t,u)^\top X(t)} \tag{10.2}$$

for all $u \in i\mathbb{R}^d$, $t \leq T$ and $x \in \mathcal{X}$.

Since the conditional characteristic function is bounded by one, the real part of the exponent $\phi(T - t, u) + \psi(T - t, u)^\top X(t)$ in (10.2) has to be negative. Note that $\phi(t, u)$ and $\psi(t, u)$ for $t \geq 0$ and $u \in i\mathbb{R}^d$ are uniquely[1] determined by (10.2), and satisfy the initial conditions $\phi(0, u) = 0$ and $\psi(0, u) = u$, in particular.

We first derive necessary and sufficient conditions for X to be affine.

---

[1] In fact, $\phi(t, u)$ may be altered by multiples of $2\pi i$. We uniquely fix the continuous function $\phi(t, u)$ by $\phi(t, 0) = 0$.

D. Filipović, *Term-Structure Models,*
Springer Finance,
DOI 10.1007/978-3-540-68015-4_10, © Springer-Verlag Berlin Heidelberg 2009

Theorem 10.1 *Suppose X is affine. Then the diffusion matrix $a(x)$ and drift $b(x)$ are affine in x. That is,*

$$a(x) = a + \sum_{i=1}^{d} x_i \alpha_i,$$

$$b(x) = b + \sum_{i=1}^{d} x_i \beta_i = b + \mathcal{B}x \tag{10.3}$$

for some $d \times d$-matrices a and α_i, and d-vectors b and β_i, where we denote by

$$\mathcal{B} = (\beta_1, \ldots, \beta_d)$$

the $d \times d$-matrix with ith column vector β_i, $1 \le i \le d$. Moreover, ϕ and $\psi = (\psi_1, \ldots, \psi_d)^\top$ solve the system of Riccati equations

$$\partial_t \phi(t, u) = \frac{1}{2} \psi(t, u)^\top a \psi(t, u) + b^\top \psi(t, u),$$

$$\phi(0, u) = 0,$$

$$\partial_t \psi_i(t, u) = \frac{1}{2} \psi(t, u)^\top \alpha_i \psi(t, u) + \beta_i^\top \psi(t, u), \quad 1 \le i \le d, \tag{10.4}$$

$$\psi(0, u) = u.$$

In particular, ϕ is determined by ψ via simple integration:

$$\phi(t, u) = \int_0^t \left(\frac{1}{2} \psi(s, u)^\top a \psi(s, u) + b^\top \psi(s, u) \right) ds.$$

Conversely, suppose the diffusion matrix $a(x)$ and drift $b(x)$ are affine of the form (10.3) and suppose there exists a solution (ϕ, ψ) of the Riccati equations (10.4) such that $\phi(t, u) + \psi(t, u)^\top x$ has a nonpositive real part for all $t \ge 0$, $u \in i\mathbb{R}^d$ and $x \in \mathcal{X}$. Then X is affine with conditional characteristic function (10.2).

Proof Suppose X is affine. For $T > 0$ and $u \in i\mathbb{R}^d$ define the complex-valued Itô process

$$M(t) = e^{\phi(T-t,u) + \psi(T-t,u)^\top X(t)}.$$

We can apply Itô's formula, separately to the real and imaginary parts of M, and obtain

$$dM(t) = I(t)\, dt + \psi(T-t, u)^\top \rho(X(t))\, dW(t), \quad t \le T,$$

with

$$I(t) = -\partial_T \phi(T-t, u) - \partial_T \psi(T-t, u)^\top X(t)$$

$$+ \psi(T-t, u)^\top b(X(t)) + \frac{1}{2} \psi(T-t, u)^\top a(X(t)) \psi(T-t, u).$$

Since M is a martingale, we have $I(t) = 0$ for all $t \leq T$ a.s. Letting $t \to 0$, by continuity of the parameters, we thus obtain

$$\partial_T \phi(T, u) + \partial_T \psi(T, u)^\top x = \psi(T, u)^\top b(x) + \frac{1}{2} \psi(T, u)^\top a(x) \psi(T, u)$$

for all $x \in \mathcal{X}$, $T \geq 0$, $u \in i\mathbb{R}^d$. Since $\psi(0, u) = u$, this implies that a and b are affine of the form (10.3). Plugging this back into the above equation and separating first-order terms in x yields (10.4).

Conversely, suppose a and b are of the form (10.3). Let (ϕ, ψ) be a solution of the Riccati equations (10.4) such that $\phi(t, u) + \psi(t, u)^\top x$ has a nonpositive real part for all $t \geq 0$, $u \in i\mathbb{R}^d$ and $x \in \mathcal{X}$. Then M, defined as above, is a uniformly bounded[2] local martingale, and hence a martingale, with $M(T) = e^{u^\top X(T)}$. Therefore $\mathbb{E}[M(T) \mid \mathcal{F}_t] = M(t)$, for all $t \leq T$, which is (10.2), and the theorem is proved. $\qquad\qquad\square$

In the sequel, we will often deal with systems of Riccati equations of the type (10.4). Therefore, we now recall an important global existence, uniqueness and regularity result for differential equations, which will be used throughout without further mention. We let K be a placeholder for either $\mathbb{R}$ or $\mathbb{C}$.

Lemma 10.1 *Consider the system of ordinary differential equations*

$$\partial_t f(t, u) = R(f(t, u)),$$
$$f(0, u) = u, \tag{10.5}$$

where $R : K^d \to K^d$ is a locally Lipschitz continuous function. Then the following holds:

(a) *For every $u \in K^d$, there exists a life time $t_+(u) \in (0, \infty]$ such that there exists a unique solution $f(\cdot, u) : [0, t_+(u)) \to K \times K^d$ of (10.5).*
(b) *The domain*

$$\mathcal{D}_K = \{(t, u) \in \mathbb{R}_+ \times K^d \mid t < t_+(u)\}$$

is open in $\mathbb{R}_+ \times K^d$ and maximal in the sense that either $t_+(u) = \infty$ or

$$\lim_{t \uparrow t_+(u)} \| f(t, u) \| = \infty,$$

respectively, for all $u \in K^d$.
(c) *For every $t \geq 0$, the t-section*

$$\mathcal{D}_K(t) = \{u \in K^d \mid (t, u) \in \mathcal{D}_K\}$$

---

[2]We note that the uniform boundedness of the local martingale M is substantial here to infer that M is a true martingale and the transform formula (10.2) holds. See also Exercise 10.4 below.

is open in K^d, and non-expanding in t in the following sense:

$$K^d = \mathcal{D}_K(0) \supseteq \mathcal{D}_K(t_1) \supseteq \mathcal{D}_K(t_2), \quad 0 \le t_1 \le t_2.$$

In fact, we have $f(s, \mathcal{D}_K(t_2)) \subseteq \mathcal{D}_K(t_1)$ for all $s \le t_2 - t_1$.
(d) *If R is analytic on K^d then f is an analytic function on $\mathcal{D}_K$.*

Proof Part (a) follows from the basic theorems for ordinary differential equations, e.g. [3, Theorem 7.4]. It is proved in [3, Theorems 7.6 and 8.3] that $\mathcal{D}_K$ is maximal and open, which is part (b). This also implies that all t-sections $\mathcal{D}_K(t)$ are open in K^d. The inclusion $\mathcal{D}_K(t_1) \supseteq \mathcal{D}_K(t_2)$ is a consequence of the maximality property from part (b), and $f(s, \mathcal{D}_K(t_2)) \subseteq \mathcal{D}_K(t_1)$ follows from the flow property $f(t_1, f(s, u)) = f(t_1 + s, u)$, whence part (c) follows. For a proof of part (d) see [55, Theorem 10.8.2]. $\qquad\square$

It is obvious to what extent Lemma 10.1 applies to the system of Riccati equations (10.4). In particular, it is easily checked that $t_+(0) = \infty$ and thus $\mathcal{D}_K(t)$ contains 0 for all $t \ge 0$. We will provide in Sect. 10.7 and Theorem 10.3 below some substantial improvements of the properties for (10.4) stated in Lemma 10.1 for the canonical state space $\mathcal{X}$ introduced in the following section.

10.2 Canonical State Space

There is an implicit trade-off between the parameters a, α_i, b, β_i in (10.3) and the state space $\mathcal{X}$:

- a, α_i, b, β_i must be such that X does not leave the set $\mathcal{X}$;
- a, α_i must be such that $a + \sum_{i=1}^d x_i \alpha_i$ is symmetric and positive semi-definite for all $x \in \mathcal{X}$.

To gain further explicit insight into this interplay, we now and henceforth assume that the state space is of the following canonical form:

$$\mathcal{X} = \mathbb{R}^m_+ \times \mathbb{R}^n$$

for some integers $m, n \ge 0$ with $m + n = d$. This canonical state space covers essentially all applications appearing in the finance literature.[3]

---

[3] Note, however, that other choices for the state space of an affine process are possible. For instance, the trivial example for $d = 1$,

$$dX = -X\,dt, \quad X(0) = x \in \mathcal{X},$$

admits as state space any closed interval $\mathcal{X} \subset \mathbb{R}$ containing 0. This degenerate diffusion process is affine, since $e^{uX(T)} = e^{ue^{-(T-t)}X(t)}$ for all $t \le T$. A non-degenerate example is provided in Exercise 10.1. See also the discussion in [61, Sect. 12]. Moreover, semi-definite matrix-valued affine processes have recently been studied and successfully applied to finance in [30, 32, 48, 49, 84, 86].

For the above canonical state space, we can give necessary and sufficient admissibility conditions on the parameters. The following terminology will be useful in the sequel. We define the index sets

$$I = \{1, \ldots, m\} \quad \text{and} \quad J = \{m+1, \ldots, m+n\}.$$

For any vector μ and matrix v, and index sets M, N, we denote by

$$\mu_M = (\mu_i)_{i \in M}, \qquad v_{MN} = (v_{ij})_{i \in M, j \in N}$$

the respective sub-vector and -matrix.

Theorem 10.2 *The process X on the canonical state space $\mathbb{R}_+^m \times \mathbb{R}^n$ is affine if and only if $a(x)$ and $b(x)$ are affine of the form (10.3) for parameters a, α_i, b, β_i which are admissible in the following sense:*

$$a, \alpha_i \text{ are symmetric positive semi-definite,}$$

$$a_{II} = 0 \quad (\text{and thus } a_{IJ} = a_{JI}^\top = 0),$$

$$\alpha_j = 0 \quad \text{for all } j \in J,$$

$$\alpha_{i,kl} = \alpha_{i,lk} = 0 \quad \text{for } k \in I \setminus \{i\}, \text{for all } 1 \le i, l \le d, \qquad (10.6)$$

$$b \in \mathbb{R}_+^m \times \mathbb{R}^n,$$

$$\mathcal{B}_{IJ} = 0,$$

$$\mathcal{B}_{II} \text{ has nonnegative off-diagonal elements.}$$

In this case, the corresponding system of Riccati equations (10.4) simplifies to

$$\partial_t \phi(t, u) = \frac{1}{2} \psi_J(t, u)^\top a_{JJ} \psi_J(t, u) + b^\top \psi(t, u),$$

$$\phi(0, u) = 0,$$

$$\partial_t \psi_i(t, u) = \frac{1}{2} \psi(t, u)^\top \alpha_i \psi(t, u) + \beta_i^\top \psi(t, u), \quad i \in I, \qquad (10.7)$$

$$\partial_t \psi_J(t, u) = \mathcal{B}_{JJ}^\top \psi_J(t, u),$$

$$\psi(0, u) = u,$$

and there exists a unique global solution $(\phi(\cdot, u), \psi(\cdot, u)) : \mathbb{R}_+ \to \mathbb{C}_- \times \mathbb{C}_-^m \times i\mathbb{R}^n$ for all initial values $u \in \mathbb{C}_-^m \times i\mathbb{R}^n$. In particular, the equation for ψ_J forms an autonomous linear system with unique global solution $\psi_J(t, u) = e^{\mathcal{B}_{JJ}^\top t} u_J$ for all $u_J \in \mathbb{C}^n$.

Before we prove the theorem, let us illustrate the admissibility conditions (10.6) for the diffusion matrix $\alpha(x)$ for dimension $d = 3$ and the corresponding cases $m =$

0, 1, 2, 3. For the first case $m = 0$ we have

$$\alpha(x) \equiv a$$

for an arbitrary positive semi-definite symmetric 3×3-matrix a. For $m = 1$, we have

$$a = \begin{pmatrix} 0 & 0 & 0 \\ & + & * \\ & & + \end{pmatrix}, \qquad \alpha_1 = \begin{pmatrix} + & * & * \\ & + & * \\ & & + \end{pmatrix},$$

for $m = 2$,

$$a = \begin{pmatrix} 0 & 0 & 0 \\ & 0 & 0 \\ & & + \end{pmatrix}, \qquad \alpha_1 = \begin{pmatrix} + & 0 & * \\ & 0 & 0 \\ & & + \end{pmatrix}, \qquad \alpha_2 = \begin{pmatrix} 0 & 0 & 0 \\ & + & * \\ & & + \end{pmatrix},$$

and for $m = 3$,

$$a = 0, \qquad \alpha_1 = \begin{pmatrix} + & 0 & 0 \\ & 0 & 0 \\ & & 0 \end{pmatrix}, \qquad \alpha_2 = \begin{pmatrix} 0 & 0 & 0 \\ & + & 0 \\ & & 0 \end{pmatrix}, \qquad \alpha_3 = \begin{pmatrix} 0 & 0 & 0 \\ & 0 & 0 \\ & & + \end{pmatrix},$$

where we leave the lower triangle of symmetric matrices blank; $+$ denotes a non-negative real number and $*$ any real number such that positive semi-definiteness holds.

Proof Suppose X is affine. That $a(x)$ and $b(x)$ are of the form (10.3) follows from Theorem 10.1. Obviously, $a(x)$ is symmetric positive semi-definite for all $x \in \mathbb{R}_+^m \times \mathbb{R}^n$ if and only if $\alpha_j = 0$ for all $j \in J$, and a and α_i are symmetric positive semi-definite for all $i \in I$.

We extend the diffusion matrix and drift continuously to $\mathbb{R}^d$ by setting

$$a(x) = a + \sum_{i \in I} x_i^+ \alpha_i \quad \text{and} \quad b(x) = b + \sum_{i \in I} x_i^+ \beta_i + \sum_{j \in J} x_j \beta_j.$$

Now let x be a boundary point of $\mathbb{R}_+^m \times \mathbb{R}^n$. That is, $x_k = 0$ for some $k \in I$. The stochastic invariance Lemma 10.11 below implies that the diffusion must be "parallel to the boundary",

$$e_k^\top \left(a + \sum_{i \in I \setminus \{k\}} x_i \alpha_i \right) e_k = 0,$$

and the drift must be "inward pointing",

$$e_k^\top \left(b + \sum_{i \in I \setminus \{k\}} x_i \beta_i + \sum_{j \in J} x_j \beta_j \right) \geq 0.$$

Since this has to hold for all $x_i \geq 0$, $i \in I \setminus \{k\}$, and $x_j \in \mathbb{R}$, $j \in J$, we obtain the following set of admissibility conditions:

$$a, \alpha_i \text{ are symmetric positive semi-definite,}$$

$$ae_k = 0 \quad \text{for all } k \in I,$$

$$\alpha_i e_k = 0 \quad \text{for all } i \in I \setminus \{k\}, \text{ for all } k \in I,$$

$$\alpha_j = 0 \quad \text{for all } j \in J,$$

$$b \in \mathbb{R}_+^m \times \mathbb{R}^n,$$

$$\beta_i^\top e_k \geq 0 \quad \text{for all } i \in I \setminus \{k\}, \text{ for all } k \in I,$$

$$\beta_j^\top e_k = 0 \quad \text{for all } j \in J, \text{ for all } k \in I,$$

which is equivalent to (10.6). The form of the system (10.7) follows by inspection.

Now suppose a, α_i, b, β_i satisfy the admissibility conditions (10.6). We show below that there exists a unique global solution $(\phi(\cdot, u), \psi(\cdot, u)) : \mathbb{R}_+ \to \mathbb{C}_- \times \mathbb{C}_-^m \times i\mathbb{R}^n$ of (10.7), for all $u \in \mathbb{C}_-^m \times i\mathbb{R}^n$. In particular, $\phi(t, u) + \psi(t, u)^\top x$ has nonpositive real part for all $t \geq 0$, $u \in i\mathbb{R}^d$ and $x \in \mathbb{R}_+^m \times \mathbb{R}^n$. Thus the first part of the theorem follows from Theorem 10.1.

It view of the admissibility conditions for a and b, it remains to show that $\psi(t, u)$ is $\mathbb{C}_-^m \times i\mathbb{R}^n$-valued and has life time $t_+(u) = \infty$ for all $u \in \mathbb{C}_-^m \times i\mathbb{R}^n$. For $i \in I$, denote the right-hand side of the equation for ψ_i by

$$R_i(u) = \frac{1}{2} u^\top \alpha_i u + \beta_i^\top u,$$

and observe that

$$\Re R_i(u) = \frac{1}{2} \Re u^\top \alpha_i \Re u - \frac{1}{2} \Im u^\top \alpha_i \Im u + \beta_i^\top \Re u.$$

Let us define $x_I^+ = (x_1^+, \ldots, x_m^+)^\top$. Since $\Re \psi_J(t, u) = 0$, it follows from the admissibility conditions (10.6) and Corollary 10.5 below, setting $f(t) = -\Re \psi(t, u)$,

$$b_i(t, x) = -\frac{1}{2} \alpha_{i,ii} \left(x_i^+ \right)^2 + \frac{1}{2} \Im \psi(t, u)^\top \alpha_i \Im \psi(t, u) + \beta_{i,I}^\top x_I^+, \quad i \in I,$$

and $b_j(t, x) = 0$ for $j \in J$, that the solution $\psi(t, u)$ of (10.7) has to take values in $\mathbb{C}_-^m \times i\mathbb{R}^n$ for all initial points $u \in \mathbb{C}_-^m \times i\mathbb{R}^n$.

Further, for $i \in I$ and $u \in \mathbb{C}^d$, one verifies that

$$\Re(\overline{u_i} R_i(u)) = \frac{1}{2} \alpha_{i,ii} |u_i|^2 \Re u_i + \Re(\overline{u_i} u_i \alpha_{i,iJ} u_J) + \frac{1}{2} \Re(\overline{u_i} u_J^\top \alpha_{i,JJ} u_J) + \Re(\overline{u_i} \beta_i^\top u)$$

$$\leq \frac{K}{2} \left(1 + \|(\Re u_I)^+\| + \|u_J\|^2 \right) \left(1 + \|u_I\|^2 \right)$$

for some finite constant K which does not depend on u. We thus obtain

$$\partial_t \|\psi_I(t, u)\|^2 = 2\Re \left(\overline{\psi_I(t, u)}^\top R_I \left(\psi_I(t, u), e^{\mathcal{B}_{JJ}^j t} u_J \right) \right)$$

$$\leq g(t) \left(1 + \|\psi_I(t, u)\|^2 \right)$$

for

$$g(t) = K \left(1 + \|(\Re\psi_I(t, u))^+\| + \|e^{\mathcal{B}_{JJ}^\top t} u_J\|^2 \right).$$

Gronwall's inequality[4] applied to $f(t) = (1 + \|\psi_I(t, u)\|^2)$ and $h(t) \equiv f(0)$, yields

$$\|\psi_I(t, u)\|^2 \leq \|u_I\|^2 + \left(1 + \|u_I\|^2 \right) \int_0^t g(s) e^{\int_s^t g(\xi) d\xi} ds. \qquad (10.8)$$

From above, for all initial points $u \in \mathbb{C}_-^m \times i\mathbb{R}^n$, we know that $(\Re\psi_I(t, u))^+ = 0$ and therefore $t_+(u) = \infty$ by (10.8). Hence the theorem is proved. $\qquad\square$

Now suppose X is affine with characteristics (10.3) satisfying the admissibility conditions (10.6). In what follows we show that not only can the functions $\phi(t, u)$ and $\psi(t, u)$, given as solutions of (10.7), be extended beyond $u \in i\mathbb{R}^d$, but also the validity of the affine transform formula (10.2) carries over. In fact, we will show that (10.2) holds for $u \in \mathbb{R}^d$ if either side is well defined. This asserts exponential moments of $X(t)$ in particular and will prove most useful for deriving pricing formulas in affine factor models.

For any set $U \subset \mathbb{R}^k$ ($k \in \mathbb{N}$), we define the strip

$$\mathcal{S}(U) = \left\{ z \in \mathbb{C}^k \mid \Re z \in U \right\}$$

in $\mathbb{C}^k$. The proof of the following theorem is postponed to Sect. 10.7.3. It builds on results that are developed in Sects. 10.6 and 10.7 below.

Theorem 10.3 *Suppose X is affine with admissible parameters as given in (10.6). Let $\mathcal{D}_K$ ($K = \mathbb{R}$ or $\mathbb{C}$) denote the maximal domain for the system of Riccati equations (10.7), and let $\tau > 0$. Then:*

---

[4]Let f, g and h be nonnegative continuous functions $[0, T] \to \mathbb{R}_+$ with

$$f(t) \leq h(t) + \int_0^t g(s) f(s) ds, \quad t \in [0, T].$$

Then

$$f(t) \leq h(t) + \int_0^t h(s) g(s) e^{\int_s^t g(\xi) d\xi} ds, \quad t \in [0, T],$$

see [55, (10.5.1.3)].

(a) $S(\mathcal{D}_{\mathbb{R}}(\tau)) \subset \mathcal{D}_{\mathbb{C}}(\tau)$.

(b) $\mathcal{D}_{\mathbb{R}}(\tau) = M(\tau)$ where

$$M(\tau) = \left\{ u \in \mathbb{R}^d \mid \mathbb{E}\left[e^{u^\top X^x(\tau)}\right] < \infty \text{ for all } x \in \mathbb{R}^m_+ \times \mathbb{R}^n \right\}.$$

(c) $\mathcal{D}_{\mathbb{R}}(\tau)$ and $\mathcal{D}_{\mathbb{R}}$ are convex sets.

Moreover, for all $0 \leq t \leq T$ and $x \in \mathbb{R}^m_+ \times \mathbb{R}^n$:

(d) (10.2) *holds for all* $u \in S(\mathcal{D}_{\mathbb{R}}(T - t))$.

(e) (10.2) *holds for all* $u \in \mathbb{C}^m_- \times i\mathbb{R}^n$.

(f) $M(t) \supseteq M(T)$.

As a corollary we may thus formulate the following key message of Theorem 10.3 parts (a), (b) and (d).

Corollary 10.1 *Suppose that either side of (10.2) is well defined for some $t \leq T$ and $u \in \mathbb{R}^d$. Then (10.2) holds, implying that both sides are well defined in particular, for u replaced by $u + iv$ for any $v \in \mathbb{R}^d$.*

Part (f) of Theorem 10.3 states that integrability of $e^{u^\top X^x(T)}$ for all $x \in \mathbb{R}^m_+ \times \mathbb{R}^n$, for some given T and $u \in \mathbb{R}^d$, implies integrability of $e^{u^\top X^x(t)}$ for all $x \in \mathbb{R}^m_+ \times \mathbb{R}^n$ and $t \leq T$. In other words, the set of exponential moment parameters $M(t)$ is non-expanding in t.

10.3 Discounting and Pricing in Affine Models

We let X be affine on the canonical state space $\mathbb{R}^m_+ \times \mathbb{R}^n$ with admissible parameters a, α_i, b, β_i as given in (10.6). Since we are interested in pricing, we interpret

$$\mathbb{P} = \mathbb{Q}$$

as risk-neutral measure and $W = W^*$ as $\mathbb{Q}$-Brownian motion in this section.[5]

A short-rate model of the form

$$r(t) = c + \gamma^\top X(t), \tag{10.9}$$

for some constant parameters $c \in \mathbb{R}$ and $\gamma \in \mathbb{R}^d$, is called an affine short-rate model. Special cases, for dimension $d = 1$, are the Vasiček and CIR short-rate models. We recall from (9.11) that an affine term-structure model always induces an affine short-rate model.

---

[5]Note, however, that the affine property of X is not preserved under an equivalent change of measure in general. Measure changes which preserve the affine structure are studied in detail in Cheridito, Filipović and Yor [42].

Now let $T > 0$, and consider a T-claim with payoff of the form $f(X(T))$ which meets the required integrability condition

$$\mathbb{E}\left[e^{-\int_0^T r(s)\,ds}\,|f(X(T))|\right] < \infty,$$

see (7.3). Its arbitrage price at time $t \le T$ is then given by

$$\pi(t) = \mathbb{E}\left[e^{-\int_t^T r(s)\,ds}\,f(X(T)) \mid \mathcal{F}_t\right]. \tag{10.10}$$

Compare this also to (5.6). A particular example is the T-bond with $f \equiv 1$. Our aim is to derive an analytic, or at least numerically tractable, pricing formula for (10.10).

As a first step, we derive a formula for the $\mathcal{F}_t$-conditional characteristic function of $X(T)$ under the T-forward measure, which equals, up to normalization with $P(t, T)$ (see Sect. 7.1),

$$\mathbb{E}\left[e^{-\int_t^T r(s)\,ds}\,e^{u^\top X(T)} \mid \mathcal{F}_t\right] \tag{10.11}$$

for $u \in i\mathbb{R}^d$. Note that the following integrability condition (a) is satisfied in particular if r is uniformly bounded from below, that is, if $\gamma \in \mathbb{R}_+^m \times \{0\}$ (see also Exercise 10.4).

Theorem 10.4 *Let $\tau > 0$. The following statements are equivalent:*

(a) $\mathbb{E}[e^{-\int_0^\tau r(s)\,ds}] < \infty$ *for all $x \in \mathbb{R}_+^m \times \mathbb{R}^n$.*
(b) *There exists a unique solution $(\Phi(\cdot, u), \Psi(\cdot, u)) : [0, \tau] \to \mathbb{C} \times \mathbb{C}^d$ of*

$$\partial_t \Phi(t, u) = \frac{1}{2}\Psi_J(t, u)^\top a_{JJ}\Psi_J(t, u) + b^\top \Psi(t, u) - c,$$

$$\Phi(0, u) = 0,$$

$$\partial_t \Psi_i(t, u) = \frac{1}{2}\Psi(t, u)^\top \alpha_i \Psi(t, u) + \beta_i^\top \Psi(t, u) - \gamma_i, \quad i \in I, \tag{10.12}$$

$$\partial_t \psi_J(t, u) = \mathcal{B}_{JJ}^\top \Psi_J(t, u) - \gamma_J,$$

$$\Psi(0, u) = u$$

for $u = 0$.

Moreover, let $\mathcal{D}_K$ $(K = \mathbb{R}$ or $\mathbb{C})$ denote the maximal domain for the system of Riccati equations (10.12). If either (a) or (b) holds then $\mathcal{D}_\mathbb{R}(S)$ is a convex open neighborhood of 0 in $\mathbb{R}^d$, and $S(\mathcal{D}_\mathbb{R}(S)) \subset \mathcal{D}_\mathbb{C}(S)$, for all $S \le \tau$. Further, (10.11) allows the following affine representation:

$$\mathbb{E}\left[e^{-\int_t^T r(s)\,ds}\,e^{u^\top X(T)} \mid \mathcal{F}_t\right] = e^{\Phi(T-t,u)+\Psi(T-t,u)^\top X(t)} \tag{10.13}$$

for all $u \in S(\mathcal{D}_\mathbb{R}(S))$, $t \le T \le t + S$ and $x \in \mathbb{R}_+^m \times \mathbb{R}^n$.

Proof We first enlarge the state space and consider the real-valued process

$$Y(t) = y + \int_0^t \left(c + \gamma^\top X(s) \right) ds, \quad y \in \mathbb{R}.$$

A moment's reflection reveals that $X' = \binom{X}{Y}$ is an $\mathbb{R}_+^m \times \mathbb{R}^{n+1}$-valued diffusion process with diffusion matrix $a' + \sum_{i \in I} x_i \alpha_i'$ and drift $b' + \mathcal{B}'x'$ where

$$a' = \begin{pmatrix} a & 0 \\ 0 & 0 \end{pmatrix}, \qquad \alpha_i' = \begin{pmatrix} \alpha_i & 0 \\ 0 & 0 \end{pmatrix}, \qquad b' = \begin{pmatrix} b \\ c \end{pmatrix}, \qquad \mathcal{B}' = \begin{pmatrix} \mathcal{B} & 0 \\ \gamma^\top & 0 \end{pmatrix}$$

form admissible parameters. We claim that X' is an affine process.

Indeed, the candidate system of Riccati equations reads, for $i \in I$:

$$\partial_t \phi'(t, u, v) = \frac{1}{2} \psi_J'(t, u, v)^\top a_{JJ} \psi_J'(t, u, v) + b^\top \psi_{\{1,\dots,d\}}'(t, u, v) + \boxed{cv},$$

$$\phi'(0, u, v) = 0,$$

$$\partial_t \psi_i'(t, u, v) = \frac{1}{2} \psi'(t, u, v)^\top \alpha_i \psi'(t, u, v) + \beta_i^\top \psi_{\{1,\dots,d\}}'(t, u, v) + \boxed{\gamma_i v},$$

$$\partial_t \psi_J'(t, u, v) = \mathcal{B}_{JJ}^\top \psi_J'(t, u, v) + \boxed{\gamma_J v},$$

$$\partial_t \psi_{d+1}'(t, u, v) = 0,$$

$$\psi'(0, u, v) = \begin{pmatrix} u \\ v \end{pmatrix}.$$

(10.14)

Here we replaced the constant solution $\psi_{d+1}'(\cdot, u, v) \equiv v$ by v in the boxes. Theorem 10.2 carries over and asserts a unique global $\mathbb{C}_- \times \mathbb{C}_-^m \times i\mathbb{R}^{n+1}$-valued solution $(\phi'(\cdot, u, v), \psi'(\cdot, u, v))$ of (10.12) for all $(u, v) \in \mathbb{C}_-^m \times i\mathbb{R}^n \times i\mathbb{R}$. The second part of Theorem 10.1 thus asserts that X' is affine with conditional characteristic function

$$\mathbb{E}\left[e^{u^\top X(T) + vY(T)} \mid \mathcal{F}_t \right] = e^{\phi'(T-t, u, v) + \psi_{\{1,\dots,d\}}'(T-t, u, v)^\top X(t) + vY(t)}$$

for all $(u, v) \in \mathbb{C}_-^m \times i\mathbb{R}^n \times i\mathbb{R}$ and $t \leq T$.

The theorem now follows from Theorem 10.3 once we set $\Phi(t, u) = \phi'(t, u, -1)$ and $\Psi(t, u) = \psi_{\{1,\dots,d\}}'(t, u, -1)$. Indeed, it is clear by inspection that $\mathcal{D}_K(S) = \{u \in K^d \mid (u, -1) \in \mathcal{D}_K'(S)\}$ where $\mathcal{D}_K'$ denotes the maximal domain for the system of Riccati equations (10.14). $\qquad\square$

As immediate consequence of Theorem 10.4, we obtain the following explicit price formulas for T-bonds in terms of Φ and Ψ.

Corollary 10.2 *For any maturity $T \leq \tau$, the T-bond price at $t \leq T$ is given as*

$$P(t, T) = e^{-A(T-t) - B(T-t)^\top X(t)}$$

where we define, in accordance with Sect. 5.3,

$$A(t) = -\Phi(t, 0), \qquad B(t) = -\Psi(t, 0).$$

Moreover, for $t \leq T \leq S \leq \tau$, the $\mathcal{F}_t$-conditional characteristic function of $X(T)$ under the S-forward measure $\mathbb{Q}^S$ is given by

$$\mathbb{E}_{\mathbb{Q}^S}\left[e^{u^\top X(T)} \mid \mathcal{F}_t\right] = \frac{e^{-A(S-T)+\Phi(T-t,u-B(S-T))+\Psi(T-t,u-B(S-T))^\top X(t)}}{P(t, S)}$$

(10.15)

for all $u \in S(\mathcal{D}_{\mathbb{R}}(T) + B(S - T))$, which contains $i\mathbb{R}^d$.

Proof The bond price formula follows from (10.13) with $u = 0$.

Now let $t \leq T \leq S \leq \tau$. In view of the flow property $\Psi(T, -B(S - T)) = -B(S)$, we know that $-B(S - T) \in \mathcal{D}_{\mathbb{R}}(T)$, and thus $S(\mathcal{D}_{\mathbb{R}}(T) + B(S - T))$ contains $i\mathbb{R}^d$. Moreover, for $u \in S(\mathcal{D}_{\mathbb{R}}(T) + B(S - T))$, we obtain from (10.13) by nested conditional expectation

$$\mathbb{E}\left[e^{-\int_t^S r(s)\,ds}e^{u^\top X(T)} \mid \mathcal{F}_t\right] = \mathbb{E}\left[e^{-\int_t^T r(s)\,ds}\mathbb{E}\left[e^{-\int_T^S r(s)\,ds} \mid \mathcal{F}_T\right]e^{u^\top X(T)} \mid \mathcal{F}_t\right]$$

$$= e^{-A(S-T)}\mathbb{E}\left[e^{-\int_t^T r(s)\,ds}e^{(u-B(S-T))^\top X(T)} \mid \mathcal{F}_t\right]$$

$$= e^{-A(S-T)+\Phi(T-t,u-B(S-T))+\Psi(T-t,u-B(S-T))^\top X(t)}.$$

Normalizing by $P(t, S)$ yields (10.15).   □

For more general payoff functions f, we can proceed as follows. Either we recognize the $\mathcal{F}_t$-conditional distribution, say $Q(t, T, dx)$, of $X(T)$ under the T-forward measure $\mathbb{Q}^T$ from its characteristic function in (10.15). Then compute the price (10.10) by (numerical) integration of f

$$\pi(t) = P(t, T)\int_{\mathbb{R}^d} f(x)Q(t, T, dx).$$

(10.16)

Or we employ Fourier transform techniques as the following two consecutive affine pricing theorems indicate.

Theorem 10.5 *Suppose either condition (a) or (b) of Theorem 10.4 is met for some $\tau \geq T$, and let $\mathcal{D}_{\mathbb{R}}$ denote the maximal domain for the system of Riccati equations (10.12). Assume that f satisfies*

$$f(x) = \int_{\mathbb{R}^q} e^{(v+iL\lambda)^\top x}\widetilde{f}(\lambda)\,d\lambda, \quad dx\text{-a.s.}$$

(10.17)

for some $v \in \mathcal{D}_{\mathbb{R}}(T)$ *and* $d \times q$*-matrix* L*, and some integrable function* $\widetilde{f} : \mathbb{R}^q \to \mathbb{C}$*, for a positive integer* $q \leq d$*. Then the price* (10.10) *is well defined and given by the formula*

$$\pi(t) = \int_{\mathbb{R}^q} e^{\Phi(T-t,v+iL\lambda)+\Psi(T-t,v+iL\lambda)^\top X(t)} \widetilde{f}(\lambda)\, d\lambda. \qquad (10.18)$$

From the Riemann–Lebesgue theorem ([156, Chap. I, Theorem 1.2]) we know that the right-hand side of (10.17) is continuous in x. Hence the equality (10.17) necessarily holds for all x if f is continuous.

Proof By assumption, we have

$$\mathbb{E}\left[e^{-\int_0^T r(s)\,ds}|f(X(T))|\right] \leq \mathbb{E}\left[\int_{\mathbb{R}^q} e^{-\int_0^T r(s)\,ds} e^{v^\top X(T)}|\widetilde{f}(\lambda)|\, d\lambda\right] < \infty.$$

Hence we may apply Fubini's theorem to change the order of integration, which gives

$$\pi(t) = \mathbb{E}\left[e^{-\int_t^T r(s)\,ds} \int_{\mathbb{R}^q} e^{(v+iL\lambda)^\top X(T)} \widetilde{f}(\lambda)\, d\lambda \mid \mathcal{F}_t\right]$$

$$= \int_{\mathbb{R}^q} \mathbb{E}\left[e^{-\int_t^T r(s)\,ds} e^{(v+iL\lambda)^\top X(T)} \mid \mathcal{F}_t\right] \widetilde{f}(\lambda)\, d\lambda$$

$$= \int_{\mathbb{R}^q} e^{\Phi(T-t,v+iL\lambda)+\Psi(T-t,v+iL\lambda)^\top X(t)} \widetilde{f}(\lambda)\, d\lambda,$$

which is (10.18). $\qquad\qquad\square$

Next, we give a more constructive and alternative approach, respectively, to the representation (10.17).

Theorem 10.6 *Suppose either condition* (a) *or* (b) *of Theorem* 10.4 *is met for some* $\tau \geq T$*, and let* $\mathcal{D}_{\mathbb{R}}$ *denote the maximal domain for the system of Riccati equations* (10.12)*. Assume that* f *is of the form*

$$f(x) = e^{v^\top x} h(L^\top x)$$

for some $v \in \mathcal{D}_{\mathbb{R}}(T)$ *and* $d \times q$*-matrix* L*, and some integrable function* $h : \mathbb{R}^q \to \mathbb{R}$*, for a positive integer* $q \leq d$*. Define the bounded function*

$$\widetilde{f}(\lambda) = \frac{1}{(2\pi)^q} \int_{\mathbb{R}^q} e^{-i\lambda^\top y} h(y)\, dy, \quad \lambda \in \mathbb{R}^q.$$

(a) *If* $\widetilde{f}(\lambda)$ *is an integrable function in* $\lambda \in \mathbb{R}^q$ *then the assumptions of Theorem* 10.5 *are met.*

(b) *If $v = Lw$, for some $w \in \mathbb{R}^q$, and $e^{\Phi(T-t,v+iL\lambda)+\Psi(T-t,v+iL\lambda)^\top X(t)}$ is an integrable function in $\lambda \in \mathbb{R}^q$ then the $\mathcal{F}_t$-conditional distribution of the $\mathbb{R}^q$-valued random variable $Y = L^\top X(T)$ under the T-forward measure $\mathbb{Q}^T$ admits the continuous density function*

$$q(t, T, y) = \frac{1}{(2\pi)^q} \int_{\mathbb{R}^q} e^{-(w+i\lambda)^\top y} \frac{e^{\Phi(T-t,v+iL\lambda)+\Psi(T-t,v+iL\lambda)^\top X(t)}}{P(t,T)} \, d\lambda.$$

(10.19)

In either case, the integral in (10.18) is well defined and the price formula (10.18) holds.

Proof We recall the fundamental inversion formula from Fourier analysis ([156, Chap. I, Corollary 1.21]): let $g : \mathbb{R}^q \to \mathbb{C}$ be an integrable function with integrable Fourier transform

$$\hat{g}(\lambda) = \int_{\mathbb{R}^q} e^{-i\lambda^\top y} g(y) \, dy.$$

Then the inversion formula

$$g(y) = \frac{1}{(2\pi)^q} \int_{\mathbb{R}^q} e^{i\lambda^\top y} \hat{g}(\lambda) \, d\lambda$$

(10.20)

holds for dy-almost all $y \in \mathbb{R}^q$.

Under the assumption of (a), the Fourier inversion formula (10.20) applied to $h(y)$ yields the representation (10.17). Hence Theorem 10.5 applies.

As for (b), we denote by $q(t, T, dy)$ the $\mathcal{F}_t$-conditional distribution of $Y = L^\top X(T)$ under the T-forward measure $\mathbb{Q}^T$. From (10.15) we infer the characteristic function of the bounded (why?) measure $e^{w^\top y} q(t, T, dy)$:

$$\int_{\mathbb{R}^q} e^{(w+i\lambda)^\top y} q(t, T, dy) = \mathbb{E}\left[e^{(w+i\lambda)^\top L^\top X(T)} \mid \mathcal{F}_t \right]$$

$$= \frac{e^{\Phi(T-t,v+iL\lambda)+\Psi(T-t,v+iL\lambda)^\top X(t)}}{P(t,T)}, \quad \lambda \in \mathbb{R}^q.$$

By assumption, this is an integrable function in λ on $\mathbb{R}^q$. The Fourier inversion formula (10.20) thus applies and the injectivity of the characteristic function (see e.g. [161, Sect. 16.6]) yields that $q(t, T, dy)$ admits the continuous density function (10.19). Moreover, we then obtain

$$P(t, T) \int_{\mathbb{R}^q} |e^{w^\top y} h(y)| q(t, T, y) \, dy$$

$$\leq \frac{1}{(2\pi)^q} \int_{\mathbb{R}^q} \int_{\mathbb{R}^q} |h(y)| \left| e^{\Phi(T-t,v+iL\lambda)+\Psi(T-t,v+iL\lambda)^\top X(t)} \right| d\lambda \, dy < \infty.$$

Hence again we can apply Fubini's theorem to change the order of integration, which gives

$$
\begin{aligned}
\pi(t) &= P(t, T) \int_{\mathbb{R}^q} e^{w^\top y} h(y) q(t, T, y) \, dy \\
&= \frac{1}{(2\pi)^q} \int_{\mathbb{R}^q} \int_{\mathbb{R}^q} e^{w^\top y} h(y) e^{-(w+i\lambda)^\top y} e^{\Phi(T-t, v+iL\lambda) + \Psi(T-t, v+iL\lambda)^\top X(t)} \, d\lambda \, dy \\
&= \frac{1}{(2\pi)^q} \int_{\mathbb{R}^q} \left(\int_{\mathbb{R}^q} h(y) e^{-i\lambda^\top y} \, dy \right) e^{\Phi(T-t, v+iL\lambda) + \Psi(T-t, v+iL\lambda)^\top X(t)} \, d\lambda,
\end{aligned}
$$

which is (10.18). $\qquad\square$

The integral in the pricing formula (10.18), as well as $\widetilde{f}$, has to be computed numerically in general. In this regard, it is remarkable that this integral is over $\mathbb{R}^q$, where q may be much smaller than the dimension d of the state process X. In fact, we will see that $\widetilde{f}$ is given in closed form and $q = 1$ for bond options.

Let us reflect for a moment on the representation (10.17): the payoff $f(X(T))$ is decomposed into a linear combination of (a continuum) of complex-valued basis "payoffs"[6] $e^{(v+iL\lambda)^\top X(T)}$ with weights $\widetilde{f}(\lambda)$. By the very nature of the affine process X, these basis claims admit closed-form complex-valued "prices"

$$
\pi_{v+iL\lambda}(t) = e^{\Phi(T-t, v+iL\lambda) + \Psi(T-t, v+iL\lambda)^\top X(t)}.
$$

Linearity of pricing thus implies that the price of $f(X(T))$ is given as linear combination of the $\pi_{v+iL\lambda}(t)$ with the same weights $\widetilde{f}(\lambda)$. But this is just formula (10.18). This reflection unfolds the power of affine diffusion processes. It suggests that we explore other types of diffusion processes that admit closed-form prices for some well specified basis of payoff functions. This approach has been pursued in e.g. [21, 40] and others. It is currently an open area of research.

Our affine pricing theorems 10.5 and 10.6 would not have much practical implications unless we find some interesting payoff functions of the form (10.17). Luckily, such representations do indeed exist for a broad range of the most important payoff functions as we shall see in the following section.

10.3.1 Examples of Fourier Decompositions

We first show that the functions $(e^y - K)^+$ and $(K - e^y)^+$ related to the European call and put option payoffs can be explicitly represented in the form (10.17).

---

[6] Admitting for the sake of reflection that there is such as a complex-valued currency.

Lemma 10.2 *Let $K > 0$. For any $y \in \mathbb{R}$ the following identities hold:*

$$
\frac{1}{2\pi} \int_{\mathbb{R}} e^{(w+i\lambda)y} \frac{K^{-(w-1+i\lambda)}}{(w+i\lambda)(w-1+i\lambda)} \, d\lambda
$$

$$
= \begin{cases} (K - e^y)^+ & \text{if } w < 0, \\ (e^y - K)^+ - e^y & \text{if } 0 < w < 1, \\ (e^y - K)^+ & \text{if } w > 1. \end{cases}
$$

The middle case $(0 < w < 1)$ obviously also equals $(K - e^y)^+ - K$.

We will give some applications of this formula for bond and stock options in Sects. 10.3.2 and 10.3.3 below.

Proof Let $w < 0$. Then the function $h(y) = e^{-wy}(K - e^y)^+$ is integrable on $\mathbb{R}$. An easy calculation shows that its Fourier transform

$$
\hat{h}(\lambda) = \int_{\mathbb{R}} e^{-(w+i\lambda)y} (K - e^y)^+ \, dy = \frac{K^{-(w-1+i\lambda)}}{(w+i\lambda)(w-1+i\lambda)} \tag{10.21}
$$

is also integrable on $\mathbb{R}$. Hence the Fourier inversion formula (10.20) applies, and we conclude that the claimed identity holds for $w < 0$. The other cases follow by similar arguments ($\rightarrow$ Exercise 10.5). $\qquad \square$

Nothing prevents us from choosing $K = e^z$ in Lemma 10.2. This way, we obtain the following useful formula related to the payoff of an exchange option.

Corollary 10.3 *For any $y, z \in \mathbb{R}$ the following identities hold:*

$$
\frac{1}{2\pi} \int_{\mathbb{R}} \frac{e^{(w+i\lambda)y - (w-1+i\lambda)z}}{(w+i\lambda)(w-1+i\lambda)} \, d\lambda = \begin{cases} (e^y - e^z)^+ & \text{if } w > 1, \\ (e^y - e^z)^+ - e^y & \text{if } 0 < w < 1. \end{cases}
$$

Suppose two asset prices are modeled as $S_i = e^{X_{m+i}}$, $i = 1, 2$, where X denotes our affine state diffusion on $\mathbb{R}_+^m \times \mathbb{R}^n$. Then the payoff of the option to exchange c_2 units of asset S_2 against c_1 units of asset S_1 at some date T is[7]

$$
f(X(T)) = \left(c_1 e^{X_{m+1}(T)} - c_2 e^{X_{m+2}(T)} \right)^+. \tag{10.22}
$$

In view of Corollary 10.3, this payoff function can be represented as (10.17) where $q = 1$, $v = we_{m+1} + (1 - w)e_{m+2}$, $L = e_{m+1} - e_{m+2}$, and

$$
\tilde{f}(\lambda) = \frac{c_1^{w+i\lambda} c_2^{-(w-1+i\lambda)}}{2\pi(w+i\lambda)(w-1+i\lambda)},
$$

for some $w > 1$.

---

[7] An exchange option is also called Margrabe option. The price formula was derived by Margrabe [121] and Fischer [76] for the case of two jointly lognormal stock price processes.

In a similar way, but now including double integration, we can find an explicit Fourier decomposition of the spread option payoff

$$f(X(T)) = \left(e^{X_{m+1}(T)} - e^{X_{m+2}(T)} - K \right)^+, \tag{10.23}$$

for some strike price $K > 0$. Indeed, Lemma 10.3 below implies that this payoff function can be represented as (10.17) where $q = 2$, $v = w_1 e_{m+1} + w_2 e_{m+2}$, $L = (e_{m+1}, e_{m+2})$, and

$$\tilde{f}(\lambda) = \frac{\Gamma(w_1 + w_2 - 1 + i(\lambda_1 + \lambda_2))\Gamma(-w_2 - i\lambda_2)}{(2\pi)^2 K^{w_1 + w_2 + i(\lambda_1 + \lambda_2)}\Gamma(w_1 + 1 + i\lambda_1)},$$

for some $w_2 < 0$ and $w_1 > 1 - w_2$.

It remains to be checked from case to case for both payoffs (10.22) and (10.23) whether $v \in \mathcal{D}_\mathbb{R}(T)$ holds for Theorem 10.5 to apply ($\to$ Exercise 10.19).

The following representation including a double Fourier integral is due to Hurd and Zhou [98].

Lemma 10.3 *Let $w = (w_1, w_2)^\top \in \mathbb{R}^2$ be such that $w_2 < 0$ and $w_1 + w_2 > 1$. Then for any $y = (y_1, y_2)^\top \in \mathbb{R}^2$ the following identity holds:*

$$\left(e^{y_1} - e^{y_2} - 1 \right)^+$$

$$= \frac{1}{(2\pi)^2} \int_{\mathbb{R}^2} e^{(w+i\lambda)^\top y} \frac{\Gamma(w_1 + w_2 - 1 + i(\lambda_1 + \lambda_2))\Gamma(-w_2 - i\lambda_2)}{\Gamma(w_1 + 1 + i\lambda_1)} \, d\lambda_1 \, d\lambda_2,$$

where the Gamma function $\Gamma(z) = \int_0^\infty t^{-1+z} e^{-t} \, dt$ is defined for all complex z with $\Re(z) > 0$.

Proof By assumption the function $h(y) = e^{-w^\top y}(e^{y_1} - e^{y_2} - 1)^+$ is integrable on $\mathbb{R}^2$. Its Fourier transform can be calculated, using (10.21) for $K = e^{y_1} - 1 > 0$ if $y_1 > 0$, as follows:

$$\hat{h}(\lambda) = \int_{\mathbb{R}^2} e^{-(w+i\lambda)^\top y} \left(e^{y_1} - e^{y_2} - 1 \right)^+ \, dy_1 \, dy_2$$

$$= \int_0^\infty e^{-(w_1 + i\lambda_1)y_1} \int_\mathbb{R} e^{-(w_2 + i\lambda_2)y_2} \left(e^{y_1} - 1 - e^{y_2} \right)^+ \, dy_2 \, dy_1$$

$$= \int_0^\infty e^{-(w_1 + i\lambda_1)y_1} \frac{(e^{y_1} - 1)^{-(w_2 - 1 + i\lambda_2)}}{(w_2 + i\lambda_2)(w_2 - 1 + i\lambda_2)} \, dy_1$$

$$= \frac{1}{(w_2 + i\lambda_2)(w_2 - 1 + i\lambda_2)}$$

$$\times \int_0^\infty (e^{-y_1})^{w_1 + w_2 - 1 + i(\lambda_1 + \lambda_2)} (1 - e^{-y_1})^{-w_2 + 1 - i\lambda_2} \, dy_1.$$

The change of variables $z = e^{-y_1}$, with $dz/z = -dy_1$, then yields

$$\widehat{h}(\lambda) = \frac{1}{(w_2 + i\lambda_2)(w_2 - 1 + i\lambda_2)} \int_0^1 z^{w_1 + w_2 - 2 + i(\lambda_1 + \lambda_2)}(1 - z)^{-w_2 + 1 - i\lambda_2} \, dz.$$

Recall that the beta function

$$B(a, b) = \frac{\Gamma(a)\Gamma(b)}{\Gamma(a + b)}$$

is defined for any complex a, b with $\Re(a), \Re(b) > 0$ by

$$B(a, b) = \int_0^1 z^{a-1}(1 - z)^{b-1} \, dz.$$

Since $w_1 + w_2 > 1$, we obtain from this and the property $\Gamma(z) = (z - 1)\Gamma(z - 1)$ that

$$
\begin{aligned}
\widehat{h}(\lambda) &= \frac{B(w_1 + w_2 - 1 + i(\lambda_1 + \lambda_2), -w_2 + 2 - i\lambda_2)}{(w_2 + i\lambda_2)(w_2 - 1 + i\lambda_2)} \\
&= \frac{\Gamma(w_1 + w_2 - 1 + i(\lambda_1 + \lambda_2))\Gamma(-w_2 + 2 - i\lambda_2)}{\Gamma(w_1 + 1 + i\lambda_1)(w_2 + i\lambda_2)(w_2 - 1 + i\lambda_2)} \\
&= \frac{\Gamma(w_1 + w_2 - 1 + i(\lambda_1 + \lambda_2))\Gamma(-w_2 - i\lambda_2)}{\Gamma(w_1 + 1 + i\lambda_1)}.
\end{aligned}
\tag{10.24}
$$

It remains to be checked whether $\widehat{h}(\lambda)$ is integrable in $\lambda \in \mathbb{R}^2$. From the definition of the beta function it follows that $|B(a, b)| \le B(\Re(a), \Re(b))$. Hence

$$|\widehat{h}(\lambda)| \le \frac{B(w_1 + w_2 - 1, -w_2 + 2)}{|(w_2 + i\lambda_2)(w_2 - 1 + i\lambda_2)|}. \tag{10.25}$$

On the other hand, factorizing the first factor in the nominator of the third line in (10.24), we can rewrite $\widehat{h}(\lambda)$ as

$$\widehat{h}(\lambda) = \frac{B(w_1 + w_2 + 1 + i(\lambda_1 + \lambda_2), -w_2 - i\lambda_2)}{(w_1 + w_2 + i(\lambda_1 + \lambda_2))(w_1 + w_2 - 1 + i(\lambda_1 + \lambda_2))}.$$

Hence

$$|\widehat{h}(\lambda)| \le \frac{B(w_1 + w_2 + 1, -w_2)}{|(w_1 + w_2 + i(\lambda_1 + \lambda_2))(w_1 + w_2 - 1 + i(\lambda_1 + \lambda_2))|}. \tag{10.26}$$

The two bounds (10.25) and (10.26) imply that $\widehat{h}(\lambda)$ is integrable in $\lambda \in \mathbb{R}^2$. The Fourier inversion formula (10.20) now yields the claim.   □

The two bounds (10.25) and (10.26) assert that the numerical integration be feasible for many models. It can be made efficient by fast Fourier transform, as outlined in [98].

10.3.2 Bond Option Pricing in Affine Models

Let us elaborate further on the pricing of bond options. We assume that either condition (a) or (b) of Theorem 10.4 is met, and fix some maturities $T < S \leq \tau$. A straightforward modification of Lemma 10.2 implies that the payoff function of the European call option on the S-bond with expiry date T and strike price K admits the integral representation

$$\left(e^{-A(S-T)-B(S-t)^{\top}x} - K\right)^{+} = \frac{1}{2\pi}\int_{\mathbb{R}} e^{-(w+i\lambda)B(S-t)^{\top}x}\, \widetilde{f}(w,\lambda)\, d\lambda$$

where we define

$$\widetilde{f}(w,\lambda) = \frac{1}{2\pi} e^{-(w+i\lambda)A(S-T)} \frac{K^{-(w-1+i\lambda)}}{(w+i\lambda)(w-1+i\lambda)}, \tag{10.27}$$

for any real $w > 1$. A similar formula results for put options. We thus obtain from Theorem 10.5 the following master pricing formula for European call and put bond options.

Corollary 10.4 *There exists some $w_- < 0$ and $w_+ > 1$ such that $-B(S-T)w \in \mathcal{D}_{\mathbb{R}}(T)$ for all $w \in (w_-, w_+)$, where $\mathcal{D}_{\mathbb{R}}$ denotes the maximal domain for the system of Riccati equations (10.12). Define $\widetilde{f}(w,\lambda)$ as in (10.27). Then the line integral*

$$\Pi(w,t) = \int_{\mathbb{R}} e^{\Phi(T-t,-(w+i\lambda)B(S-T))+\Psi(T-t,-(w+i\lambda)B(S-T))^{\top}X(t)}\, \widetilde{f}(w,\lambda)\, d\lambda$$

is well defined for all $w \in (w_-, w_+) \setminus \{0,1\}$ and $t \leq T$. Moreover, the time t prices of the European call and put option on the S-bond with expiry date T and strike price K are given by any of the following identities:

$$\pi_{call}(t) = \begin{cases} \Pi(w,t), & \text{if } w \in (1, w_+), \\ \Pi(w,t) + P(t,S), & \text{if } w \in (0,1) \end{cases}$$

$$= P(t,S)q(t,S,\mathcal{I}) - KP(t,T)q(t,T,\mathcal{I}),$$

$$\pi_{put}(t) = \begin{cases} \Pi(w,t) + KP(t,T), & \text{if } w \in (0,1), \\ \Pi(w,t), & \text{if } w \in (w_-, 0) \end{cases} \tag{10.28}$$

$$= KP(t,T)q(t,T,\mathbb{R}\setminus\mathcal{I}) - P(t,S)q(t,S,\mathbb{R}\setminus\mathcal{I}),$$

where $\mathcal{I} = (A(S-T) + \log K, \infty)$, and $q(t,S,dy)$ and $q(t,T,dy)$ denote the $\mathcal{F}_t$-conditional distributions of the real-valued random variable $Y = -B(S-T)^{\top}X(T)$ under the S- and T-forward measure, respectively.

Proof From the flow property $\Psi(T, -B(S-T)) = -B(S)$, we know that $-B(S-T) \in \mathcal{D}_{\mathbb{R}}(T)$. Since $\mathcal{D}_{\mathbb{R}}(T)$ is a convex open neighborhood of 0 in $\mathbb{R}^d$, we obtain that $-B(S-T)w \in \mathcal{D}_{\mathbb{R}}(T)$ for all $w \in (w_-, w_+)$, for some $w_- < 0$ and $w_+ > 1$.

It then follows by inspection that $e^{\Phi(T-t,-(w+i\lambda)B(S-T))+\Psi(T-t,-(w+i\lambda)B(S-T))^{\top}X(t)}$ is uniformly bounded and $\widetilde{f}(w,\lambda)$ is integrable in $\lambda \subset \mathbb{R}$, for any fixed $w \in (w_-, w_+) \setminus \{0, 1\}$. Hence the line integral $\Pi(w,t)$ is well defined for all $w \in (w_-, w_+) \setminus \{0, 1\}$ and $t \leq T$.

Further, we recall from Chap. 7 that we can decompose (10.10), according to (7.7). For the call option we thus obtain

$$\pi(t) = P(t, S)\mathbb{Q}^S[E \mid \mathcal{F}_t] - KP(t, T)\mathbb{Q}^T[E \mid \mathcal{F}_t],$$

for the exercise event $E = \{-B(S-T)^{\top}X(T) > A(S-T) + \log K\}$, and similarly for the put option.

The price formulas (10.28) now follow from the above discussion, and Theorem 10.5 and Lemma 10.2 ($\rightarrow$ Exercise 10.6). This proves the corollary. $\square$

Thus the pricing of European call and put bond options in the present d-dimensional affine factor model boils down to the computation of a line integral $\Pi(w,t)$, which is a simple numerical task. Moreover, in case the distributions $q(t, S, dy)$ and $q(t, T, dy)$ are explicitly known, the pricing is reduced to the computation of the respective probabilities in (10.28) of the exercise events $\mathcal{I}$ and $\mathbb{R} \setminus \mathcal{I}$.

In the following two subsections, we illustrate this approach for the Vasiček and CIR short-rate models.

10.3.2.1 Example: Vasiček Short-Rate Model

The state space is $\mathbb{R}$, and we set $r = X$ for the Vasiček short-rate model

$$dr = (b + \beta r)\, dt + \sigma\, dW.$$

The system (10.12) reads

$$\Phi(t, u) = \frac{1}{2}\sigma^2 \int_0^t \Psi^2(s, u)\, ds + b \int_0^t \Psi(s, u)\, ds,$$

$$\partial_t \Psi(t, u) = \beta\Psi(t, u) - 1,$$

$$\Psi(0, u) = u,$$

which admits a unique global solution with

$$\Phi(t, u) = \frac{1}{2}\sigma^2 \left(\frac{u^2}{2\beta}(e^{2\beta t} - 1) + \frac{1}{2\beta^3}(e^{2\beta t} - 4e^{\beta t} + 2\beta t + 3) \right.$$
$$\left. - \frac{u}{\beta^2}(e^{2\beta t} - 2e^{\beta t} + 2\beta) \right) + b \left(\frac{e^{\beta t} - 1}{\beta}u - \frac{e^{\beta t} - 1 - \beta t}{\beta^2} \right),$$

$$\Psi(t, u) = e^{\beta t}u - \frac{e^{\beta t} - 1}{\beta}$$

for all $u \in \mathbb{C}$. Hence (10.13) holds for all $u \in \mathbb{C}$ and $t \leq T$.

Moreover, we see that the exponent of the $\mathcal{F}_t$-conditional characteristic function of $r(T)$ under the S-forward measure (10.15) is a quadratic polynomial in u. Hence, under the S-forward measure, $r(T)$ is $\mathcal{F}_t$-conditionally Gaussian distributed with variance $\sigma^2 \frac{e^{2\beta(T-t)}-1}{2\beta}$. This is in line with (5.11) (why?). A straightforward calculation yields the $\mathcal{F}_t$-conditional $\mathbb{Q}^S$-mean of $r(T)$. The bond option price formula for the Vasiček short-rate model from Proposition 7.2 and Sect. 7.2.1 can now be derived via (10.28) ($\rightarrow$ Exercise 10.7).

10.3.2.2 Example: CIR Short-Rate Model

The state space is $\mathbb{R}_+$, and we set $r = X$ for the CIR short-rate model

$$dr = (b + \beta r)\,dt + \sigma\sqrt{r}\,dW.$$

The system (10.12) reads

$$\Phi(t, u) = b \int_0^t \Psi(s, u)\,ds,$$

$$\partial_t \Psi(t, u) = \frac{1}{2}\sigma^2\Psi^2(t, u) + \beta\Psi(t, u) - 1, \qquad (10.29)$$

$$\Psi(0, u) = u.$$

By Lemma 10.12 below, there exists a unique solution $(\Phi(\cdot, u), \Psi(\cdot, u)) : \mathbb{R}_+ \to \mathbb{C}_- \times \mathbb{C}_-$, and thus identity (10.13) holds, for all $u \in \mathbb{C}_-$ and $t \leq T$. In fact, the solution is given explicitly as

$$\Phi(t, u) = \frac{2b}{\sigma^2} \log\left(\frac{2\theta e^{\frac{(\theta - \beta)t}{2}}}{L_3(t) - L_4(t)u}\right),$$

$$\Psi(t, u) = -\frac{L_1(t) - L_2(t)u}{L_3(t) - L_4(t)u},$$

where $\theta = \sqrt{\beta^2 + 2\sigma^2}$ and

$$L_1(t) = 2\left(e^{\theta t} - 1\right),$$

$$L_2(t) = \theta\left(e^{\theta t} + 1\right) + \beta\left(e^{\theta t} - 1\right),$$

$$L_3(t) = \theta\left(e^{\theta t} + 1\right) - \beta\left(e^{\theta t} - 1\right),$$

$$L_4(t) = \sigma^2\left(e^{\theta t} - 1\right).$$

Some tedious but elementary algebraic manipulations show that the $\mathcal{F}_t$-conditional characteristic function of $r(T)$ under the S-forward measure $\mathbb{Q}^S$ is given by

$$\mathbb{E}_{\mathbb{Q}^S}\left[e^{ur(T)} \mid \mathcal{F}_t\right] = \frac{e^{\frac{C_2(t,T,S)r(t)C_1(t,T,S)u}{1-C_1(t,T,S)u}}}{(1 - C_1(t,T,S)u)^{\frac{2b}{\sigma^2}}},$$

where

$$C_1(t,T,S) = \frac{L_3(S-T)L_4(T-t)}{2\theta L_3(S-t)}, \qquad C_2(t,T,S) = \frac{L_2(T-t)}{L_4(T-t)} - \frac{L_1(S-t)}{L_3(S-t)}.$$

Comparing this with Lemma 10.4 below, we conclude that the $\mathcal{F}_t$-conditional distribution of the random variable

$$Z(t,T) = \frac{2r(T)}{C_1(t,T,S)}$$

under the S-forward measure $\mathbb{Q}^S$ is noncentral χ^2 with $\frac{4b}{\sigma^2}$ degrees of freedom and parameter of noncentrality $2C_2(t,T,S)r(t)$. The noncentral χ^2-distribution is a generalization of the distribution of the sum of the squares of independent normal distributed random variables ($\to$ Exercise 10.8). It is good to know that the noncentral χ^2-distribution is hard coded in most statistical software packages.[8] Combining this with Corollary 10.4, we obtain explicit European bond option price formulas for the CIR model.

Lemma 10.4 (Noncentral χ^2-Distribution) *The noncentral χ^2-distribution with $\delta > 0$ degrees of freedom and noncentrality parameter $\zeta > 0$ has density function*

$$f_{\chi^2(\delta,\zeta)}(x) = \frac{1}{2}e^{-\frac{x+\zeta}{2}}\left(\frac{x}{\zeta}\right)^{\frac{\delta}{4}-\frac{1}{2}} I_{\frac{\delta}{2}-1}(\sqrt{\zeta x}), \quad x \geq 0$$

and characteristic function

$$\int_{\mathbb{R}_+} e^{ux} f_{\chi^2(\delta,\zeta)}(x)\, dx = \frac{e^{\frac{\zeta u}{1-2u}}}{(1-2u)^{\frac{\delta}{2}}}, \quad u \in \mathbb{C}_-.$$

Here $I_\nu(x) = \sum_{j\geq 0} \frac{1}{j!\Gamma(j+\nu+1)}\left(\frac{x}{2}\right)^{2j+\nu}$ denotes the modified Bessel function of the first kind of order $\nu > -1$.

Proof See e.g. [104, Chap. 29].   □

For illustration, we now fix the following CIR model parameters

$$\sigma^2 = 0.033, \qquad b = 0.08, \qquad \beta = -0.9, \qquad r_0 = 0.08. \qquad (10.30)$$

---

[8]The sampling from a noncentral χ^2-distribution is described in [79, Sect. 3.4.1].

Fig. 10.1 Line integral $\Pi(w, 0)$ as a function of w

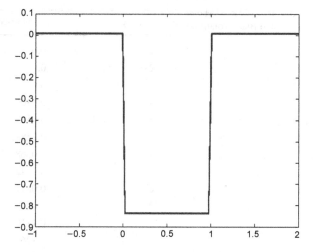

Fig. 10.2 Real part of the integrand of $\Pi(w, 0)$, for $w = -0.5, 0.5, 1.5$, as a function of λ

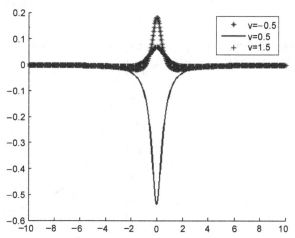

Moreover, we set $t = 0$, $T = 1$ and $S = 2$. Using any software capable of numerical integration, we see that the line integral $\Pi(w, 0)$ in Corollary 10.4 behaves numerically stable for w ranging between $(-1, 2) \setminus \{0, 1\}$ (see Fig. 10.1). On the other hand, we know that $\Pi(w, 0)$ diverges for $w \to +\infty$ (why?). The real part of the integrand of $\Pi(w, 0)$, for $w = -0.5, 0.5, 1.5$, is plotted as function of λ in Fig. 10.2. The resulting ATM call and put option strike price is $K = 0.9180$. The call and put option prices, $\pi_{call}(0) = \pi_{put}(0) = 0.0078$, can now be computed by any of the formulas in (10.28) ($\to$ Exercise 10.13).

As an application, we next compute ATM cap prices and implied Black volatilities ($\to$ Exercise 10.14). The tenor is as follows: $t = 0$ (today), $T_0 = 1/4$ (first reset date), and $T_i - T_{i-1} \equiv 1/4$, $i = 1, \ldots, 119$ (the maturity of the last cap is $T_{119} = 30$). Table 10.1 and Fig. 10.3 show the ATM cap prices and implied Black volatilities for a range of maturities. Like the Vasiček model (see Fig. 7.1), the CIR model seems incapable of producing humped volatility curves.

Table 10.1 CIR ATM cap prices and Black volatilities

Maturity	ATM prices	ATM vols
1	0.0073	0.4506
2	0.0190	0.3720
3	0.0302	0.3226
4	0.0406	0.2890
5	0.0501	0.2647
6	0.0588	0.2462
7	0.0668	0.2316
8	0.0742	0.2198
10	0.0871	0.2017
12	0.0979	0.1886
15	0.1110	0.1744
20	0.1265	0.1594
30	0.1430	0.1442

Fig. 10.3 CIR ATM cap Black volatilities

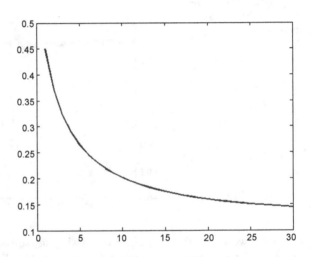

10.3.3 Heston Stochastic Volatility Model

This affine model, proposed by Heston [91], generalizes the Black–Scholes model (see Exercise 4.7 and Sect. 7.3) by assuming a stochastic volatility.

Interest rates are assumed to be constant $r(t) \equiv r \geq 0$, and there is one risky asset (stock) $S = e^{X_2}$, where $X = (X_1, X_2)$ is the affine process with state space $\mathbb{R}_+ \times \mathbb{R}$ and dynamics

$$dX_1 = (k + \kappa X_1)\,dt + \sigma\sqrt{2X_1}\,dW_1,$$

$$dX_2 = (r - X_1)\,dt + \sqrt{2X_1}\left(\rho dW_1 + \sqrt{1 - \rho^2}dW_2\right)$$

for some constant parameters $k, \sigma \geq 0$, $\kappa \in \mathbb{R}$, and some $\rho \in [-1, 1]$.

The implied risk-neutral stock dynamics read

$$dS = Sr \, dt + S\sqrt{2X_1} \, dW$$

for the Brownian motion $W = \rho W_1 + \sqrt{1 - \rho^2} W_2$. We see that $\sqrt{2X_1}$ is the stochastic volatility of the price process S. They have possibly non-zero covariation

$$d\langle S, X_1 \rangle = 2\rho\sigma S X_1 \, dt.$$

The corresponding system of Riccati equations (10.7) is equivalent to ($\rightarrow$ Exercise 10.15)

$$\phi(t, u) = k \int_0^t \psi_1(s, u) \, ds + r u_2 t,$$

$$\partial_t \psi_1(t, u) = \sigma^2 \psi_1^2(t, u) + (2\rho\sigma u_2 + \kappa)\psi_1(t, u) + u_2^2 - u_2, \qquad (10.31)$$

$$\psi_1(0, u) = u_1,$$

$$\psi_2(t, u) = u_2,$$

which, in view of Lemma 10.12(b) below admits an explicit global solution if $u_1 \in \mathbb{C}_-$ and $0 \leq \Re u_2 \leq 1$. In particular, for $u = (0, 1)$, we obtain

$$\phi(t, 0, 1) = rt, \qquad \psi(t, 0, 1) = (0, 1)^\top.$$

Theorem 10.3 thus implies that $S(T)$ has a finite first moment, for any $T \in \mathbb{R}_+$, and

$$\mathbb{E}[e^{-rT} S(T) \mid \mathcal{F}_t] = e^{-rT} \mathbb{E}[e^{X_2(T)} \mid \mathcal{F}_t] = e^{-rT} e^{r(T-t)+X_2(t)} = e^{-rt} S(t),$$

for $t \leq T$, which is just the martingale property of S.

We now want to compute the price

$$\pi(t) = e^{-r(T-t)} \mathbb{E}\left[(S(T) - K)^+ \mid \mathcal{F}_t\right]$$

of a European call option on $S(T)$ with maturity T and strike price K. Fix some $w > 1$ small enough with $(0, w) \in \mathcal{D}_\mathbb{R}(T)$, where $\mathcal{D}_\mathbb{R}$ denotes the maximal domain for the system of Riccati equations (10.31). Formula (10.18) combined with Lemma 10.2 then yields ($\rightarrow$ Exercise 10.16)

$$\pi(t) = e^{-r(T-t)} \int_\mathbb{R} e^{\phi(T-t,0,w+i\lambda)+\psi_1(T-t,0,w+i\lambda)X_1(t)+(w+i\lambda)X_2(t)} \tilde{f}(\lambda) \, d\lambda$$

$$(10.32)$$

with

$$\tilde{f}(\lambda) = \frac{1}{2\pi} \frac{K^{-(w-1+i\lambda)}}{(w + i\lambda)(w - 1 + i\lambda)}.$$

Table 10.2 Call option prices in the Heston model

T–K	0.8	0.9	1.0	1.1	1.2
0.2	0.2016	0.1049	0.0348	0.0074	0.0012
0.4	0.2037	0.1120	0.0478	0.0168	0.0053
0.6	0.2061	0.1183	0.0571	0.0245	0.0100
0.8	0.2088	0.1239	0.0646	0.0310	0.0144
1.0	0.2115	0.1291	0.0711	0.0368	0.0186

Table 10.3 Black–Scholes implied volatilities for the call option prices in the Heston model

T–K	0.8	0.9	1.0	1.1	1.2
0.2	0.1715	0.1786	0.1899	0.2017	0.2126
0.4	0.1641	0.1712	0.1818	0.1930	0.2033
0.6	0.1585	0.1656	0.1755	0.1858	0.1954
0.8	0.1544	0.1612	0.1704	0.1799	0.1889
1.0	0.1513	0.1579	0.1664	0.1751	0.1835

Alternatively, we may fix any $0 < w < 1$ and then

$$\pi(t) = S(t) + e^{-r(T-t)} \int_{\mathbb{R}} e^{\phi(T-t,0,w+i\lambda) + \psi_1(T-t,0,w+i\lambda)X_1(t) + (w+i\lambda)X_2(t)} \, \widetilde{f}(\lambda) \, d\lambda. \tag{10.33}$$

For illustration, we choose the model parameters

$$X_1(0) = 0.02, \qquad X_2(0) = 0, \qquad \sigma = 0.1, \qquad \kappa = -2.0,$$

$$k = 0.02, \qquad r = 0.01, \qquad \rho = 0.5.$$

Table 10.2 shows European call option prices at $t = 0$ for various strikes K and maturities T. The corresponding implied Black–Scholes volatilities are shown in Table 10.3 and Fig. 10.4 ($\rightarrow$ Exercise 10.17).

10.4 Affine Transformations and Canonical Representation

As in the beginning of Sect. 10.3, we let X be affine on the canonical state space $\mathbb{R}_+^m \times \mathbb{R}^n$ with admissible parameters a, α_i, b, β_i. Hence, in view of (10.1), for any $x \in \mathbb{R}_+^m \times \mathbb{R}^n$ the process $X = X^x$ satisfies

$$dX = (b + \mathcal{B}X) \, dt + \rho(X) \, dW,$$
$$X(0) = x, \tag{10.34}$$

and $\rho(x)\rho(x)^\top = a + \sum_{i \in I} x_i \alpha_i$.

Fig. 10.4 Implied volatility
surface for the Heston model

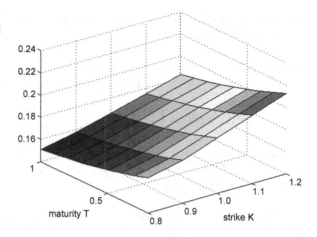

It can easily be checked ($\rightarrow$ Exercise 10.20) that for every invertible $d \times d$-matrix Λ, the linear transform

$$Y = \Lambda X$$

satisfies

$$dY = \left(\Lambda b + \Lambda \mathcal{B} \Lambda^{-1} Y\right) dt + \Lambda \rho \left(\Lambda^{-1} Y\right) dW, \quad Y(0) = \Lambda x. \tag{10.35}$$

Hence, Y has again an affine drift and diffusion matrix

$$\Lambda b + \Lambda \mathcal{B} \Lambda^{-1} y \quad \text{and} \quad \Lambda \alpha (\Lambda^{-1} y) \Lambda^\top, \tag{10.36}$$

respectively.

On the other hand, the affine short-rate model (10.9) can be expressed in terms of $Y(t)$ as

$$r(t) = c + \gamma^\top \Lambda^{-1} Y(t). \tag{10.37}$$

This shows that Y and (10.37) specify an affine short-rate model producing the same short rates, and thus bond prices, as X and (10.9). That is, an invertible linear transformation of the state process changes the particular form of the stochastic differential equation (10.34). But it leaves observable quantities, such as short rates and bond prices invariant.

This motivates the question whether there exists a classification method ensuring that affine short-rate models with the same observable implications have a unique canonical representation. This topic has been addressed in [43, 44, 50, 105], see also the notes section. We now elaborate on this issue and show that the diffusion matrix $\alpha(x)$ can always be brought into block-diagonal form by a regular linear transform Λ with $\Lambda(\mathbb{R}^m_+ \times \mathbb{R}^n) = \mathbb{R}^m_+ \times \mathbb{R}^n$.

We denote by

$$\text{diag}(z_1, \ldots, z_m)$$

the diagonal matrix with diagonal elements $z_1, \ldots, z_m$, and we write I_m for the $m \times m$-identity matrix.

Lemma 10.5 *There exists some invertible $d \times d$-matrix Λ with $\Lambda(\mathbb{R}_+^m \times \mathbb{R}^n) = \mathbb{R}_+^m \times \mathbb{R}^n$ such that $\Lambda\alpha(\Lambda^{-1}y)\Lambda^\top$ is block-diagonal of the form*

$$\Lambda\alpha(\Lambda^{-1}y)\Lambda^\top = \begin{pmatrix} \mathrm{diag}(y_1, \ldots, y_q, 0, \ldots, 0) & 0 \\ 0 & p + \sum_{i \in I} y_i \pi_i \end{pmatrix}$$

for some integer $0 \le q \le m$ and symmetric positive semi-definite $n \times n$ matrices $p, \pi_1, \ldots, \pi_m$. Moreover, Λb and $\Lambda\mathcal{B}\Lambda^{-1}$ meet the respective admissibility conditions (10.6) in lieu of b and $\mathcal{B}$.

Proof From (10.3) we know that $\Lambda\alpha(x)\Lambda^\top$ is block-diagonal for all $x = \Lambda^{-1}y$ if and only if $\Lambda a \Lambda^\top$ and $\Lambda\alpha_i\Lambda^\top$ are block-diagonal for all $i \in I$. By permutation and scaling of the first m coordinate axes (this is a linear bijection from $\mathbb{R}_+^m \times \mathbb{R}^n$ onto itself, which preserves the admissibility of the transformed b and $\mathcal{B}$), we may assume that there exists some integer $0 \le q \le m$ such that $\alpha_{1,11} = \cdots = \alpha_{q,qq} = 1$ and $\alpha_{i,ii} = 0$ for $q < i \le m$. Hence a and α_i for $q < i \le m$ are already block-diagonal of the special form

$$a = \begin{pmatrix} 0 & 0 \\ 0 & a_{JJ} \end{pmatrix}, \qquad \alpha_i = \begin{pmatrix} 0 & 0 \\ 0 & \alpha_{i,JJ} \end{pmatrix}.$$

For $1 \le i \le q$, we may have non-zero off-diagonal elements in the ith row $\alpha_{i,iJ}$. We thus define the $n \times m$-matrix $D = (\delta_1, \ldots, \delta_m)$ with ith column $\delta_i = -\alpha_{i,iJ}$ and set

$$\Lambda = \begin{pmatrix} I_m & 0 \\ D & I_n \end{pmatrix}.$$

One checks by inspection that D is invertible and maps $\mathbb{R}_+^m \times \mathbb{R}^n$ onto $\mathbb{R}_+^m \times \mathbb{R}^n$. Moreover,

$$D\alpha_{i,II} = -\alpha_{i,JI}, \quad i \in I.$$

From here we easily verify that

$$\Lambda\alpha_i = \begin{pmatrix} \alpha_{i,II} & \alpha_{i,IJ} \\ 0 & D\alpha_{i,IJ} + \alpha_{i,JJ} \end{pmatrix},$$

and thus

$$\Lambda\alpha_i \Lambda^\top = \begin{pmatrix} \alpha_{i,II} & 0 \\ 0 & D\alpha_{i,IJ} + \alpha_{i,JJ} \end{pmatrix}.$$

Since $\Lambda a \Lambda^\top = a$, the first assertion is proved.

The admissibility conditions for Λb and $\Lambda\mathcal{B}\Lambda^{-1}$ can easily be checked as well. $\qquad\square$

In view of (10.36), (10.37) and Lemma 10.5 we thus obtain the following result.

Theorem 10.7 (Canonical Representation) *Any affine short-rate model* (10.9), *after some modification of γ if necessary, admits an $\mathbb{R}_+^m \times \mathbb{R}^n$-valued affine state process X with block-diagonal diffusion matrix of the form*

$$\alpha(x) = \begin{pmatrix} \operatorname{diag}(x_1, \ldots, x_q, 0, \ldots, 0) & 0 \\ 0 & a + \sum_{i \in I} x_i \alpha_{i,JJ} \end{pmatrix} \qquad (10.38)$$

for some integer $0 \le q \le m$.

10.5 Existence and Uniqueness of Affine Processes

All we said about the affine process X so far was under the premise that there exists a unique solution $X = X^x$ of the stochastic differential equation (10.1) on some appropriate state space $\mathcal{X} \subset \mathbb{R}^d$. However, if the diffusion matrix $\rho(x)\rho(x)^\top$ is affine then $\rho(x)$ cannot be Lipschitz continuous in x in general. This raises the question whether (10.1) admits a solution at all.

In this section, we show how X can always be realized as unique solution of the stochastic differential equation (10.1), which is (10.34), in the canonical affine framework $\mathcal{X} = \mathbb{R}_+^m \times \mathbb{R}^n$ and for particular choices of $\rho(x)$.

We recall from Theorem 10.1 that the affine property of X imposes explicit conditions on $\rho(x)\rho(x)^\top$, but not on $\rho(x)$ as such. Indeed, for any orthogonal $d \times d$-matrix D, the function $\rho(x)D$ yields the same diffusion matrix, $\rho(x)DD^\top\rho(x)^\top = \rho(x)\rho(x)^\top$, as $\rho(x)$ (see also Remark 4.2).

On the other hand, from Theorem 10.2 we know that any admissible parameters a, α_i, b, β_i in (10.3) uniquely determine the functions $(\phi(\cdot, u), \psi(\cdot, u)) : \mathbb{R}_+ \to \mathbb{C}_- \times \mathbb{C}_-^m \times i\mathbb{R}^n$ as solution of the Riccati equations (10.7), for all $u \in \mathbb{C}_-^m \times i\mathbb{R}^n$. These in turn uniquely determine the law of the process X. Indeed, for any $0 \le t_1 < t_2$ and $u_1, u_2 \in \mathbb{C}_-^m \times i\mathbb{R}^n$, we infer by iteration of (10.2)

$$\mathbb{E}\left[e^{u_1^\top X(t_1) + u_2^\top X(t_2)}\right] = \mathbb{E}\left[e^{u_1^\top X(t_1)} \mathbb{E}\left[e^{u_2^\top X(t_2)} \mid \mathcal{F}_{t_1}\right]\right]$$

$$= \mathbb{E}\left[e^{u_1^\top X(t_1)} e^{\phi(t_2 - t_1, u_2) + \psi(t_2 - t_1, u_2)^\top X(t_1)}\right]$$

$$= e^{\phi(t_2 - t_1, u_2) + \phi(t_1, u_1 + \psi(t_2 - t_1, u_2)) + \psi(t_1, u_1 + \psi(t_2 - t_1, u_2))^\top x}.$$

Hence the joint distribution of $(X(t_1), X(t_2))$ is uniquely determined by the functions ϕ and ψ. By further iteration of this argument, we conclude that every finite-dimensional distribution, and thus the law, of X is uniquely determined by the parameters a, α_i, b, β_i.

We conclude that the law of an affine process X, while uniquely determined by its characteristics (10.3), can be realized by infinitely many variants of the stochastic differential equation (10.34) by replacing $\rho(x)$ by $\rho(x)D$, for any orthogonal $d \times d$-matrix D. We now propose a canonical choice of $\rho(x)$ as follows:

- In view of (10.35) and Lemma 10.5, every affine process X on $\mathbb{R}^m_+ \times \mathbb{R}^n$ can be written as $X = \Lambda^{-1}Y$ for some invertible $d \times d$-matrix Λ and some affine process Y on $\mathbb{R}^m_+ \times \mathbb{R}^n$ with block-diagonal diffusion matrix. It is thus enough to consider such $\rho(x)$ where $\rho(x)\rho(x)^\top$ is of the form (10.38). Obviously, $\rho(x) \equiv \rho(x_I)$ is a function of x_I only.

- Set $\rho_{IJ}(x) \equiv 0$, $\rho_{JI}(x) \equiv 0$, and

$$\rho_{II}(x_I) = \mathrm{diag}(\sqrt{x_1}, \ldots, \sqrt{x_q}, 0, \ldots, 0).$$

Choose for $\rho_{JJ}(x_I)$ any measurable $n \times n$-matrix-valued function satisfying

$$\rho_{JJ}(x_I)\rho_{JJ}(x_I)^\top = a + \sum_{i \in I} x_i \alpha_{i,JJ}. \tag{10.39}$$

In practice, one would determine $\rho_{JJ}(x_I)$ via Cholesky factorization, see e.g. [129, Theorem 2.2.5]. If $a + \sum_{i \in I} x_i \alpha_{i,JJ}$ is positive definite, then $\rho_{JJ}(x_I)$ turns out to be the unique lower triangular matrix with positive diagonal elements and which satisfies (10.39). If $a + \sum_{i \in I} x_i \alpha_{i,JJ}$ is merely positive semi-definite, then the algorithm becomes more involved. In any case, $\rho_{JJ}(x_I)$ will depend measurably on x_I.

- The stochastic differential equation (10.34) now reads

$$dX_I = (b_I + \mathcal{B}_{II}X_I)\,dt + \rho_{II}(X_I)\,dW_I,$$

$$dX_J = (b_J + \mathcal{B}_{JI}X_I + \mathcal{B}_{JJ}X_J)\,dt + \rho_{JJ}(X_I)\,dW_J, \tag{10.40}$$

$$X(0) = x.$$

Lemma 10.6 below asserts the existence and uniqueness of an $\mathbb{R}^m_+ \times \mathbb{R}^n$-valued solution $X = X^x$, for any $x \in \mathbb{R}^m_+ \times \mathbb{R}^n$.

We thus have shown:

Theorem 10.8 *Let a, α_i, b, β_i be admissible parameters. Then there exists a measurable function $\rho : \mathbb{R}^m_+ \times \mathbb{R}^n \to \mathbb{R}^{d \times d}$ with $\rho(x)\rho(x)^\top = a + \sum_{i \in I} x_i \alpha_i$, and such that, for any $x \in \mathbb{R}^m_+ \times \mathbb{R}^n$, there exists a unique $\mathbb{R}^m_+ \times \mathbb{R}^n$-valued solution $X = X^x$ of (10.34).*

Moreover, the law of X is uniquely determined by a, α_i, b, β_i, and does not depend on the particular choice of ρ.

The proof of the following lemma uses the concept of a weak solution, which is beyond the scope of this book and therefore mentioned without further explanation. The interested reader will find detailed background in e.g. [106, Sect. 5.3]. At first reading, the following result may simply be taken for granted.

Lemma 10.6 *For any $x \in \mathbb{R}^m_+ \times \mathbb{R}^n$, there exists a unique $\mathbb{R}^m_+ \times \mathbb{R}^n$-valued solution $X = X^x$ of (10.40).*

Proof First, we extend ρ continuously to $\mathbb{R}^d$ by setting $\rho(x) = \rho(x_1^+, \ldots, x_m^+)$, where we define $x_i^+ = \max(0, x_i)$.

Now observe that X_I solves the autonomous equation

$$dX_I = (b_I + \mathcal{B}_{II} X_I) \, dt + \rho_{II}(X_I) \, dW_I, \quad X_I(0) = x_I. \tag{10.41}$$

Obviously, there exists a finite constant K such that the linear growth condition

$$\|b_I + \mathcal{B}_{II} x_I\|^2 + \|\rho(x_I)\|^2 \leq K(1 + \|x_I\|^2)$$

is satisfied for all $x \in \mathbb{R}^m$. By [99, Theorems 2.3 and 2.4] there exists a weak solution[9] of (10.41). On the other hand, (10.41) is exactly of the form as assumed in [162, Theorem 1], which implies that pathwise uniqueness[10] holds for (10.41). The Yamada–Watanabe theorem, see [162, Corollary 3] or [106, Corollary 5.3.23], thus implies that there exists a unique solution $X_I = X_I^{x_I}$ of (10.41), for all $x_I \in \mathbb{R}^m$.

Given $X_I^{x_I}$, it is then easily seen that

$$X_J(t) = e^{\mathcal{B}_{JJ} t} \left(x_J + \int_0^t e^{-\mathcal{B}_{JJ} s} (b_J + \mathcal{B}_{JI} X_I(s)) \, ds \right.$$

$$\left. + \int_0^t e^{-\mathcal{B}_{JJ} s} \rho_{JJ}(X_I(s)) \, dW_J(s) \right)$$

is the unique solution to the second equation in (10.40).

Admissibility of the parameters b and β_i and the stochastic invariance Lemma 10.11 eventually imply that $X_I = X_I^{x_I}$ is $\mathbb{R}_+^m$-valued for all $x_I \in \mathbb{R}_+^m$. Whence the lemma is proved. $\qquad\square$

10.6 On the Regularity of Characteristic Functions

This auxiliary section provides some analytic regularity results for characteristic functions, which are of independent interest. These results enter the main text only via the proof of Theorem 10.3 in Sect. 10.7.3 below. This section may thus be skipped at the first reading.

---

[9] A weak solution consists of a filtered probability space $(\Omega, \mathcal{F}, (\mathcal{F}_t), \mathbb{P})$ carrying a continuous adapted process X_I and a Brownian motion W_I such that (10.41) is satisfied. The crux of a weak solution is that X_I is not necessarily adapted to the filtration generated by the Brownian motion W_I. See [162, Definition 1] or [106, Definition 5.3.1].

[10] Pathwise uniqueness holds if, for any two weak solutions (X_I, W_I) and (X_I', W_I) of (10.41) defined on the same probability space $(\Omega, \mathcal{F}, \mathbb{P})$ with common Brownian motion W_I and with common initial value $X_I(0) = X_I'(0)$, the two processes are indistinguishable: $\mathbb{P}[X_I(t) = X_I'(t)$ for all $t \geq 0] = 1$. See [162, Definition 2] or [106, Sect. 5.3].

Let v be a bounded measure on $\mathbb{R}^d$, and denote by

$$G(z) = \int_{\mathbb{R}^d} e^{z^\top x} v(dx)$$

its characteristic function[11] for $z \in i\mathbb{R}^d$. Note that $G(z)$ is actually well defined for $z \in \mathcal{S}(V)$ where

$$V = \left\{ y \in \mathbb{R}^d \mid \int_{\mathbb{R}^d} e^{y^\top x} v(dx) < \infty \right\}.$$

We first investigate the interplay between the (marginal) moments of v and the corresponding (partial) regularity of G.

Lemma 10.7 *Define $g(y) = G(iy)$ for $y \in \mathbb{R}^d$, and let $k \in \mathbb{N}$ and $1 \leq i \leq d$. If $\partial_{y_i}^{2k} g(0)$ exists then*

$$\int_{\mathbb{R}^d} |x_i|^{2k} v(dx) < \infty.$$

On the other hand, if $\int_{\mathbb{R}^d} \|x\|^k v(dx) < \infty$ then $g \in C^k$ and

$$\partial_{y_{i_1}} \cdots \partial_{y_{i_l}} g(y) = i^l \int_{\mathbb{R}^d} x_{i_1} \cdots x_{i_l} e^{iy^\top x} v(dx)$$

for all $y \in \mathbb{R}^d$, $1 \leq i_1, \ldots, i_l \leq d$ and $1 \leq l \leq k$.

Proof As usual, let e_i denote the ith standard basis vector in $\mathbb{R}^d$. Observe that $s \mapsto g(se_i)$ is the characteristic function of the image measure of v on $\mathbb{R}$ by the mapping $x \mapsto x_i$. Since $\partial_s^{2k} g(se_i)|_{s=0} = \partial_{y_i}^{2k} g(0)$, the assertion follows from the one-dimensional case, see [120, Theorem 2.3.1].

The second part of the lemma follows by differentiating under the integral sign, which is allowed by dominated convergence.   □

Lemma 10.8 *The set V is convex. Moreover, if $U \subset V$ is an open set in $\mathbb{R}^d$, then G is analytic on the open strip $\mathcal{S}(U)$ in $\mathbb{C}^d$.*

Proof Since $G : \mathbb{R}^d \to [0, \infty]$ is a convex function, its domain $V = \{y \in \mathbb{R}^d \mid G(y) < \infty\}$ is convex, and so is every level set $V_l = \{y \in \mathbb{R}^d \mid G(y) \leq l\}$ for $l \geq 0$.

Now let $U \subset V$ be an open set in $\mathbb{R}^d$. Since any convex function on $\mathbb{R}^d$ is continuous on the open interior of its domain, see [136, Theorem 10.1], we infer that G is continuous on U. We may thus assume that $U_l = \{y \in \mathbb{R}^d \mid G(y) < l\} \cap U \subset V_l$ is open in $\mathbb{R}^d$ and non-empty for $l > 0$ large enough.

---

[11] This is a slight abuse of terminology, since the characteristic function $g(y) = G(iy)$ of v is usually defined on real arguments $y \in \mathbb{R}^d$. However, it facilitates the subsequent notation.

Let $z \in \mathcal{S}(U_l)$ and (z_n) be a sequence in $\mathcal{S}(U_l)$ with $z_n \to z$. For n large enough, there exists some $p > 1$ such that $pz_n \in \mathcal{S}(U_l)$. This implies $p\Re z_n \in V_l$ and hence

$$\int_{\mathbb{R}^d} \left| e^{z_n^\top x} \right|^p \nu(dx) \le l.$$

Hence the class of functions $\{ e^{z_n^\top x} \mid n \in \mathbb{N} \}$ is uniformly integrable with respect to ν, see [161, 13.3]. Since $e^{z_n^\top x} \to e^{z^\top x}$ for all x, we conclude by Lebesgue's convergence theorem that

$$|G(z_n) - G(z)| \le \int_{\mathbb{R}^d} \left| e^{z_n^\top x} - e^{z^\top x} \right| \nu(dx) \to 0.$$

Hence G is continuous on $\mathcal{S}(U_l)$.

It thus follows from the Cauchy formula, see [55, Sect. IX.9], that G is analytic on $\mathcal{S}(U_l)$ if and only if, for every $z \in \mathcal{S}(U_l)$ and $1 \le i \le d$, the function $\zeta \mapsto G(z + \zeta e_i)$ is analytic on $\{ \zeta \in \mathbb{C} \mid z + \zeta e_i \in \mathcal{S}(U_l) \}$. Here, as usual, we denote by e_i the ith standard basis vector in $\mathbb{R}^d$.

We thus let $z \in \mathcal{S}(U_l)$ and $1 \le i \le d$. Then there exists some $\varepsilon_- < 0 < \varepsilon_+$ such that $z + \zeta e_i \in \mathcal{S}(U_l)$ for all $\zeta \in \mathcal{S}([\varepsilon_-, \varepsilon_+])$. In particular, $|e^{(z+\varepsilon_- e_i)^\top x}| \nu(dx)$ and $|e^{(z+\varepsilon_+ e_i)^\top x}| \nu(dx)$ are bounded measures on $\mathbb{R}^d$. By dominated convergence, it follows that the two summands

$$G(z + \zeta e_i) = \int_{\{x_i < 0\}} e^{(\zeta - \varepsilon_-)x_i} e^{(z+\varepsilon_- e_i)^\top x} \nu(dx)$$

$$+ \int_{\{x_i \ge 0\}} e^{(\zeta - \varepsilon_+)x_i} e^{(z+\varepsilon_+ e_i)^\top x} \nu(dx),$$

are complex differentiable, and thus G is analytic, in $\zeta \in \mathcal{S}((\varepsilon_-, \varepsilon_+))$. Whence G is analytic on $\mathcal{S}(U_l)$. Since $\mathcal{S}(U) = \bigcup_{l>0} \mathcal{S}(U_l)$, the lemma follows. $\qquad \square$

In general, V does not have an open interior in $\mathbb{R}^d$. The next lemma provides sufficient conditions for the existence of an open set $U \subset V$ in $\mathbb{R}^d$.

Lemma 10.9 *Let U' be an open neighborhood of 0 in $\mathbb{C}^d$ and h an analytic function on U'. Suppose that $U = U' \cap \mathbb{R}^d$ is star-shaped around 0 and $G(z) = h(z)$ for all $z \in U' \cap i\mathbb{R}^d$. Then $U \subset V$ and $G = h$ on $U' \cap \mathcal{S}(U)$.*

Proof We first suppose that $U' = P_\rho$ for the open polydisc

$$P_\rho = \left\{ z \in \mathbb{C}^d \mid |z_i| < \rho_i, \ 1 \le i \le d \right\},$$

for some $\rho = (\rho_1, \ldots, \rho_d) \in \mathbb{R}^d_{++}$. Note the symmetry $iP_\rho = P_\rho$.

As in Lemma 10.7, we define $g(y) = G(iy)$ for $y \in \mathbb{R}^d$. By assumption, $g(y) = h(iy)$ for all $y \in P_\rho \cap \mathbb{R}^d$. Hence g is analytic on $P_\rho \cap \mathbb{R}^d$, and the Cauchy formula,

[55, Sect. IX.9], yields

$$g(y) = \sum_{i_1,\ldots,i_d \in \mathbb{N}_0} c_{i_1,\ldots,i_d} y_1^{i_1} \cdots y_d^{i_d} \quad \text{for } y \in P_\rho \cap \mathbb{R}^d$$

where $\sum_{i_1,\ldots,i_d \in \mathbb{N}_0} c_{i_1,\ldots,i_d} z_1^{i_1} \cdots z_d^{i_d} = h(iz)$ for all $z \in P_\rho$. This power series is absolutely convergent on P_ρ, that is,

$$\sum_{i_1,\ldots,i_d \in \mathbb{N}_0} |c_{i_1,\ldots,i_d}| |z_1^{i_1} \cdots z_d^{i_d}| < \infty \quad \text{for all } z \in P_\rho.$$

From the first part of Lemma 10.7, we infer that ν possesses all moments, that is, $\int_{\mathbb{R}^d} \|x\|^k \nu(dx) < \infty$ for all $k \in \mathbb{N}$. From the second part of Lemma 10.7 thus

$$c_{i_1,\ldots,i_d} = \frac{i^{i_1+\cdots+i_d}}{i_1! \cdots i_d!} \int_{\mathbb{R}^d} x_1^{i_1} \cdots x_d^{i_d} \nu(dx).$$

From the inequality $|x_i|^{2k-1} \le (x_i^{2k} + x_i^{2k-2})/2$, for $k \in \mathbb{N}$, and the above properties, we infer that for all $z \in P_\rho$,

$$\int_{\mathbb{R}^d} e^{\sum_{i=1}^d |z_i||x_i|} \nu(dx) = \sum_{i_1,\ldots,i_d \in \mathbb{N}_0} \frac{|z_1^{i_1} \cdots z_d^{i_d}|}{i_1! \cdots i_d!} \int_{\mathbb{R}^d} |x_1^{i_1} \cdots x_d^{i_d}| \nu(dx) < \infty.$$

Hence $P_\rho \cap \mathbb{R}^d \subset V$, and Lemma 10.8 implies that G is analytic on $\mathcal{S}(P_\rho \cap \mathbb{R}^d)$. Since the power series for G and h coincide on $P_\rho \cap i\mathbb{R}^d$, we conclude that $G = h$ on P_ρ, and the lemma is proved for $U' = P_\rho$.

Now let U' be an open neighborhood of 0 in $\mathbb{C}^d$. Then there exists some open polydisc $P_\rho \subset U'$ with $\rho \in \mathbb{R}_{++}^d$. By the preceding case, we have $P_\rho \cap \mathbb{R}^d \subset V$ and $G = h$ on P_ρ. In view of Lemma 10.8 it thus remains to show that $U = U' \cap \mathbb{R}^d \subset V$.

To this end, let $a \in U$. Since U is star-shaped around 0 in $\mathbb{R}^d$, there exists some $s_1 > 1$ such that $sa \in U$ for all $s \in [0, s_1]$ and $h(sa)$ is analytic in $s \in (0, s_1)$. On the other hand, there exists some $0 < s_0 < s_1$ such that $sa \in P_\rho \cap \mathbb{R}^d$ for all $s \in [0, s_0]$, and $G(sa) = h(sa)$ for $s \in (0, s_0)$. This implies

$$\int_{\{a^\top x \ge 0\}} e^{sa^\top x} \nu(dx) = h(sa) - \int_{\{a^\top x < 0\}} e^{sa^\top x} \nu(dx)$$

for $s \in (0, s_0)$. By Lemma 10.8, the right-hand side is an analytic function in $s \in (0, s_1)$. We conclude by Lemma 10.10 below, for μ defined as the image measure of ν on $\mathbb{R}_+$ by the mapping $x \mapsto a^\top x$, that $a \in V$. Hence the lemma is proved. $\square$

Lemma 10.10 *Let μ be a bounded measure on $\mathbb{R}_+$, and h an analytic function on $(0, s_1)$, such that*

$$\int_{\mathbb{R}_+} e^{sx} \mu(dx) = h(s) \tag{10.42}$$

for all $s \in (0, s_0)$, for some numbers $0 < s_0 < s_1$. Then (10.42) *also holds for $s \in (0, s_1)$.*

Proof Define $f(s) = \int_{\mathbb{R}_+} e^{sx} \mu(dx)$ and $s_\infty = \sup\{s > 0 \mid f(s) < \infty\} \geq s_0$, such that

$$f(s) = +\infty \quad \text{for } s > s_\infty. \tag{10.43}$$

We assume, by contradiction, that $s_\infty < s_1$. Then there exists some $s_* \in (0, s_\infty)$ and $\varepsilon > 0$ such that $s_* < s_\infty < s_* + \varepsilon$ and such that h can be developed in an absolutely convergent power series

$$h(s) = \sum_{k \geq 0} \frac{c_k}{k!} (s - s_*)^k \quad \text{for } s \in (s_* - \varepsilon, s_* + \varepsilon).$$

In view of Lemma 10.8, f is analytic, and thus $f = h$, on $(0, s_\infty)$. Hence we obtain, by dominated convergence,

$$c_k = \frac{d^k}{ds^k} h(s) \bigg|_{s=s_*} = \frac{d^k}{ds^k} f(s) \bigg|_{s=s_*} = \int_{\mathbb{R}_+} x^k e^{s_* x} \mu(dx) \geq 0.$$

By monotone convergence, we conclude

$$h(s) = \sum_{k \geq 0} \int_{\mathbb{R}_+} \frac{x^k}{k!} (s - s_*)^k e^{s_* x} \mu(dx) = \int_{\mathbb{R}_+} \sum_{k \geq 0} \frac{x^k}{k!} (s - s_*)^k e^{s_* x} \mu(dx)$$

$$= \int_{\mathbb{R}_+} e^{sx} \mu(dx)$$

for all $s \in (s_*, s_* + \varepsilon)$. But this contradicts (10.43). Whence $s_\infty \geq s_1$, and the lemma is proved. $\qquad \square$

10.7 Auxiliary Results for Differential Equations

In this section we deliver invariance and comparison results for stochastic and ordinary differential equations, which are used in the proofs of the main Theorems 10.2, 10.3 and 10.4 and Lemma 10.6. This section can be skipped at the first reading.

10.7.1 Some Invariance Results

We start with an invariance result for the stochastic differential equation (10.1).

Lemma 10.11 *Suppose b and ρ in (10.1) admit a continuous and measurable extension to $\mathbb{R}^d$, respectively, and such that a is continuous on $\mathbb{R}^d$. Let $u \in \mathbb{R}^d \setminus \{0\}$ and define the half space*

$$H = \{x \in \mathbb{R}^d \mid u^\top x \geq 0\},$$

its interior $H^0 = \{x \in \mathbb{R}^d \mid u^\top x > 0\}$, and its boundary $\partial H = \{x \in H \mid u^\top x = 0\}$.

(a) *Fix $x \in \partial H$ and let $X = X^x$ be a solution of (10.1). If $X(t) \in H$ for all $t \geq 0$, then necessarily*

$$u^\top a(x)u = 0, \tag{10.44}$$

$$u^\top b(x) \geq 0. \tag{10.45}$$

(b) *Conversely, if (10.44) and (10.45) hold for all $x \in \mathbb{R}^d \setminus H^0$, then any solution X of (10.1) with $X(0) \in H$ satisfies $X(t) \in H$ for all $t \geq 0$.*

Intuitively speaking, (10.44) means that the diffusion must be "parallel to the boundary", and (10.45) says that the drift must be "inward pointing" at the boundary of H.

Proof Fix $x \in \partial H$ and let $X = X^x$ be a solution of (10.1). Hence

$$u^\top X(t) = \int_0^t u^\top b(X(s))\,ds + \int_0^t u^\top \rho(X(s))\,dW(s).$$

Since a and b are continuous, there exists a stopping time $\tau_1 > 0$ and a finite constant K such that

$$|u^\top b(X(t \wedge \tau_1))| \leq K$$

and

$$\|u^\top \rho(X(t \wedge \tau_1))\|^2 = u^\top a(X(t \wedge \tau_1))u \leq K$$

for all $t \geq 0$. In particular, the stochastic integral part of $u^\top X(t \wedge \tau_1)$ is a martingale. Hence

$$\mathbb{E}\left[u^\top X(t \wedge \tau_1)\right] = \mathbb{E}\left[\int_0^{t \wedge \tau_1} u^\top b(X(s))\,ds\right], \quad t \geq 0.$$

We now argue by contradiction, and assume first that $u^\top b(x) < 0$. By continuity of b and $X(t)$, there exists some $\varepsilon > 0$ and a stopping time $\tau_2 > 0$ such that $u^\top b(X(t)) \leq -\varepsilon$ for all $t \leq \tau_2$. In view of the above this implies

$$\mathbb{E}\left[u^\top X(\tau_2 \wedge \tau_1)\right] < 0.$$

This contradicts $X(t) \in H$ for all $t \geq 0$, whence (10.45) holds.

As for (10.44), let $C > 0$ be a finite constant and define the stochastic exponential $Z_t = \mathcal{E}(-C \int_0^t u^\top \rho(X) \, dW)$. Then Z is a positive local martingale. Integration by parts yields

$$u^\top X(t) Z(t) = \int_0^t Z(s) \left(u^\top b(X(s)) - C u^\top a(X(s)) u \right) ds + M(t)$$

where M is a local martingale. Hence there exists a stopping time $\tau_3 > 0$ such that for all $t \geq 0$,

$$\mathbb{E}\left[u^\top X(t \wedge \tau_3) Z(t \wedge \tau_3) \right] = \mathbb{E}\left[\int_0^{t \wedge \tau_3} Z(s) \left(u^\top b(X(s)) - C u^\top a(X(s)) u \right) ds \right].$$

Now assume that $u^\top a(x) u > 0$. By continuity of a and $X(t)$, there exists some $\varepsilon > 0$ and a stopping time $\tau_4 > 0$ such that $u^\top a(X(t)) u \geq \varepsilon$ for all $t \leq \tau_4$. For $C > K/\varepsilon$, this implies

$$\mathbb{E}\left[u^\top X(\tau_4 \wedge \tau_3 \wedge \tau_1) Z(\tau_4 \wedge \tau_3 \wedge \tau_1) \right] < 0.$$

This contradicts $X(t) \in H$ for all $t \geq 0$. Hence (10.44) holds, and part (a) is proved.

As for part (b), suppose (10.44) and (10.45) hold for all $x \in \mathbb{R}^d \setminus H^0$, and let X be a solution of (10.1) with $X(0) \in H$. For $\delta, \varepsilon > 0$ define the stopping time

$$\tau_{\delta,\varepsilon} = \inf \left\{ t \mid u^\top X(t) \leq -\varepsilon \text{ and } u^\top X(s) < 0 \text{ for all } s \in [t - \delta, t] \right\}.$$

Then on $\{\tau_{\delta,\varepsilon} < \infty\}$ we have $u^\top \rho(X(s)) = 0$ for $\tau_{\delta,\varepsilon} - \delta \leq s \leq \tau_{\delta,\varepsilon}$ and thus

$$0 > u^\top X(\tau_{\delta,\varepsilon}) - u^\top X(\tau_{\delta,\varepsilon} - \delta) = \int_{\tau_{\delta,\varepsilon}-\delta}^{\tau_{\delta,\varepsilon}} u^\top b(X(s)) \, ds \geq 0,$$

a contradiction. Hence $\tau_{\delta,\varepsilon} = \infty$. Since $\delta, \varepsilon > 0$ were arbitrary, we conclude that $u^\top X(t) \geq 0$ for all $t \geq 0$, as desired. Whence the lemma is proved. $\qquad\square$

It is straightforward to extend Lemma 10.11 towards a polyhedral convex set $\bigcap_{i=1}^k H_i$ with half-spaces $H_i = \{x \in \mathbb{R}^d \mid u_i^\top x \geq 0\}$, for some elements $u_1, \ldots, u_k \in \mathbb{R}^d \setminus \{0\}$ and some $k \in \mathbb{N}$. This holds in particular for the canonical state space $\mathbb{R}_+^m \times \mathbb{R}^n$. Moreover, Lemma 10.11 includes time-inhomogeneous[12] ordinary differential equations as special case. The proofs of the following two corollaries are left to the reader ($\rightarrow$ Exercise 10.22).

Corollary 10.5 *Let $H_i = \{x \in \mathbb{R}^d \mid x_i \geq 0\}$ denote the ith canonical half space in $\mathbb{R}^d$, for $i = 1, \ldots, m$. Let $b : \mathbb{R}_+ \times \mathbb{R}^d \rightarrow \mathbb{R}^d$ be a continuous map satisfying, for*

---

[12]Time-inhomogeneous differential equations can be made homogeneous by enlarging the state space.

all t ≥ 0,

$$b(t, x) = b(t, x_1^+, \ldots, x_m^+, x_{m+1}, \ldots, x_d) \quad \text{for all } x \in \mathbb{R}^d, \text{ and}$$

$$b_i(t, x) \geq 0 \quad \text{for all } x \in \partial H_i, i = 1, \ldots, m.$$

Then any solution f of

$$\partial_t f(t) = b(t, f(t))$$

with $f(0) \in \mathbb{R}_+^m \times \mathbb{R}^n$ satisfies $f(t) \in \mathbb{R}_+^m \times \mathbb{R}^n$ for all $t \geq 0$.

Corollary 10.6 *Let $B(t)$ and $C(t)$ be continuous $\mathbb{R}^{m \times m}$- and $\mathbb{R}_+^m$-valued parameters, respectively, such that $B_{ij}(t) \geq 0$ whenever $i \neq j$. Then the solution f of the linear differential equation in $\mathbb{R}^m$*

$$\partial_t f(t) = B(t) f(t) + C(t)$$

with $f(0) \in \mathbb{R}_+^m$ satisfies $f(t) \in \mathbb{R}_+^m$ for all $t \geq 0$.

Here and subsequently, we let $\succeq$ denote the partial order on $\mathbb{R}^m$ induced by the cone $\mathbb{R}_+^m$. That is, $x \succeq y$ if $x - y \in \mathbb{R}_+^m$. Then Corollary 10.6 may be rephrased, for $C(t) \equiv 0$, by saying that the operator $e^{\int_0^t B(s)\, ds}$ is $\succeq$-order preserving, i.e. $e^{\int_0^t B(s)\, ds} \mathbb{R}_+^m \subseteq \mathbb{R}_+^m$.

10.7.2 Some Results on Riccati Equations

We first provide the explicit solution for the one-dimensional Riccati equation.

Lemma 10.12 *Consider the Riccati differential equation*

$$\partial_t G = AG^2 + BG - C, \qquad G(0, u) = u, \tag{10.46}$$

where $A, B, C \in \mathbb{C}$ and $u \in \mathbb{C}$, with $A \neq 0$ and $B^2 + 4AC \in \mathbb{C} \setminus \mathbb{R}_-$. Let $\sqrt{\cdot}$ denote the analytic extension of the real square root to $\mathbb{C} \setminus \mathbb{R}_-$, and define $\theta = \sqrt{B^2 + 4AC}$.

(a) *The function*

$$G(t, u) = -\frac{2C(e^{\theta t} - 1) - (\theta(e^{\theta t} + 1) + B(e^{\theta t} - 1))u}{\theta(e^{\theta t} + 1) - B(e^{\theta t} - 1) - 2A(e^{\theta t} - 1)u} \tag{10.47}$$

is the unique solution of equation (10.46) on its maximal interval of existence $[0, t_+(u))$. Moreover,

$$\int_0^t G(s, u)\, ds = \frac{1}{A} \log \left(\frac{2\theta e^{\frac{\theta - B}{2} t}}{\theta(e^{\theta t} + 1) - B(e^{\theta t} - 1) - 2A(e^{\theta t} - 1)u} \right). \tag{10.48}$$

(b) *If, moreover, $A > 0$, $B \in \mathbb{R}$, $\Re(C) \geq 0$ and $u \in \mathbb{C}_-$ then $t_+(u) = \infty$ and $G(t, u)$ is $\mathbb{C}_-$-valued.*

Proof Recall that the square root $\sqrt{z} = e^{1/2 \log(z)}$ is the well-defined analytic extension of the real square root to $\mathbb{C} \setminus \mathbb{R}_-$, through the main branch of the logarithm which can be written in the form $\log(z) = \int_{[0,z]} \frac{dz}{z}$. Hence we may write (10.46) as

$$\partial_t G = A(G - \theta_+)(G - \theta_-), \qquad G(0, u) = u,$$

where $\theta_\pm = \frac{-B \pm \sqrt{B^2 + 4AC}}{2A}$, and it follows that

$$G(t, u) = \frac{\theta_+(u - \theta_-) - \theta_-(u - \theta_+)e^{\theta t}}{(u - \theta_-) - (u - \theta_+)e^{\theta t}},$$

which can be seen to be equivalent to (10.47). As $\theta_+ \neq \theta_-$, numerator and denominator cannot vanish at the same time t, and certainly not for t near zero. Hence, by the maximality of $t_+(u)$, (10.47) is the solution of (10.46) for $t \in [0, t_+(u))$. Finally, the integral (10.48) is checked by differentiation. This proves (a).

As for (b), we show along the lines of the proof of Theorem 10.2, that for this choice of coefficients global solutions exist for initial data $u \in \mathbb{C}_-$ and stay in $\mathbb{C}_-$. To this end, write $R(G) = AG^2 + BG - C$, then

$$\Re(R(G)) = A(\Re(G))^2 - A(\Im(G))^2 + B\Re(G) - \Re(C) \leq A(\Re(G))^2 + B\Re(G)$$

and since $A, B \in \mathbb{R}$ we have that $\Re(G(t, u)) \leq 0$ for all times $t \in [0, t_+(u))$, see Corollary 10.5 below. Furthermore, we see that $\Re(\overline{G}R(G)) \leq (1 + |G|^2)(|B| + |C|)$, hence $\partial_t |G(t, u)|^2 \leq 2(1 + |G(t, u)|^2)(|B| + |C|)$. This implies, by Gronwall's inequality ([55, (10.5.1.3)]), that $t_+(u) = \infty$. Hence the lemma is proved. $\qquad \square$

Next, we consider time-inhomogeneous Riccati equations in $\mathbb{R}^m$ of the special form

$$\partial_t f_i(t) = A_i f_i(t)^2 + B_i^\top f(t) + C_i(t), \quad i = 1, \ldots, m, \tag{10.49}$$

for some parameters $A, B, C(t)$ satisfying the following admissibility conditions:

$$A = (A_1, \ldots, A_m) \in \mathbb{R}^m,$$

$$B_{i,j} \geq 0 \quad \text{for } 1 \leq i \neq j \leq m, \tag{10.50}$$

$$C(t) = (C_1(t), \ldots, C_m(t)) \quad \text{continuous } \mathbb{R}^m\text{-valued.}$$

The following lemma provides a comparison result for (10.49). It shows, in particular, that the solution of (10.49) is uniformly bounded from below on compacts with respect to $\succeq$ if $A \succeq 0$.

Lemma 10.13 *Let $A^{(k)}, B, C^{(k)}, k = 1, 2$, be parameters satisfying the admissibility conditions (10.50), and*

$$A^{(1)} \preceq A^{(2)}, \qquad C^{(1)}(t) \preceq C^{(2)}(t). \tag{10.51}$$

Let $\tau > 0$ and $f^{(k)} : [0, \tau) \to \mathbb{R}^m$ be solutions of (10.50) *with A and C replaced by* $A^{(k)}$ *and* $C^{(k)}$, *respectively,* $k = 1, 2$. *If* $f^{(1)}(0) \preceq f^{(2)}(0)$ *then* $f^{(1)}(t) \preceq f^{(2)}(t)$ *for all* $t \in [0, \tau)$. *If, moreover,* $A^{(1)} = 0$ *then*

$$e^{Bt}\left(f^{(1)}(0) + \int_0^t e^{-Bs} C^{(1)}(s)\, ds\right) \preceq f^{(2)}(t)$$

for all $t \in [0, \tau)$.

Proof The function $f = f^{(2)} - f^{(1)}$ solves

$$\partial_t f_i = A_i^{(2)}\left(f_i^{(2)}\right)^2 - A_i^{(1)}\left(f_i^{(1)}\right)^2 + B_i^\top f + C_i^{(2)} - C_i^{(1)}$$

$$= \left(A_i^{(2)} - A_i^{(1)}\right)\left(f_i^{(2)}\right)^2 + A_i^{(1)}\left(f_i^{(2)} + f_i^{(1)}\right) f_i + B_i^\top f + C_i^{(2)} - C_i^{(1)}$$

$$= \tilde{B}_i^\top f + \tilde{C}_i,$$

where we write

$$\tilde{B}_i = \tilde{B}_i(t) = B_i + A_i^{(1)}\left(f_i^{(2)}(t) + f_i^{(1)}(t)\right) e_i,$$

$$\tilde{C}_i = \tilde{C}_i(t) = \left(A_i^{(2)} - A_i^{(1)}\right)\left(f_i^{(2)}(t)\right)^2 + C_i^{(2)}(t) - C_i^{(1)}(t).$$

Note that $\tilde{B} = (\tilde{B}_{i,j})$ and $\tilde{C}$ satisfy the assumptions of Corollary 10.6 in lieu of B and C, and $f(0) \in \mathbb{R}_+^m$. Hence Corollary 10.6 implies $f(t) \in \mathbb{R}_+^m$ for all $t \in [0, \tau)$, as desired. The last statement of the lemma follows by the variation of constants formula for $f^{(1)}(t)$. $\qquad\square$

After these preliminary comparison results for the Riccati equation (10.49), we now can state and prove an important result for the system of Riccati equations (10.7).

Lemma 10.14 *Let $\mathcal{D}_\mathbb{R}$ denote the maximal domain for the system of Riccati equations* (10.7). *Let $(\tau, u) \in \mathcal{D}_\mathbb{R}$. Then:*

(a) $\mathcal{D}_\mathbb{R}(\tau)$ *is star-shaped around zero.*
(b) $\theta^* = \sup\{\theta \geq 0 \mid \theta u \in \mathcal{D}_\mathbb{R}(\tau)\}$ *satisfies either $\theta^* = \infty$ or*

$$\lim_{\theta \uparrow \theta^*} \|\psi_I(t, \theta u)\| = \infty.$$

In the latter case, there exists some $x^* \in \mathbb{R}_+^m \times \mathbb{R}^n$ such that

$$\lim_{\theta \uparrow \theta^*} \phi(\tau, \theta u) + \psi(\tau, \theta u)^\top x^* = \infty.$$

Proof We first assume that the matrices α_i are block-diagonal, such that $\alpha_{i,iJ} = 0$, for all $i = 1, \ldots, m$.

Fix $\theta \in (0, 1]$. We claim that $\theta u \in \mathcal{D}_\mathbb{R}(\tau)$. It follows by inspection that $f^{(\theta)}(t) = \frac{\psi_I(t, \theta u)}{\theta}$ solves (10.49) with

$$A_i^{(\theta)} = \frac{1}{2}\theta\alpha_{i,ii}, \qquad B = \mathcal{B}_{II}^\top,$$

$$C_i^{(\theta)}(t) = \beta_{i,J}^\top \psi_J(t, u) + \frac{1}{2}\psi_J(t, u)^\top \theta\alpha_{i,JJ}\psi_J(t, u),$$

and $f(0) = u$. Lemma 10.13 thus implies that $f^{(\theta)}(t)$ is nicely behaved, as

$$e^{\mathcal{B}_{II}^\top t}\left(u + \int_0^t e^{-\mathcal{B}_{II}^\top s}C^{(0)}(s)\,ds\right) \preceq f^{(\theta)}(t) \preceq \psi_I(t, u), \tag{10.52}$$

for all $t \in [0, t_+(\theta u)) \cap [0, \tau]$. By the maximality of $\mathcal{D}_\mathbb{R}$ we conclude that $\tau < t_+(\theta u)$, which implies $\theta u \in \mathcal{D}_\mathbb{R}(\tau)$, as desired. Hence $\mathcal{D}_\mathbb{R}(\tau)$ is star-shaped around zero, which is part (a).

Next suppose that $\theta^* < \infty$. Since $\mathcal{D}_\mathbb{R}(\tau)$ is open, this implies $\theta^* u \notin \mathcal{D}_\mathbb{R}(\tau)$ and thus $t_+(\theta^* u) \le \tau$. From part (a) we know that $(t, \theta u) \in \mathcal{D}_\mathbb{R}$ for all $t < t_+(\theta^* u)$ and $0 \le \theta \le \theta^*$. On the other hand, there exists a sequence $t_n \uparrow t_+(\theta^* u)$ such that $\|\psi_I(t_n, \theta^* u)\| > n$ for all $n \in \mathbb{N}$. By continuity of ψ on $\mathcal{D}_\mathbb{R}$, we conclude that there exists some sequence $\theta_n \uparrow \theta^*$ with $\|\psi_I(t_n, \theta_n u) - \psi_I(t_n, \theta^* u)\| \le 1/n$ and hence

$$\lim_n \|\psi_I(t_n, \theta_n u)\| = \infty. \tag{10.53}$$

Applying Lemma 10.13 as above, where initial time $t = 0$ is shifted to t_n, yields

$$g_n = e^{\mathcal{B}_{II}^\top(\tau - t_n)}\left(f^{(\theta_n)}(t_n) + \int_{t_n}^\tau e^{\mathcal{B}_{II}^\top(t_n - s)}C^{(0)}(s)\,ds\right) \preceq f^{(\theta_n)}(\tau).$$

Corollary 10.6 implies that $e^{\mathcal{B}_{II}^\top(\tau - t_n)}$ is $\succeq$-order preserving. That is, $e^{\mathcal{B}_{II}^\top(\tau - t_n)}\mathbb{R}_+^m \subseteq \mathbb{R}_+^m$. Hence, in view of (10.52) for $f^{(\theta_n)}(t_n)$,

$$g_n \succeq e^{\mathcal{B}_{II}^\top(\tau - t_n)}\left(e^{\mathcal{B}_{II}^\top t_n}\left(u + \int_0^{t_n} e^{-\mathcal{B}_{II}^\top s}C^{(0)}(s)\,ds\right) + \int_{t_n}^\tau e^{\mathcal{B}_{II}^\top(t_n - s)}C^{(0)}(s)\,ds\right)$$

$$= e^{\mathcal{B}_{II}^\top \tau}\left(u + \int_0^\tau e^{-\mathcal{B}_{II}^\top s}C^{(0)}(s)\,ds\right).$$

On the other hand, elementary operator norm inequalities yield

$$\|g_n\| \ge e^{-\|\mathcal{B}_{II}\|\tau}\|f^{(\theta_n)}(t_n)\| - e^{\|\mathcal{B}_{II}\|\tau}\tau \sup_{s \in [0,\tau]} \|C^{(0)}(s)\|.$$

Together with (10.53), this implies $\|g_n\| \to \infty$. From Lemma 10.15 below we conclude that $\lim_n f^{(\theta_n)}(\tau)^\top y^* = \infty$ for some $y^* \in \mathbb{R}_+^m$. Moreover, in view of Lemma 10.13, we know that $f^{(\theta)}(\tau)^\top y^*$ is nondecreasing in θ. Therefore $\lim_{\theta \uparrow \theta^*} f^{(\theta)}(\tau)^\top y^* = \infty$. Applying (10.52) and Lemma 10.15 below again, this also implies that

$$\lim_{\theta \uparrow \theta^*} \| f^{(\theta)}(\tau) \| = \infty.$$

It remains to set $x^* = (y^*, 0)$ and observe that $b_I \in \mathbb{R}_+^m$ and thus

$$\phi(\tau, \theta u) = \int_0^\tau \left(\frac{1}{2} \psi_J(t, \theta u)^\top a_{JJ} \, \psi_J(t, \theta u) + b_I^\top \psi_I(t, \theta u) + b_J^\top \psi_J(t, \theta u) \right) dt$$

is uniformly bounded from below for all $\theta \in [0, \theta^*)$. Thus the lemma is proved under the premise that the matrices α_i are block-diagonal for all $i = 1, \ldots, m$.

The general case of admissible parameters a, α_i, b, β_i is reduced to the preceding block-diagonal case by a linear transformation along the lines of Lemma 10.5. Indeed, define the invertible $d \times d$-matrix Λ

$$\Lambda = \begin{pmatrix} I_m & 0 \\ D & I_n \end{pmatrix}, \tag{10.54}$$

where the $n \times m$-matrix $D = (\delta_1, \ldots, \delta_m)$ has ith column vector

$$\delta_i = \begin{cases} -\dfrac{\alpha_{i,iJ}}{\alpha_{i,ii}}, & \text{if } \alpha_{i,ii} > 0, \\ 0, & \text{else.} \end{cases}$$

It is then not hard to see ($\to$ Exercise 10.23) that $\Lambda(\mathbb{R}_+^m \times \mathbb{R}^n) = \mathbb{R}_+^m \times \mathbb{R}^n$, and

$$\widetilde{\phi}(t, u) = \phi(t, \Lambda^\top u), \qquad \widetilde{\psi}(t, u) = \left(\Lambda^\top \right)^{-1} \psi(t, \Lambda^\top u) \tag{10.55}$$

satisfy the system of Riccati equations (10.7) with a, α_i, b, and $\mathcal{B} = (\beta_1, \ldots, \beta_d)$ replaced by the admissible parameters

$$\widetilde{a} = \Lambda a \Lambda^\top, \qquad \widetilde{\alpha}_i = \Lambda \alpha_i \Lambda^\top, \qquad \widetilde{b} = \Lambda b, \qquad \widetilde{\mathcal{B}} = \Lambda \mathcal{B} \Lambda^{-1}. \tag{10.56}$$

Moreover, $\widetilde{\alpha}_i$ are block-diagonal, for all $i = 1, \ldots, m$.

By the first part of the proof, the corresponding maximal domain $\widetilde{\mathcal{D}}_\mathbb{R}(\tau)$, and hence also $\mathcal{D}_\mathbb{R}(\tau) = \Lambda^\top \widetilde{\mathcal{D}}_\mathbb{R}(\tau)$, is star-shaped around zero. Moreover, if $\theta^* < \infty$, then

$$\lim_{\theta \uparrow \theta^*} \| \psi_I(\tau, \theta u) \| = \lim_{\theta \uparrow \theta^*} \left\| \widetilde{\psi}_I \left(\tau, \theta \left(\Lambda^\top \right)^{-1} u \right) \right\| = \infty,$$

and there exists some $x^* \in \mathbb{R}_+^m \times \mathbb{R}^n$ such that

$$\lim_{\theta \uparrow \theta^*} \phi(\tau, \theta u) + \psi(\tau, \theta u)^\top x^*$$

$$= \lim_{\theta \uparrow \theta^*} \widetilde{\phi} \left(\tau, \theta \left(\Lambda^\top \right)^{-1} u \right) + \widetilde{\psi} \left(\tau, \theta \left(\Lambda^\top \right)^{-1} u \right)^\top \Lambda x^* = \infty.$$

Hence the lemma is proved. $\square$

Lemma 10.15 *Let $c \in \mathbb{R}^m$, and (c_n) and (d_n) be sequences in $\mathbb{R}^m$ such that*

$$c \preceq c_n \preceq d_n$$

for all $n \in \mathbb{N}$. Then the following are equivalent:

(a) $\|c_n\| \to \infty$.
(b) $c_n^\top y^* \to \infty$ *for some $y^* \in \mathbb{R}_+^m \setminus \{0\}$.*

In either case, $\|d_n\| \to \infty$ and $d_n^\top y^* \to \infty$.

Proof (a) $\Rightarrow$ (b): since $\|c_n\|^2 = \sum_{i=1}^m (c_n^\top e_i)^2$ and $c_n^\top e_i \geq c^\top e_i$, we conclude that $c_n^\top e_i \to \infty$ for some $i = 1, \ldots, m$.

(b) $\Rightarrow$ (a): this follows from $\|c_n^\top y^*\| \leq \|c_n\| \|y^*\|$.

The last statement now follows since $d_n^\top y^* \geq c_n^\top y^*$. □

We now have all the ingredients needed for the proof of Theorem 10.3.

10.7.3 Proof of Theorem 10.3

We first claim that, for every $u \in \mathbb{C}^d$ with $t_+(u) < \infty$, there exists some $i \in I$ and some sequence $t_n \uparrow t_+(u)$ such that

$$\lim_n (\Re \psi_i(t_n, u))^+ = \infty. \tag{10.57}$$

Indeed, otherwise we would have $\sup_{t \in [0, t_+(u))} \|(\Re \psi_I(t, u))^+\| < \infty$. But then (10.8) would imply $\sup_{t \in [0, t_+(u))} \|\psi_I(t, u)\| < \infty$, which is absurd. Whence (10.57) is proved.

In the following, we write

$$G(u, t, x) = \mathbb{E}\left[e^{u^\top X^x(t)}\right], \qquad V(t, x) = \left\{u \in \mathbb{R}^d \mid G(u, t, x) < \infty\right\}.$$

Since X is affine, by definition we have $\mathbb{R}_+ \times i\mathbb{R}^d \subset \mathcal{D}_{\mathbb{C}}$ and (10.2) implies

$$G(u, t, x) = e^{\phi(t, u) + \psi(t, u)^\top x} \tag{10.58}$$

for all $u \in i\mathbb{R}^d$, $t \in \mathbb{R}_+$ and $x \in \mathbb{R}_+^m \times \mathbb{R}^n$. Moreover, by Lemma 10.14, $\mathcal{D}_{\mathbb{R}}(t) = \mathcal{D}_{\mathbb{C}}(t) \cap \mathbb{R}^d$ is open and star-shaped around 0 in $\mathbb{R}^d$. Hence Lemma 10.9 implies that $\mathcal{D}_{\mathbb{R}}(t) \subset V(t, x)$ and (10.58) holds for all $u \in \mathcal{D}_{\mathbb{C}}(t) \cap S(\mathcal{D}_{\mathbb{R}}(t))$, for all $x \in \mathbb{R}_+^m \times \mathbb{R}^n$ and $t \in [0, \tau]$.

Now let $u \in \mathcal{D}_{\mathbb{R}}(\tau)$ and $v \in \mathbb{R}^d$, and define

$$\theta^* = \inf\{\theta \in \mathbb{R}_+ \mid u + i\theta v \notin \mathcal{D}_{\mathbb{C}}(\tau)\}.$$

We claim that $\theta^* = \infty$. Arguing by contradiction, assume that $\theta^* < \infty$. Since $\mathcal{D}_{\mathbb{C}}(\tau)$ is open, this implies $u + i\theta^* v \notin \mathcal{D}_{\mathbb{C}}(\iota)$, and thus

$$t_+(u + i\theta^* v) \le \tau. \tag{10.59}$$

On the other hand, since $\mathcal{D}_{\mathbb{R}}(\tau)$ is open, $(1 + \varepsilon)u \in \mathcal{D}_{\mathbb{R}}(\tau)$ for some $\varepsilon > 0$. Hence (10.58) holds and $G(t, (1 + \varepsilon)u, x)$ is uniformly bounded in $t \in [0, \tau]$, by continuity of $\phi(t, (1 + \varepsilon)u)$ and $\psi(t, (1 + \varepsilon)u)$ in t. We infer that the class of random variables $\{e^{(u + i\theta^* v)^\top X(t)} \mid t \in [0, \tau]\}$ is uniformly integrable, see [161, 13.3]. Since $X(t)$ is continuous in t, we conclude by Lebesgue's convergence theorem that $G(t, u + i\theta^* v, x)$ is continuous in $t \in [0, \tau]$, for all $x \in \mathbb{R}_+^m \times \mathbb{R}^n$. But for all $t < t_+(u + i\theta^* v)$ we have $(t, u + i\theta^* v) \in \mathcal{D}_{\mathbb{C}}(t) \cap \mathcal{S}(\mathcal{D}_{\mathbb{R}}(t))$, and thus (10.58) holds for all $x \in \mathbb{R}_+^m \times \mathbb{R}^n$. In view of (10.57), this contradicts (10.59). Whence $\theta^* = \infty$ and thus $u + iv \in \mathcal{D}_{\mathbb{C}}(\tau)$. This proves (a).

Applying the above arguments to[13] $\mathbb{E}[e^{u^\top X(T)} \mid \mathcal{F}_t] = G(T - t, u, X(t))$ with $T = t + \tau$ yields (d). Part (e) follows, since, by Theorem 10.2, $\mathbb{C}_-^m \times i\mathbb{R}^n \subset \mathcal{S}(\mathcal{D}_{\mathbb{R}}(t))$ for all $t \in \mathbb{R}_+$.

As for (b), we first let $u \in \mathcal{D}_{\mathbb{R}}(\tau)$. From part (d) it follows that $u \in M(\tau)$. Conversely, let $u \in M(\tau)$, and define $\theta^* = \sup\{\theta \ge 0 \mid \theta u \in \mathcal{D}_{\mathbb{R}}(\tau)\}$. We have to show that $\theta^* > 1$. Assume, by contradiction, that $\theta^* \le 1$. From Lemma 10.14, we know that there exists some $x^* \in \mathbb{R}_+^m \times \mathbb{R}^n$ such that

$$\lim_{\theta \uparrow \theta^*} \phi(\tau, \theta u) + \psi(\tau, \theta u)^\top x^* = \infty. \tag{10.60}$$

On the other hand, from part (d) and Jensen's inequality, we obtain

$$e^{\phi(\tau, \theta u) + \psi(\tau, \theta u)^\top x^*} = G(\tau, \theta u, x^*) \le G(\tau, u, x^*)^\theta \le G(\tau, u, x^*) < \infty$$

for all $\theta < \theta^*$. But this contradicts (10.60), hence $u \in \mathcal{D}_{\mathbb{R}}(\tau)$, and part (b) is proved. Since $M(\tau)$ is convex, this also implies (c). Finally, part (f) follows from part (b) and the respective inclusion property $\mathcal{D}_{\mathbb{R}}(t) \supseteq \mathcal{D}_{\mathbb{R}}(T)$. Whence Theorem 10.3 is proved.

10.8 Exercises

Exercise 10.1 This exercise provides an example for a multivariate affine process defined on a state space which is not of the form $\mathbb{R}_+^m \times \mathbb{R}^n$.

Consider the epigraph $\mathcal{X} = \{x \in \mathbb{R}^2 \mid x_1 \ge x_2^2\}$ of the parabola $x_1 = x_2^2$ in $\mathbb{R}^2$. Let $W = (W_1, W_2)^\top$ be a two-dimensional standard Brownian motion. For every $y \ge 0$, there exists a unique nonnegative affine diffusion process $Y = Y^y$ satisfying

$$dY = 2\sqrt{Y}\, dW_1, \quad Y(0) = y$$

---

[13] Here we use the Markov property of X, see Theorem 4.5.

(you do not have to prove this fact, it follows from Lemma 10.6). For every $x \in \mathcal{X}$ we define the $\mathcal{X}$-valued diffusion process $X = X^x$ by

$$X_1(t) = (W_2(t) + x_2)^2 + Y^y(t),$$
$$X_2(t) = W_2(t) + x_2,$$

where $y = y(x)$ is the unique nonnegative number with $x_1 = x_2^2 + y$.

(a) Show that X satisfies

$$dX_1 = dt + 2\sqrt{X_1 - X_2^2}\, dW_1 + 2X_2\, dW_2,$$

$$dX_2 = dW_2.$$

Conclude that the drift and diffusion matrix of X are affine functions of x. Verify that the diffusion matrix is positive semi-definite on $\mathcal{X}$.

(b) Verify by solving the corresponding Riccati equations that

$$\mathbb{E}\left[e^{u^\top X(T)} \mid \mathcal{F}_t\right] = \frac{1}{\sqrt{1 - 2u_1(T - t)}} e^{\frac{(T-t)u_2^2 + 2u^\top X(t)}{2(1 - 2u_2(T-t))}} \quad \text{for } u = (u_1, u_2)^\top \in i\mathbb{R}^2.$$

Conclude that X is an affine process.

Exercise 10.2 Let B be a Brownian motion and define the $\mathbb{R}_+^2$-valued process X by $X_i(t) = (\sqrt{x_i} + B(t))^2$, for $i = 1, 2$, for some $x \in \mathbb{R}_+^2$.

(a) Show that X satisfies

$$dX_1 = dt + 2\sqrt{X_1}\, dW,$$

$$dX_2 = dt + 2\sqrt{X_2}\, dW,$$

$$X(0) = x,$$

for some Brownian motion W. Is X an affine process? Why (not)?

(b) Compute the characteristic function of $X(t)$ and verify your finding concerning the (supposed) affine property of X.

Exercise 10.3 Let W be a Brownian motion. The aim of this exercise is to find some $\gamma \in \mathcal{L}$ such that the stochastic exponential $\mathcal{E}(\gamma \bullet W)$ is a martingale, while γ does not satisfy Novikov's condition (Theorem 4.7).

(a) Let $c > 0$ be some real constant. Show that

$$\mathbb{E}\left[e^{c \int_0^t W(s)^2\, ds}\right] = \begin{cases} \dfrac{1}{\sqrt{\cos(t\sqrt{2c})}}, & t < \dfrac{\pi}{2\sqrt{2c}}, \\ \infty, & t \geq \dfrac{\pi}{2\sqrt{2c}}. \end{cases}$$

Hint: show that $X_1(t) = (\sqrt{x_1} + W(t))^2$ and $X_2(t) = x_2 + \int_0^t X_1(s)\, ds$ define an affine process in $\mathbb{R}_+^2$.

(b) Define the positive local martingale $M = \mathcal{E}(-W \bullet W)$. Prove that M is a martingale, that is, $\mathbb{E}[M(t)] = 1$ for all finite $t \geq 0$. Hint: show that $\int_0^t W(s) \, dW(s) = W(t)^2/2 - t/2$ and use the affine process (X_1, X_2) from part (a).

(c) Conclude that you have just found a $\gamma \in \mathcal{L}$ such that the stochastic exponential $\mathcal{E}(\gamma \bullet W)$ is a true martingale while γ does not satisfy Novikov's condition.

(d) Finally show that, for any finite time horizon T, $d\mathbb{Q}/d\mathbb{P} = M(T)$ defines an equivalent measure such that W has a mean reverting drift under $\mathbb{Q}$:

$$dW(t) = -W(t) \, dt + dW^*(t),$$

where $W^*(t)$ denotes the Girsanov transformed Brownian motion, for $t \leq T$.

Exercise 10.4 The aim of this exercise is to give a direct proof for the validity of (10.13) in the case where $\gamma \in \mathbb{R}_+^m \times \{0\}$. Let $u \in \mathbb{C}_-^m \times i\mathbb{R}^n$ and $T \in \mathbb{R}_+$.

(a) Along the lines of the proof of Theorem 10.2 show that there exists a unique solution $(\Phi(\cdot, u), \Psi(\cdot, u)) : \mathbb{R}_+ \to \mathbb{C} \times \mathbb{C}_-^m \times i\mathbb{R}^n$ of (10.12) with

$$\Re(\Phi(t, u)) = -ct.$$

(b) Now argue as in the proof of Theorem 10.1, and show that

$$M(t) = e^{-\int_0^t r(s) \, ds} e^{\Phi(T-t, u) + \Psi(T-t, u)^\top X(t)}, \quad t \leq T,$$

is a martingale with $M(T) = e^{-\int_0^T r(s) \, ds} e^{u^\top X(T)}$.

(c) Conclude that (10.13) holds.

Exercise 10.5 Complete the proof of Lemma 10.2.

Exercise 10.6 Complete the proof of Corollary 10.4 by deriving the price formulas (10.28).

Exercise 10.7 Derive explicit call and put bond option price formulas for the Vasiček short-rate model and the results from Sect. 7.2.1 using the approach outlined in Sect. 10.3.2.1.

Exercise 10.8 The aim of this exercise is to derive an intuition for the noncentral χ^2-distribution, and its interplay with affine processes. Fix $\delta \in \mathbb{N}$ and some real numbers $v_1, \ldots, v_\delta$, and define $\zeta = \sum_{i=1}^\delta v_i^2$.

(a) Let $N_1, \ldots, N_\delta$ be independent standard normal distributed random variables. Define $Z = \sum_{i=1}^\delta (N_i + v_i)^2$. Show by direct integration that the characteristic function of Z equals

$$\mathbb{E}[e^{uZ}] = \frac{e^{\frac{\zeta u}{1-2u}}}{(1-2u)^{\frac{\delta}{2}}}, \quad u \in \mathbb{C}_-.$$

Conclude by Lemma 10.4 that Z is noncentral χ^2-distributed with δ degrees of freedom and noncentrality parameter ζ.

(b) Now let $W_1, \ldots, W_\delta$ be independent standard Brownian motions with respect to some filtration $(\mathcal{F}_t)$, and define the process $X(t) = \sum_{i=1}^{\delta}(W_i(t) + v_i)^2$, $t \geq 0$. Using (a), show that the $\mathcal{F}_t$-conditional characteristic function of $X(T)$ equals

$$\mathbb{E}[e^{uX(T)} \mid \mathcal{F}_t] = \frac{e^{\frac{u}{1-2(T-t)u}X(t)}}{(1 - 2(T-t)u)^{\frac{\delta}{2}}}, \quad u \in \mathbb{C}_-, \ t < T. \tag{10.61}$$

Conclude that the $\mathcal{F}_t$-conditional distribution of $X(T)$ is noncentral χ^2 with δ degrees of freedom and noncentrality parameter $\frac{X(t)}{T-t}$.

(c) Along the lines of Exercise 5.4, show that X satisfies the stochastic differential equation

$$dX(t) = \delta \, dt + 2\sqrt{X(t)} \, dB(t), \quad X(0) = \zeta,$$

for the Brownian motion $dB = \sum_{i=1}^{\delta} \frac{W_i + v_i}{\sqrt{X}} \, dW_i$.

(d) Conclude by either (b) or (c) that X is an affine process with state space $\mathbb{R}_+$.

(e) Find the explicit solutions of the corresponding Riccati equations

$$\partial_t \phi(t, u) = \cdots, \qquad \partial_t \psi(t, u) = \cdots,$$

for X and verify (10.61) by applying the affine transform formula

$$\mathbb{E}[e^{uX(T)} \mid \mathcal{F}_t] = e^{\phi(T-t,u) + \psi(T-t,u)X(t)}.$$

Exercise 10.9 Let $b, \sigma > 0$ and $\beta \in \mathbb{R}$, and consider the affine process

$$dX = (b + \beta X) \, dt + \sigma \sqrt{X} \, dW, \quad X(0) = x \in \mathbb{R}_+,$$

with state space $\mathbb{R}_+$.

(a) Use Lemma 10.12 for finding the explicit solutions ϕ and ψ of the corresponding Riccati equations.

(b) Define $C(\tau) = \frac{\sigma^2(e^{\beta\tau}-1)}{4\beta}$ $(= \frac{\sigma^2\tau}{4}$ if $\beta = 0)$ for $\tau > 0$. Check that $C(\tau)$ is positive and increasing in τ. Let $t < T$, and show that the $\mathcal{F}_t$-conditional distribution of $\frac{X(T)}{C(T-t)}$ is noncentral χ^2 with $\frac{4b}{\sigma^2}$ degrees of freedom and noncentrality parameter $\frac{e^{\beta(T-t)}X(t)}{C(T-t)}$.

(c) Verify the findings of Exercise 10.8(b).

Exercise 10.10 Let $\sigma > 0$ and $\beta \in \mathbb{R}$, and consider the affine process

$$dX = \beta X \, dt + \sigma \sqrt{X} \, dW, \quad X(0) = x \in \mathbb{R}_+,$$

with state space $\mathbb{R}_+$. Define $C(t)$ as in Exercise 10.9(b).

(a) Show that the characteristic function of $Z(t) = \frac{X(t)}{C(t)}$ equals

$$\mathbb{E}[e^{uZ(t)}] = \exp\left[\frac{\frac{e^{\beta t}x}{C(t)}u}{1 - 2u}\right], \quad u \in \mathbb{C}_-. \tag{10.62}$$

This is the characteristic function of the so-called noncentral χ^2 distribution with zero degrees of freedom[14] and noncentrality parameter $\frac{e^{\beta t}x}{C(t)}$.

(b) Show that

$$\mathbb{P}[X(t) = 0] = \exp\left[-\frac{e^{\beta t}x}{2C(t)}\right] > 0, \quad t > 0.$$

Conclude that the noncentral χ^2 distribution with zero degrees of freedom admits no density on $\mathbb{R}_+$ (hint: take limit $u \to -\infty$ in (10.62)).

(c) Derive the asymptotic result

$$\lim_{t\to\infty} \mathbb{P}[X(t) = 0] = \begin{cases} 1, & \beta \leq 0, \\ e^{-\frac{2\beta}{\sigma^2}}, & \beta > 0. \end{cases}$$

(d) Now suppose $\beta = 0$. Show that, for every $T > 0$, there exists some reals M, $p > 0$ such that

$$\mathbb{E}[e^{pX(t)}] \leq M, \quad t \leq T.$$

Conclude that X is a true martingale.

Exercise 10.11 Fix a constant interest rate $r > 0$ and $\sigma > 0$, and consider the affine stock model with risk-neutral dynamics ($\mathbb{P} = \mathbb{Q}$)

$$dS = rS\,dt + \sigma\sqrt{S}\,dW, \quad S(0) = s_0 \geq 0. \tag{10.63}$$

(a) Using Exercise 10.10(d), show that the discounted stock price process $e^{-rt}S(t)$ is a martingale ≥ 0.

(b) Check, by Exercise 10.10, that $\mathbb{P}[S(t) = 0] > 0$ and argue that S may be interpreted as defaultable stock price: once the price hits zero, it remains zero (hint: the solution of (10.63) is unique).

(c) Define $C(t) = \frac{\sigma^2(e^{rt}-1)}{4r}$, and derive from Exercise 10.10 that $\frac{S(t)}{C(t)}$ has a noncentral χ^2 distribution with zero degrees of freedom and parameter of noncentrality $\frac{e^{rt}s_0}{C(t)}$.

(d) For the parameters $s_0 = 100$, $r = 0.01$ and $\sigma = 4$ derive the European call option prices and implied volatilities, by inverting the Black–Scholes option price formula in Proposition 7.3, as shown in Tables 10.4–10.5 and Fig. 10.5. Show that the risk-neutral default probability is $\mathbb{P}[S(1) = 0] = 0.3501 \times 10^{-5}$.

---

[14] The noncentral χ^2 distribution with zero degrees of freedom has been defined by Siegel [152].

Table 10.4 Call option prices in the affine stock model (10.63)

T–K	80	90	100	110	120
0.2	21.1236	13.1847	7.2226	3.4253	1.3983
0.4	22.9365	15.8324	10.2530	6.2162	3.5274
0.6	24.6300	17.9697	12.5878	8.4644	5.4662
0.8	26.1726	19.8052	14.5570	10.3953	7.2166
1.0	27.5739	21.4231	16.2776	12.1002	8.8050

Table 10.5 Black–Scholes implied volatilities for the affine stock model (10.63)

T–K	80	90	100	110	120
0.2	0.4229	0.4108	0.4001	0.3907	0.3822
0.4	0.4230	0.4109	0.4003	0.3908	0.3823
0.6	0.4232	0.4110	0.4004	0.3909	0.3824
0.8	0.4232	0.4111	0.4004	0.3909	0.3824
1.0	0.4226	0.4106	0.4001	0.3906	0.3821

Fig. 10.5 Implied volatility surface for the affine stock model (10.63)

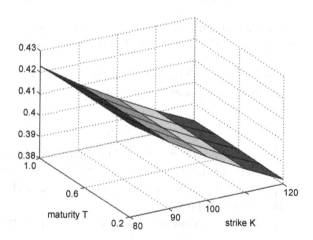

Exercise 10.12 Let $b \geq 0$, $\beta \in \mathbb{R}$ and $\sigma > 0$, and consider the affine process

$$dX = (b + \beta X)\,dt + \sigma\sqrt{X}\,dW, \quad X(0) = x_0 > 0,$$

with state space $\mathbb{R}_+$. For any $c \geq 0$ we define the stopping time

$$\tau_c = \inf\{t \geq 0 \mid X(t) = c\} \leq \infty.$$

Thus, $\{\tau_0 = \infty\}$ is the event that X never hits zero. The aim of this exercise is to prove the following claims:

(a) If $b \geq \frac{\sigma^2}{2}$, then $\mathbb{P}[\tau_0 = \infty] = 1$.

(b) If $b < \frac{\sigma^2}{2}$ and $\beta \leq 0$, then $\mathbb{P}[\tau_0 < \infty] = 1$.

(c) If $b < \frac{\sigma^2}{2}$ and $\beta > 0$, then $\mathbb{P}[\tau_0 < \infty] \in (0, 1)$.

Note that Exercise 10.10(b) and (c) is a special case of the claims (b) and (c) (why?).

- Define the function

$$f(x) = \int_1^x e^{-\frac{2\beta}{\sigma^2}y} y^{-\frac{2b}{\sigma^2}} \, dy, \quad x \geq 0,$$

and show that $f(X)$ is a local martingale (hint: Itô's formula).

- Let $0 < r < x_0 < R$, and define the stopping time $\tau_{r,R} = \tau_r \wedge \tau_R$. Show that

$$f(X(t \wedge \tau_{r,R})) - f(x_0) = \int_0^t f'(X(s))\sigma\sqrt{X(s)}\mathbf{1}_{\{s \leq \tau_{r,R}\}} \, dW(s), \quad t \geq 0.$$

- Taking the second moment on both sides (hint: Itô isometry), derive

$$M_1 \geq \mathbb{E}\left[\left(f(X(t \wedge \tau_{r,R})) - f(x_0)\right)^2\right] \geq M_2\mathbb{E}[t \wedge \tau_{r,R}],$$

for some real constant $M_1, M_2 > 0$ which do not depend on t (hint: show that $\sigma^2 x f'(x)^2 \geq M_2$ for all $x \geq r$). Conclude that $\mathbb{E}[\tau_{r,R}] < \infty$, and hence $\tau_{r,R} < \infty$ a.s.

- Show that $f(x_0) = \mathbb{E}[f(X(t \wedge \tau_{r,R}))] = \mathbb{E}[f(X(\tau_{r,R}))]$ (hint: show that $f(X(t \wedge \tau_{r,R}))$ is a martingale and use dominated convergence). Derive from this the identity

$$f(x_0) = f(r)\mathbb{P}[\tau_r < \tau_R] + f(R)\mathbb{P}[\tau_r > \tau_R].$$

- Using monotone convergence and the continuity of X, show that $\lim_{r \to 0} \mathbb{P}[\tau_r < \tau_R] = \mathbb{P}[\tau_0 < \tau_R]$, $\lim_{r \to 0} \mathbb{P}[\tau_r > \tau_R] = \mathbb{P}[\tau_0 > \tau_R]$, and $\tau_R \uparrow \infty$ for $R \uparrow \infty$.

- Suppose $b \geq \frac{\sigma^2}{2}$. Show that $\lim_{r \to 0} f(r) = -\infty$, and infer that $\mathbb{P}[\tau_0 = \infty] = 1$. Thus claim (a) is proved.

- Suppose $b < \frac{\sigma^2}{2}$. Show that $f(0) = \lim_{r \to 0} f(r)$ exists in $\mathbb{R}$, and

$$\lim_{R \to \infty} f(R) \begin{cases} = \infty, & \beta \leq 0, \\ \in \mathbb{R}, & \beta > 0. \end{cases}$$

Deduce from this that claims (b) and (c) hold.

Exercise 10.13 For the CIR model with parameters (10.30), compute the at-the-money European call and put option prices $\pi_{call}(t = 0)$ and $\pi_{put}(t = 0)$ on the $(S = 2)$-bond with expiry date $T = 1$, using each of the formulas in (10.28). Compare the results.

Exercise 10.14 Derive the ATM cap prices and Black volatilities in Table 10.1.

Exercise 10.15 Derive (10.31), and check how (10.12) would look like in the Heston stochastic volatility model.

Exercise 10.16 Derive the call option price formulas (10.32) and (10.33) in the Heston model.

Exercise 10.17

(a) Compute the call option prices in Table 10.2, using either formula (10.32) or (10.33) and Lemma 10.12, in the Heston model.
(b) Then derive the corresponding implied volatilities from Table 10.3 by inverting the Black–Scholes option price formula in Proposition 7.3.
(c) Show that the implied volatilities are decreasing with increasing strike price K if the stock price S and the volatility process X_1 have negative covariation $d\langle S, X_1 \rangle$, that is if $\rho < 0$.

Exercise 10.18 Find the representation of the affine process X underlying the Heston stochastic volatility model in Sect. 10.3.3 in the form (10.34) in terms of parameters $a, \alpha_1, \alpha_2, b, \mathcal{B} = (\beta_1, \beta_2)$.

(a) Show that $\mathcal{B}$ is singular, and hence cannot have negative eigenvalues (hence X is not strictly mean reverting).
(b) Verify that the diffusion matrix is not of block-diagonal form. Find an invertible 2×2-matrix Λ with $\Lambda(\mathbb{R}_+ \times \mathbb{R}) = \mathbb{R}_+ \times \mathbb{R}$ and an affine process Y on $\mathbb{R}_+ \times \mathbb{R}$ with block-diagonal diffusion matrix such that $X = \Lambda^{-1}Y$. How do you have to adjust the stock price process S to be expressed as a function of Y?

Exercise 10.19 Consider the following multivariate extension of Heston's stochastic volatility model. Interest rates are assumed to be constant $r(t) \equiv r \geq 0$, and there are two risky assets $S_i = e^{X_{i+1}}$, $i = 1, 2$, where $X = (X_1, X_2, X_3)^{\top}$ is the affine process with state space $\mathbb{R}_+ \times \mathbb{R}^2$ and dynamics

$$dX_1 = (k + \kappa X_1)\,dt + \sigma\sqrt{2X_1}\,dW_1,$$

$$dX_2 = (r - \sigma_2 X_1)\,dt + \sigma_2\sqrt{2X_1}\left(\rho_1\,dW_1 + \sqrt{1 - \rho_1^2}\,dW_2\right),$$

$$dX_3 = (r - \sigma_3 X_1)\,dt + \sigma_2\sqrt{2X_1}\left(\rho_2\,dW_1 + \rho_3\,dW_2 + \sqrt{1 - \rho_2^2 - \rho_3^2}\,dW_3\right)$$

for some constant parameters $k, \sigma \geq 0$, $\kappa \in \mathbb{R}$, $\sigma_1, \sigma_2 > 0$, and some $\rho_i \in [-1, 1]$ satisfying $\rho_2^2 + \rho_3^2 \leq 1$.

(a) Verify that this is an arbitrage-free model.
(b) Find and solve the corresponding Riccati equations.
(c) Compute the price of the exchange option with payoff $(c_1 S_1(T) - c_2 S_2(T))^+$ for various parameter specifications.

(d) Compute the price of the spread option with payoff $(S_1(T) - S_2(T) - K)^+$ for various parameter specifications.

Exercise 10.20 Let X be the affine process given in (10.34), and let Λ be a regular $d \times d$-matrix and $\lambda \in \mathbb{R}^d$. Show that the affine transform $Y = \Lambda X + \lambda$ satisfies

$$dY = \left(\Lambda b - \Lambda \beta^\top \Lambda^{-1} \lambda + \Lambda \beta^\top \Lambda^{-1} Y \right) dt + \Lambda \rho \left(\Lambda^{-1}(Y - \lambda) \right) dW.$$

Verify that the drift and diffusion matrix of Y are affine in $Y(t)$.

Exercise 10.21 Derive the corresponding system of Riccati equations (10.7) for the block-diagonal diffusion matrix (10.38).

Exercise 10.22 Derive Corollaries 10.5 and 10.6 as special cases from the invariance Lemma 10.11.

Exercise 10.23 Finish the proof of Lemma 10.14 by showing that:

(a) Λ in (10.54) satisfies $\Lambda(\mathbb{R}^m_+ \times \mathbb{R}^n) = \mathbb{R}^m_+ \times \mathbb{R}^n$.
(b) $\tilde{a}, \tilde{\alpha}_i, \tilde{b}, \tilde{B}$, given by (10.56), are admissible and $\tilde{\alpha}_i$ are block-diagonal, for all $i = 1, \ldots, m$.
(c) $\tilde{\phi}$ and $\tilde{\psi}$ in (10.55) satisfy the system of Riccati equations (10.7) with a, α_i, b, and $B = (\beta_1, \ldots, \beta_d)$ replaced by $\tilde{a}, \tilde{\alpha}_i, \tilde{b}, \tilde{B}$.

10.9 Notes

Affine Markov models have been employed in finance for decades, and they have found growing interest due to their computational tractability as well as their capability to capture empirical evidence from financial time series. Their main applications lie in the theory of term-structure of interest rates, stochastic volatility option pricing and the modeling of credit risk (see [61] and the references therein). There is a vast literature on affine models. We mention here explicitly just the few articles [4, 29, 43, 50, 58, 60, 80, 91, 114] and [61] for a broader overview. The generalizations to time-inhomogeneous affine processes have been studied in detail in [71].

A preliminary version of this chapter has been published as a review article [72]. Theorem 10.3(b) and (d) was first proved by Glasserman and Kim [80] for strictly mean reverting affine diffusion processes, which, however, excludes the Heston stochastic volatility model (see Exercise 10.18). The strict mean reversion assumption was subsequently relaxed in [72]. The blow-up property of ψ_I stated in Lemma 10.14(b) is crucial for the proof of Theorem 10.3(b). Its proof is inspired by the line of arguments in [80]. It is yet unclear whether it holds for the class of affine jump-diffusion processes in general. The convexity property of the maximal domain stated in Theorem 10.3(c) represents a non-trivial result for ordinary differential equations. Only in the mid 1990s were corresponding convexity results derived in the analysis literature, see Lakshmikantham et al. [109].

The Fourier transform of an option payoff as shown in Lemma 10.2 was first proposed and utilized for option pricing via fast Fourier transform methods by Carr and Madan [36]. The formula for the exchange option in Corollary 10.3 is obviously related, but seems to be new in the financial literature. The Fourier decomposition of the spread option payoff in Lemma 10.3 has been found and explored by Hurd and Zhou [98]. More examples of payoff functions with explicit Fourier decomposition, including the one in Lemma 10.2, can be found in Hubalek et al. [94].

The classification problem for affine term-structure models raised in Sect. 10.4 has been addressed by the specification analysis in Dai and Singleton [50], which was subsequently extended by Collin-Dufresne et al. [44]. However, it is shown in Cheridito et al. [43] that the Dai–Singleton classification is not exhaustive for state space dimension $d \geq 4$.

The existence and uniqueness proof of affine diffusion processes given in Sect. 10.5 builds on the seminal result by Yamada and Watanabe [162]. We note that this approach is different from the one used in [61], which uses infinite divisibility on the canonical state space and the Markov semigroup theory, thereby asserting existence of weak solutions only.

Section 10.6 is a more elaborated version from the Appendix in [61]. Exercise 10.12 is adapted from [110, Exercise 34].

Chapter 11
Market Models

Instantaneous forward rates are not so simple to estimate, as we have seen. One may want to model other rates, such as LIBOR, directly. There has been some effort in the years after the publication of HJM [90] in 1992 to develop arbitrage-free models of other than instantaneous, continuously compounded rates. The breakthrough came 1997, when the LIBOR market models were introduced by Miltersen et al. [124] and Brace et al. [23] who succeeded in finding a HJM-type model inducing lognormal LIBOR rates. At the same time, Jamshidian [102] developed a framework for arbitrage-free LIBOR and swap rate models not based on HJM. The principal idea of these approaches is to choose a different numeraire than the risk-free account (the latter does not even necessarily have to exist). Both approaches lead to Black's formula for either caps (LIBOR models) or swaptions (swap rate models). Because of this they are usually referred to as "market models".

11.1 Heuristic Derivation

To start with we consider the HJM setup from Chap. 7. Recall that, for a fixed $\delta > 0$, the forward δ-period LIBOR for the future date T prevailing at time t is the simple forward rate

$$L(t, T) = F(t; T, T + \delta) = \frac{1}{\delta} \left(\frac{P(t, T)}{P(t, T + \delta)} - 1 \right).$$

We have seen in Chap. 7 that $P(t, T)/P(t, T + \delta)$ is a martingale for the $(T + \delta)$-forward measure $\mathbb{Q}^{T+\delta}$. In particular, by Lemma 7.1,

$$d \left(\frac{P(t, T)}{P(t, T + \delta)} \right) = \frac{P(t, T)}{P(t, T + \delta)} \sigma_{T, T+\delta}(t) \, dW^{T+\delta}(t),$$

where $\sigma_{T, T+\delta}(t) = \int_T^{T+\delta} \sigma(t, u) \, du$ was defined in (7.2). Hence

$$dL(t, T) = \frac{1}{\delta} d \left(\frac{P(t, T)}{P(t, T + \delta)} \right) = \frac{1}{\delta} \frac{P(t, T)}{P(t, T + \delta)} \sigma_{T, T+\delta}(t) \, dW^{T+\delta}(t)$$

$$= \frac{1}{\delta} (\delta L(t, T) + 1) \sigma_{T, T+\delta}(t) \, dW^{T+\delta}(t).$$

Now suppose there exists an $\mathbb{R}^d$-valued deterministic function $\lambda(t, T)$ such that

$$\sigma_{T, T+\delta}(t) = \frac{\delta L(t, T)}{\delta L(t, T) + 1} \lambda(t, T). \tag{11.1}$$

D. Filipović, *Term-Structure Models*,
Springer Finance,
DOI 10.1007/978-3-540-68015-4_11, © Springer-Verlag Berlin Heidelberg 2009

Plugging this in the above formula, we get

$$dL(t, T) = L(t, T)\lambda(t, T) dW^{T+\delta}(t),$$

which is equivalent to

$$L(t, T) = L(s, T) \exp\left(\int_s^t \lambda(u, T) dW^{T+\delta}(u) - \frac{1}{2}\int_s^t \|\lambda(u, T)\|^2 du\right),$$

for $s \leq t \leq T$. Hence the $\mathbb{Q}^{T+\delta}$-distribution of $\log L(T, T)$ conditional on $\mathcal{F}_t$ is Gaussian with mean

$$\log L(t, T) - \frac{1}{2}\int_t^T \|\lambda(s, T)\|^2 ds$$

and variance

$$\int_t^T \|\lambda(s, T)\|^2 ds.$$

The time t price of a caplet with reset date T, settlement date $T + \delta$ and strike rate κ is thus

$$\mathbb{E}_{\mathbb{Q}}\left[e^{-\int_t^{T+\delta} r(s) ds}\delta(L(T, T) - \kappa)^+ \mid \mathcal{F}_t\right]$$

$$= P(t, T + \delta)\mathbb{E}_{\mathbb{Q}^{T+\delta}}\left[\delta(L(T, T) - \kappa)^+ \mid \mathcal{F}_t\right]$$

$$= \delta P(t, T + \delta)\left(L(t, T)\Phi(d_1(t, T)) - \kappa\Phi(d_2(t, T))\right),$$

where

$$d_{1,2}(t, T) = \frac{\log(\frac{L(t,T)}{\kappa}) \pm \frac{1}{2}\int_t^T \|\lambda(s, T)\|^2 ds}{(\int_t^T \|\lambda(s, T)\|^2 ds)^{\frac{1}{2}}},$$

and Φ is the standard Gaussian cumulative distribution function. This is just Black's formula for the caplet price with $\sigma(t)^2$ set equal to

$$\frac{1}{T - t}\int_t^T \|\lambda(s, T)\|^2 ds,$$

as introduced in Sect. 2.6.

We have thus shown that any HJM model satisfying (11.1) yields Black's formula for caplet prices. The question remains, whether such HJM models exist. The answer is yes, but the construction and proof are not easy. The idea is to rewrite (11.1), using the definition of $\sigma_{T,T+\delta}(t)$, as ($\rightarrow$ Exercise 11.1)

$$\int_T^{T+\delta} \sigma(t, u) du = \left(1 - e^{-\int_T^{T+\delta} f(t,u) du}\right)\lambda(t, T). \tag{11.2}$$

Differentiating in T gives

$$\sigma(t, T+\delta) = \sigma(t, T) + (f(t, T+\delta) - f(t, T))e^{-\int_T^{T+\delta} f(t,u)\,du}\lambda(t, T)$$
$$+ \left(1 - e^{-\int_T^{T+\delta} f(t,u)\,du}\right)\partial_T\lambda(t, T).$$

This is a recurrence relation that can be solved by forward induction, once $\sigma(t, \cdot)$ is determined on $[0, \delta)$ (typically, $\sigma(t, T) = 0$ for $T \in [0, \delta)$). This gives a complicated dependence of σ on the forward curve. Now it has to be proved that the corresponding HJM equations for the forward rates have a unique and well-behaved solution. All of this has been carried out by [23], see also [68, Sect. 5.6].

11.2 LIBOR Market Model

There is a more direct approach to LIBOR models without making reference to continuously compounded forward and short rates. In a sense, we place ourselves outside of the HJM framework (although HJM is often implicitly adopted). Instead of the risk-neutral martingale measure we will work under forward measures, the numeraires accordingly being bond price processes.

We fix a finite time horizon $T_M = M\delta$, for some $M \in \mathbb{N}$, and a filtered probability space

$$(\Omega, \mathcal{F}, (\mathcal{F}_t)_{t\in[0,T_M]}, \mathbb{Q}^{T_M}),$$

which carries a d-dimensional Brownian motion $W^{T_M}(t)$, $t \in [0, T_M]$. The notation already suggests that $\mathbb{Q}^{T_M}$ will play the role of the T_M-forward measure. Write

$$T_m = m\delta, \quad m = 0, \dots, M.$$

We are going to construct a model for the M forward LIBOR rates with maturities $T_0, \dots, T_{M-1}$. We take as given:

- for every $m \le M - 1$, an $\mathbb{R}^d$-valued deterministic bounded measurable function $\lambda(t, T_m)$, $t \in [0, T_m]$, which represents the volatility of $L(t, T_m)$;
- an initial positive and nonincreasing discrete term-structure

$$P(0, T_m), \quad m = 0, \dots, M,$$

and hence nonnegative initial forward LIBOR rates

$$L(0, T_m) = \frac{1}{\delta}\left(\frac{P(0, T_m)}{P(0, T_{m+1})} - 1\right), \quad m = 0, \dots, M - 1. \tag{11.3}$$

We proceed by backward induction and *postulate* first that

$$dL(t, T_{M-1}) = L(t, T_{M-1})\lambda(t, T_{M-1})\,dW^{T_M}(t), \quad t \in [0, T_{M-1}],$$

$$L(0, T_{M-1}) = \frac{1}{\delta} \left(\frac{P(0, T_{M-1})}{P(0, T_M)} - 1 \right)$$

which is of course equivalent to

$$L(t, T_{M-1}) = \frac{1}{\delta} \left(\frac{P(0, T_{M-1})}{P(0, T_M)} - 1 \right) \mathcal{E}_t \left(\lambda(\cdot, T_{M-1}) \bullet W^{T_M} \right).$$

Motivated by (11.1), we now define the $\mathbb{R}^d$-valued bounded progressive process

$$\sigma_{T_{M-1}, T_M}(t) = \frac{\delta L(t, T_{M-1})}{\delta L(t, T_{M-1}) + 1} \lambda(t, T_{M-1}), \quad t \in [0, T_{M-1}].$$

This induces an equivalent probability measure $\mathbb{Q}^{T_{M-1}} \sim \mathbb{Q}^{T_M}$ on $\mathcal{F}_{T_{M-1}}$ via

$$\frac{d\mathbb{Q}^{T_{M-1}}}{d\mathbb{Q}^{T_M}} = \mathcal{E}_{T_{M-1}} \left(\sigma_{T_{M-1}, T_M} \bullet W^{T_M} \right),$$

and by Girsanov's theorem

$$W^{T_{M-1}}(t) = W^{T_M}(t) - \int_0^t \sigma_{T_{M-1}, T_M}(s)^\top \, ds, \quad t \in [0, T_{M-1}],$$

is a $\mathbb{Q}^{T_{M-1}}$-Brownian motion.

Hence we can postulate

$$dL(t, T_{M-2}) = L(t, T_{M-2}) \lambda(t, T_{M-2}) \, dW^{T_{M-1}}(t), \quad t \in [0, T_{M-2}],$$

$$L(0, T_{M-2}) = \frac{1}{\delta} \left(\frac{P(0, T_{M-2})}{P(0, T_{M-1})} - 1 \right),$$

that is,

$$L(t, T_{M-2}) = \frac{1}{\delta} \left(\frac{P(0, T_{M-2})}{P(0, T_{M-1})} - 1 \right) \mathcal{E}_t \left(\lambda(\cdot, T_{M-2}) \bullet W^{T_{M-1}} \right),$$

and define the $\mathbb{R}^d$-valued bounded progressive process

$$\sigma_{T_{M-2}, T_{M-1}}(t) = \frac{\delta L(t, T_{M-2})}{\delta L(t, T_{M-2}) + 1} \lambda(t, T_{M-2}), \quad t \in [0, T_{M-2}],$$

yielding an equivalent probability measure $\mathbb{Q}^{T_{M-2}} \sim \mathbb{Q}^{T_{M-1}}$ on $\mathcal{F}_{T_{M-2}}$ via

$$\frac{d\mathbb{Q}^{T_{M-2}}}{d\mathbb{Q}^{T_{M-1}}} = \mathcal{E}_{T_{M-2}} \left(\sigma_{T_{M-2}, T_{M-1}} \bullet W^{T_{M-1}} \right),$$

and the $\mathbb{Q}^{T_{M-2}}$-Brownian motion

$$W^{T_{M-2}}(t) = W^{T_{M-1}}(t) - \int_0^t \sigma_{T_{M-2}, T_{M-1}}(s)^\top \, ds, \quad t \in [0, T_{M-2}].$$

Repeating this procedure leads to a family of lognormal martingales $(L(t, T_m))_{t \in [0, T_m]}$ under their respective measures $\mathbb{Q}^{T_{m+1}}$.

11.2.1 LIBOR Dynamics Under Different Measures

Next we are interested in finding the dynamics of $L(t, T_m)$ under any of the forward measures $\mathbb{Q}^{T_{n+1}}$.

Lemma 11.1 *Let $0 \leq m, n \leq M - 1$. Then the dynamics of $L(t, T_m)$ under $\mathbb{Q}^{T_{n+1}}$ is given according to the three cases*

$$\frac{dL(t, T_m)}{L(t, T_m)} = \begin{cases} -\lambda(t, T_m) \sum_{l=m+1}^{n} \sigma_{T_l, T_{l+1}}(t)^\top dt + \lambda(t, T_m) dW^{T_{n+1}}(t), & m < n, \\ \lambda(t, T_m) dW^{T_{n+1}}(t), & m = n, \\ \lambda(t, T_m) \sum_{l=n+1}^{m} \sigma_{T_l, T_{l+1}}(t)^\top dt + \lambda(t, T_m) dW^{T_{n+1}}(t), & m > n, \end{cases}$$

for $t \in [0, T_m \wedge T_{n+1}]$.

Proof This follows from the obvious equality

$$W^{T_{i+1}}(t) = W^{T_{j+1}}(t) - \sum_{l=i+1}^{j} \int_0^t \sigma_{T_l, T_{l+1}}(s)^\top ds,$$

$$t \in [0, T_{i+1}], \ 0 \leq i < j \leq M - 1 \tag{11.4}$$

($\rightarrow$ Exercise 11.2). $\qquad\qquad\qquad\qquad\qquad\qquad\qquad\qquad\qquad\qquad\qquad$ $\square$

11.3 Implied Bond Market

What can be said about bond prices? First, for all $m = 1, \ldots, M$, we can *define* the forward price process

$$\frac{P(t, T_{m-1})}{P(t, T_m)} = \delta L(t, T_{m-1}) + 1, \quad t \in [0, T_{m-1}].$$

Since

$$d\left(\frac{P(t, T_{m-1})}{P(t, T_m)}\right) = \delta \, dL(t, T_{m-1}) = \delta L(t, T_{m-1})\lambda(t, T_{m-1}) \, dW^{T_m}(t)$$

$$= \frac{P(t, T_{m-1})}{P(t, T_m)} \sigma_{T_{m-1}, T_m}(t) \, dW^{T_m}(t)$$

we get that

$$\frac{P(t, T_{m-1})}{P(t, T_m)} = \frac{P(0, T_{m-1})}{P(0, T_m)} \mathcal{E}_t \left(\sigma_{T_{m-1}, T_m} \bullet W^{T_m} \right), \quad t \in [0, T_{m-1}], \tag{11.5}$$

which is a $\mathbb{Q}^{T_m}$-martingale.

We now extend this further and define the T_m-forward price processes for all T_k-bonds via

$$\frac{P(t, T_k)}{P(t, T_m)} = \begin{cases} \frac{P(t, T_k)}{P(t, T_{k+1})} \cdots \frac{P(t, T_{m-1})}{P(t, T_m)}, & k < m, \\ \left(\frac{P(t, T_m)}{P(t, T_{m+1})} \right)^{-1} \cdots \left(\frac{P(t, T_{k-1})}{P(t, T_k)} \right)^{-1}, & k > m, \end{cases}$$

for $t \in [0, T_k \wedge T_m]$. The following result is, formally, in accordance with Lemma 7.1.

Lemma 11.2 *For every $1 \le k \ne m \le M$, the forward price process satisfies*

$$\frac{P(t, T_k)}{P(t, T_m)} = \frac{P(0, T_k)}{P(0, T_m)} \mathcal{E}_t \left(\sigma_{T_k, T_m} \bullet W^{T_m} \right), \quad t \in [0, T_k \wedge T_m],$$

for the $\mathbb{R}^d$-valued bounded progressive process

$$\sigma_{T_k, T_m} = \begin{cases} \sum_{l=k}^{m-1} \sigma_{T_l, T_{l+1}}, & k < m, \\ -\sum_{l=m}^{k-1} \sigma_{T_l, T_{l+1}}, & k > m. \end{cases} \tag{11.6}$$

Hence $\frac{P(t, T_k)}{P(t, T_m)}, t \in [0, T_k \wedge T_m]$, is a positive $\mathbb{Q}^{T_m}$-martingale.

Proof Suppose first $k < m$. Then (11.5) and (11.4) imply

$$\frac{P(t, T_k)}{P(t, T_m)} = \prod_{l=k}^{m-1} \frac{P(t, T_l)}{P(t, T_{l+1})} = \frac{P(0, T_k)}{P(0, T_m)} \prod_{l=k}^{m-1} \mathcal{E}_t \left(\sigma_{T_l, T_{l+1}} \bullet W^{T_{l+1}} \right)$$

$$= \frac{P(0, T_k)}{P(0, T_m)} \exp \left[\int_0^t \sum_{l=k}^{m-1} \sigma_{T_l, T_{l+1}}(s) \left(dW^{T_m}(s) - \sum_{i=l+1}^{m-1} \sigma_{T_i, T_{i+1}}(s)^\top ds \right) \right.$$

$$\left. - \frac{1}{2} \int_0^t \sum_{l=k}^{m-1} \| \sigma_{T_l, T_{l+1}}(s) \|^2 ds \right]$$

$$= \frac{P(0, T_k)}{P(0, T_m)} \exp \left[\int_0^t \sum_{l=k}^{m-1} \sigma_{T_l, T_{l+1}}(s) dW^{T_m}(s) \right.$$

$$\left. - \frac{1}{2} \int_0^t \left\| \sum_{l=k}^{m-1} \sigma_{T_l, T_{l+1}}(s) \right\|^2 ds \right].$$

Hence

$$\frac{P(t, T_k)}{P(t, T_m)} = \frac{P(0, T_k)}{P(0, T_m)} \mathcal{E}_t \left(\left(\sum_{l=k}^{m-1} \sigma_{T_l, T_{l+1}} \right) \bullet W^{T_m} \right),$$

as desired. The case $k > m$ follows by similar argumentation ($\rightarrow$ Exercise 11.3). $\square$

From the above we can derive, for $0 \leq m < n \leq M$, the nominal T_n-bond prices

$$P(T_m, T_n) = \prod_{k=m+1}^{n} \frac{P(T_m, T_k)}{P(T_m, T_{k-1})} = \prod_{k=m+1}^{n} \frac{1}{\delta L(T_m, T_{k-1}) + 1}$$

at dates $t = T_m$. However, it is not possible to uniquely determine the continuous time dynamics of the bond price $P(t, T_n)$ in the discrete-tenor model of forward LIBOR rates. The knowledge of forward LIBOR rates for all maturities $T \in [0, T_{M-1}]$ would be necessary. This will be tackled in Sect. 11.8 below.

Notwithstanding, we have defined an arbitrage-free market model for the bonds with maturities $T_0, \ldots, T_M$, since we have shown that $\mathbb{Q}^{T_m}$ is a martingale measure for the T_m-bond as numeraire. In view of Proposition 7.1, any T_m-contingent claim X with $\mathbb{E}_{\mathbb{Q}^{T_m}}[|X|] < \infty$ can thus consistently be priced at $t \leq T_m$ in terms of the T_m-bond as numeraire via

$$\frac{\pi(t)}{P(t, T_m)} = \mathbb{E}_{\mathbb{Q}^{T_m}}[X \mid \mathcal{F}_t].$$

We can express this price relative to any future T_n-bond, as the following lemma indicates.

Lemma 11.3 *The T_m-bond discounted T_m-contingent claim price satisfies*

$$\frac{\pi(t)}{P(t, T_m)} = \frac{P(t, T_n)}{P(t, T_m)} \mathbb{E}_{\mathbb{Q}^{T_n}} \left[\frac{X}{P(T_m, T_n)} \,\Big|\, \mathcal{F}_t \right],$$

for all $m < n \leq M$.

Proof Notice that, by Lemma 11.2,

$$\frac{d\mathbb{Q}^{T_k}}{d\mathbb{Q}^{T_{k+1}}} \bigg|_{\mathcal{F}_t} = \mathcal{E}_t \left(\sigma_{T_k, T_{k+1}} \bullet W^{T_{k+1}} \right) = \frac{P(0, T_{k+1})}{P(0, T_k)} \frac{P(t, T_k)}{P(t, T_{k+1})}, \quad t \in [0, T_k].$$

Hence

$$\frac{d\mathbb{Q}^{T_m}}{d\mathbb{Q}^{T_n}} \bigg|_{\mathcal{F}_t} = \prod_{k=m}^{n-1} \frac{d\mathbb{Q}^{T_k}}{d\mathbb{Q}^{T_{k+1}}} \bigg|_{\mathcal{F}_t} = \prod_{k=m}^{n-1} \frac{P(0, T_{k+1})}{P(0, T_k)} \frac{P(t, T_k)}{P(t, T_{k+1})}$$

$$= \frac{P(0, T_n)}{P(0, T_m)} \frac{P(t, T_m)}{P(t, T_n)}.$$

Bayes' rule now yields the assertion. $\square$

As a corollary, we may now restate the caplet pricing formula derived in Sect. 11.1.

Corollary 11.1 *Let $m + 1 < n \leq M$. The time T_m price of the nth caplet with reset date T_{n-1}, settlement date T_n and strike rate κ is*

$$Cpl(T_m; T_{n-1}, T_n) = \delta P(T_m, T_n) \left(L(T_m, T_{n-1}) \Phi(d_1(n; T_m)) - \kappa \Phi(d_2(n; T_m)) \right),$$

where

$$d_{1,2}(n; T_m) = \frac{\log(\frac{L(T_m, T_{n-1})}{\kappa}) \pm \frac{1}{2} \int_{T_m}^{T_{n-1}} \|\lambda(s, T_{n-1})\|^2 \, ds}{(\int_{T_m}^{T_{n-1}} \|\lambda(s, T_{n-1})\|^2 \, ds)^{\frac{1}{2}}},$$

and Φ is the standard Gaussian cumulative distribution function.

This is exactly Black's formula (2.6) for the caplet price with

$$\sigma(T_m)^2 = \frac{1}{T_{n-1} - T_m} \int_{T_m}^{T_{n-1}} \|\lambda(s, T_{n-1})\|^2 \, ds.$$

11.4 Implied Money-Market Account

Given the LIBOR $L(T_{i-1}, T_{i-1})$ for period $[T_{i-1}, T_i]$, for all $i = 1, \ldots, M$, we can define the discrete-time implied money-market account process

$$B^*(0) = 1,$$
$$B^*(T_m) = (1 + \delta L(T_{m-1}, T_{m-1})) B^*(T_{m-1}), \quad m = 1, \ldots, M,$$

that is,

$$B^*(T_n) = B^*(T_m) \prod_{k=m}^{n-1} \frac{1}{P(T_k, T_{k+1})}, \quad m < n \leq M.$$

Hence $B^*(T_m)$ can be interpreted as the cash amount accumulated up to time T_m by rolling over a series of zero-coupon bonds with the shortest maturities available.

By construction, B^* is an nondecreasing and progressive process with respect to the discrete-time filtration $(\mathcal{F}_{T_m})$, that is,

$$B^*(T_m) \text{ is } \mathcal{F}_{T_{m-1}}\text{-measurable}, \quad \text{for all } m = 1, \ldots, M.$$

Define the right-continuous[1] integer-valued function η on $[0, T_{M-1}]$ by

$$T_{\eta(t)-1} \leq t < T_{\eta(t)}, \quad t \geq 0. \tag{11.7}$$

---

[1] We could as well define η by $T_{\eta(t)-1} < t \leq T_{\eta(t)}$ to be left-continuous. But for the discrete approximation in (11.11) below this would make a difference.

In line with Lemma 11.3 one can show the following result.

Lemma 11.4 *For all $t \in [0, T_{M-1}]$ we have*

$$\mathbb{E}_{\mathbb{Q}^{T_M}} \left[B^*(T_M) P(0, T_M) \mid \mathcal{F}_t \right] = \mathcal{E}_t \left(\sigma_{T_0, T_{M-1}} \bullet W^{T_M} \right),$$

where we define, in accordance with (11.6), the $\mathbb{R}^d$-valued bounded progressive process

$$\sigma_{T_0, T_{M-1}}(t) = \sum_{k=\eta(t)}^{M-1} \sigma_{T_k, T_{k+1}}(t).$$

In particular, for all $0 \le m \le M$ we have

$$\mathbb{E}_{\mathbb{Q}^{T_M}} \left[B^*(T_M) \mid \mathcal{F}_{T_m} \right] = \frac{B^*(T_m)}{P(T_m, T_M)}.$$

Proof Follows from Lemma 11.2 ($\rightarrow$ Exercise 11.4). $\qquad\square$

In view of Lemma 11.4, we can now define the equivalent probability measure $\mathbb{Q}^* \sim \mathbb{Q}^{T_M}$ on $\mathcal{F}_{T_M}$ by

$$\frac{d\mathbb{Q}^*}{d\mathbb{Q}^{T_M}} = B^*(T_M) P(0, T_M).$$

Moreover, we have

$$\frac{d\mathbb{Q}^*}{d\mathbb{Q}^{T_M}} \bigg|_{\mathcal{F}_t} = \begin{cases} \mathcal{E}_t(\sigma_{T_0, T_{M-1}} \bullet W^{T_M}), & t \in [0, T_{M-1}], \\ B^*(T_m) \frac{P(0, T_M)}{P(T_m, T_M)}, & \text{if } t = T_m, \text{ for } m \le M - 1, \text{ in particular.} \end{cases}$$

Hence, on the one hand, $\mathbb{Q}^*$ can be interpreted as risk-neutral martingale measure. Following Jamshidian [102], it is also called the "spot LIBOR measure". Indeed, an application of Bayes' rule implies the following result ($\rightarrow$ Exercise 11.5).

Lemma 11.5 *The time T_k price of the T_m-contingent claim X from Lemma 11.3 satisfies*

$$\pi(T_k) = B^*(T_k) \mathbb{E}_{\mathbb{Q}^*} \left[\frac{X}{B^*(T_m)} \bigg| \mathcal{F}_{T_k} \right],$$

for all $k \le m$.

Since $\pi(T_k) = P(T_k, T_m)$ for $X = 1$, this implies that for any $0 \le m \le M$ the discrete-time process

$$\left(\frac{P(T_k, T_m)}{B^*(T_k)} \right)_{k=0,\dots,m}$$

is a $\mathbb{Q}^*$-martingale with respect to $(\mathcal{F}_{T_k})$. We have thus constructed a full discrete-time interest rate model.

On the other hand, Girsanov's theorem tells us that

$$W^*(t) = W^{T_M}(t) - \int_0^t \sigma_{T_0, T_{M-1}}(s)\, ds, \quad t \in [0, T_{M-1}],$$

is a $\mathbb{Q}^*$-Brownian motion. Thus we find the following useful extension of Lemma 11.1:

Lemma 11.6 *Let $0 \le m \le M - 1$. Then the dynamics of $L(t, T_m)$ under $\mathbb{Q}^*$ is given according to*

$$\frac{dL(t, T_m)}{L(t, T_m)} = \lambda(t, T_m) \sum_{k=\eta(t)}^m \sigma_{T_k, T_{k+1}}(t)^\top dt + \lambda(t, T_m)\, dW^*(t), \quad t \in [0, T_m].$$

Proof Follows from Lemma 11.1 for $n = M - 1$ and the definition of W^*. $\qquad\square$

11.5 Swaption Pricing

Consider a payer swaption with nominal 1, strike rate K, maturity T_μ and underlying tenor $T_\mu, T_{\mu+1}, \ldots, T_\nu$ (T_μ is the first reset date and T_ν the maturity of the underlying swap), for some $\mu < \nu \le M$. We recall from (2.7) that its payoff at maturity T_μ is

$$\delta \left(\sum_{m=\mu+1}^\nu P(T_\mu, T_m)(L(T_\mu, T_{m-1}) - K) \right)^+.$$

In view of Lemmas 11.3 and 11.5, the swaption price at $t = 0$ therefore is

$$\pi = \delta P(0, T_\mu) \mathbb{E}_{\mathbb{Q}^{T_\mu}} \left[\left(\sum_{m=\mu+1}^\nu P(T_\mu, T_m)(L(T_\mu, T_{m-1}) - K) \right)^+ \right]$$

$$= \delta \mathbb{E}_{\mathbb{Q}^*} \left[\frac{1}{B^*(T_\mu)} \left(\sum_{m=\mu+1}^\nu P(T_\mu, T_m)(L(T_\mu, T_{m-1}) - K) \right)^+ \right].$$

To compute π we thus need to know the joint distribution of

$$L(T_\mu, T_\mu), \ L(T_\mu, T_{\mu+1}), \ldots, L(T_\mu, T_{\nu-1})$$

under either forward measure $\mathbb{Q}^{T_\mu}$ or $\mathbb{Q}^*$. It turns out that this cannot be done exactly analytically in the context of the LIBOR market model.

11.5.1 Forward Swap Measure

We now describe a pricing attempt via a suitable change of numeraire. We consider the above payer swap. The corresponding forward swap rate at time $t \leq T_\mu$ is

$$R_{swap}(t) = \frac{P(t, T_\mu) - P(t, T_\nu)}{\delta \sum_{k=\mu+1}^{\nu} P(t, T_k)} = \frac{1 - \frac{P(t, T_\nu)}{P(t, T_\mu)}}{\delta \sum_{k=\mu+1}^{\nu} \frac{P(t, T_k)}{P(t, T_\mu)}}. \tag{11.8}$$

In view of Lemma 11.2, $R_{swap}(t)$ is thus given in terms of the above-constructed LIBOR rates.

Define the positive $\mathbb{Q}^{T_\mu}$-martingale

$$D(t) = \sum_{k=\mu+1}^{\nu} \frac{P(t, T_k)}{P(t, T_\mu)}, \quad t \in [0, T_\mu].$$

This induces an equivalent probability measure $\mathbb{Q}^{swap} \sim \mathbb{Q}^{T_\mu}$, the *forward swap measure*, on $\mathcal{F}_{T_\mu}$ by

$$\frac{d\mathbb{Q}^{swap}}{d\mathbb{Q}^{T_\mu}} = \frac{D(T_\mu)}{D(0)}.$$

Lemma 11.7 *The forward swap rate process $R_{swap}(t)$, $t \in [0, T_\mu]$, is a positive $\mathbb{Q}^{swap}$-martingale.*

Moreover, there exists some d-dimensional $\mathbb{Q}^{swap}$-Brownian motion W^{swap} and an $\mathbb{R}^d$-valued progressive swap volatility process ρ^{swap} such that

$$dR_{swap}(t) = R_{swap}(t)\rho^{swap}(t)\, dW^{swap}(t), \quad t \in [0, T_\mu].$$

Proof Let $0 \leq m \leq M$ and $0 \leq s \leq t \leq T_m \wedge T_\mu$. Then

$$\mathbb{E}_{\mathbb{Q}^{swap}}\left[\frac{P(t, T_m)}{P(t, T_\mu)D(t)}\bigg|\mathcal{F}_s\right] = \frac{1}{D(s)}\mathbb{E}_{\mathbb{Q}^{T_\mu}}\left[\frac{P(t, T_m)}{P(t, T_\mu)D(t)}D(t)\bigg|\mathcal{F}_s\right]$$

$$= \frac{1}{D(s)}\frac{P(s, T_m)}{P(s, T_\mu)}.$$

On the other hand, (11.8) implies

$$R_{swap}(t) = \frac{1}{\delta D(t)} - \frac{P(t, T_\nu)}{\delta P(t, T_\mu)D(t)}.$$

Hence $R_{swap}(t)$ is a positive $\mathbb{Q}^{swap}$-martingale. The representation of $R_{swap}(t)$ in terms of W^{swap} and ρ^{swap} follows from Lemma 11.2 and Girsanov's theorem. $\square$

Recall from (2.8) that the payoff at maturity of the above swaption can be written as

$$\delta D(T_\mu)\left(R_{swap}(T_\mu) - K\right)^+.$$

Hence the price equals

$$\pi = \delta P(0, T_\mu) \mathbb{E}_{\mathbb{Q}^{T_\mu}} \left[D(T_\mu) \left(R_{swap}(T_\mu) - K \right)^+ \right]$$

$$= \delta P(0, T_\mu) D(0) \mathbb{E}_{\mathbb{Q}^{swap}} \left[\left(R_{swap}(T_\mu) - K \right)^+ \right]$$

$$= \delta \sum_{k=\mu+1}^{\nu} P(0, T_k) \mathbb{E}_{\mathbb{Q}^{swap}} \left[\left(R_{swap}(T_\mu) - K \right)^+ \right].$$

Under the hypothesis:

(H) $\rho^{swap}(t)$ is deterministic,

we would have that $\log R_{swap}(T_\mu)$ is Gaussian distributed under $\mathbb{Q}^{swap}$ with mean

$$\log R_{swap}(0) - \frac{1}{2} \int_0^{T_\mu} \|\rho^{swap}(t)\|^2 \, dt$$

and variance

$$\int_0^{T_\mu} \|\rho^{swap}(t)\|^2 \, dt.$$

Hence the swaption price would then be

$$\pi = \delta \sum_{k=\mu+1}^{\nu} P(0, T_k) \left(R_{swap}(0)\Phi(d_1) - K\Phi(d_2) \right),$$

with

$$d_{1,2} = \frac{\log(\frac{R_{swap}(0)}{K}) \pm \frac{1}{2} \int_0^{T_\mu} \|\rho^{swap}(t)\|^2 \, dt}{(\int_0^{T_\mu} \|\rho^{swap}(t)\|^2 \, dt)^{\frac{1}{2}}}.$$

This is Black's formula (2.9) with volatility σ^2 given by

$$\frac{1}{T_\mu} \int_0^{T_\mu} \|\rho^{swap}(t)\|^2 \, dt.$$

However, it can be shown that ρ^{swap} cannot be deterministic, and hence hypothesis **(H)** does not hold, in our lognormal LIBOR setup. For swaption pricing it would be natural to model the forward swap rates directly and postulate that they are lognormal under the forward swap measures (the so-called swap market model). This approach has been carried out by Jamshidian [102], and computationally improved by Pelsser [131]. It could be shown, however, that then the forward LIBOR rate volatility cannot be deterministic. So either one gets Black's formula for caps or for swaptions, but not simultaneously for both. Put in other words, when we insist on lognormal forward LIBOR rates then swaption prices have to be approximated. One possibility is to use Monte Carlo methods, as outlined below. Another approach is via analytic approximation, which we now sketch in the following section.

11.5.2 Analytic Approximations

We have seen in Sect. 2.4.3 that the forward swap rate can be written as weighted sum of forward LIBOR rates

$$R_{swap}(t) = \sum_{m=\mu+1}^{\nu} w_m(t) L(t, T_{m-1}),$$

with weights

$$w_m(t) = \frac{P(t, T_m)}{D(t) P(t, T_\mu)} = \frac{\frac{1}{1+\delta L(t,T_\mu)} \cdots \frac{1}{1+\delta L(t,T_{m-1})}}{\sum_{k=\mu+1}^{\nu} \frac{1}{1+\delta L(t,T_\mu)} \cdots \frac{1}{1+\delta L(t,T_{k-1})}}.$$

According to empirical studies, the variability of the w_m's is small compared to the variability of the forward LIBOR rates. We thus approximate $w_m(t)$ by its deterministic initial value $w_m(0)$, so that

$$R_{swap}(t) \approx \sum_{m=\mu+1}^{\nu} w_m(0) L(t, T_{m-1}),$$

and hence, under the T_μ-forward measure $\mathbb{Q}^{T_\mu}$

$$dR_{swap}(t) \approx (\cdots) dt + \sum_{m=\mu+1}^{\nu} w_m(0) L(t, T_{m-1}) \lambda(t, T_{m-1}) dW^{T_\mu}, \quad t \in [0, T_\mu],$$

for some appropriate drift term. We obtain that the forward swap volatility satisfies

$$\|\rho^{swap}(t)\|^2 = \frac{d \langle \log R_{swap}, \log R_{swap} \rangle_t}{dt}$$

$$\approx \sum_{k,l=\mu+1}^{\nu} \frac{w_k(0) w_l(0) L(t, T_{k-1}) L(t, T_{l-1}) \lambda(t, T_{k-1}) \lambda(t, T_{l-1})^\top}{R_{swap}^2(t)}.$$

In a further approximation we replace all random variables by their time 0 values, such that the quadratic variation of $\log R_{swap}(t)$ becomes approximatively deterministic

$$\|\rho^{swap}(t)\|^2 \approx \sum_{k,l=\mu+1}^{\nu} \frac{w_k(0) w_l(0) L(0, T_{k-1}) L(0, T_{l-1}) \lambda(t, T_{k-1}) \lambda(t, T_{l-1})^\top}{R_{swap}^2(0)}.$$

Denote the square root of the right-hand side by $\tilde{\rho}^{swap}(t)$. By Lévy's characterization theorem, the following is a $\mathbb{Q}^{swap}$-Brownian motion:

$$\mathcal{W}(t) = \int_0^t \sum_{j=1}^{d} \frac{\rho_j^{swap}(s)}{\|\rho^{swap}(s)\|} dW_j^{swap}(s), \quad t \in [0, T_\mu].$$

We then have

$$dR_{swap}(t) = R_{swap}(t) \|\rho^{swap}(t)\| \, dW(t)$$

$$\approx R_{swap}(t) \tilde{\rho}^{swap}(t) \, dW(t).$$

Hence we can approximate the swaption price in our lognormal forward LIBOR model by Black's swaption price formula (2.9) where σ^2 is to be replaced by

$$\frac{1}{T_\mu} \int_0^{T_\mu} \sum_{k,l=\mu+1}^{\nu} \frac{w_k(0)w_l(0)L(0, T_{k-1})L(0, T_{l-1})\lambda(t, T_{k-1})\lambda(t, T_{l-1})^\top}{R_{swap}^2(0)} \, dt.$$

$$(11.9)$$

This is "Rebonato's formula", since it originally appears in his book [134]. The goodness of this approximation has been numerically tested by several authors, see [27, Chap. 8]. They conclude that "the approximation is satisfactory in general".

11.6 Monte Carlo Simulation of the LIBOR Market Model

As we have seen in the swaption case above, pricing the claims in Lemmas 11.3 and 11.5 typically requires Monte Carlo simulation. There are countless ways to simulate forward LIBOR rates. We will sketch here a particular Euler scheme, and refer to Glasserman [79] for a thorough discussion of the topic.

Let us focus on the risk-neutral martingale measure $\mathbb{Q}^*$, albeit the following can be carried out under any forward measure $\mathbb{Q}^{T_n}$. We aim at simulating the entire M-vector of forward LIBOR rates $(L(t, T_0), \ldots, L(t, T_{M-1}))^\top$. But instead of discretizing the system of stochastic differential equations in Lemma 11.6, we consider the transforms $H_m(t) = \log L(t, T_m)$. An application of Itô's formula and inserting the definition of $\sigma_{T_k, T_{k+1}}(t)$ implies

$$dH_m(t) = \alpha_m(t) \, dt + \lambda(t, T_m) \, dW^*(t), \quad t \leq T_m \qquad (11.10)$$

with the respective drift term

$$\alpha_m(t) = \lambda(t, T_m) \sum_{k=\eta(t)}^{m} \frac{\delta e^{H_k(t)}}{1 + \delta e^{H_k(t)}} \lambda(t, T_k)^\top - \frac{1}{2} \|\lambda(t, T_m)\|^2.$$

Simulating the logarithm of $L(t, T_m)$ has the advantage that it keeps the simulated rate $L(t, T_m)$ nonnegative. Moreover, H_m has Gaussian increments, a fact which improves the convergence of the Euler scheme.

Now suppose we want to price a T_n-claim with payoff of the form

$$f(H_n(T_n), \ldots, H_{M-1}(T_n)).$$

According to Lemma 11.5, this price at $t = 0$ is given by

$$\pi = \mathbb{E}_{\mathbb{Q}^*}\left[\frac{f\left(H_n(T_n), \ldots, H_{M-1}(T_n)\right)}{B^*(T_n)}\right].$$

Let us fix a time grid $t_i = i\,\Delta t$, $i = 0, \ldots, N$, with $\Delta t = T_n/N$ for N large enough over which to simulate. The corresponding Euler approximation of (11.10) is[2]

$$H_m(t_i) = H_m(t_{i-1}) + \alpha_m(t_{i-1})\,\Delta t + \lambda(t_{i-1}, T_m)\,Z(i)\,\sqrt{\Delta t}, \quad 1 \le i \le N, \quad (11.11)$$

where $Z(1), \ldots, Z(N)$ is a sequence of independent standard normal random vectors in $\mathbb{R}^d$.

The principle of Monte Carlo is to simulate via the Euler scheme (11.11) a number K of independent copies $\Pi^{(1)}, \ldots, \Pi^{(K)}$ of the random variable $\Pi = \frac{f(H_n(T_n), \ldots, H_{M-1}(T_n))}{B^*(T_n)}$, and then to estimate π via averaging

$$\overline{\Pi} = \frac{1}{K} \sum_{j=1}^{K} \Pi^{(j)}.$$

Three considerations are important for the efficiency of this simulation estimator: bias, variance, and computing time. This will now briefly be discussed.

First, a bias is introduced via the Euler approximation (11.11). It means that $\mathbb{E}_{\mathbb{Q}^*}[\overline{\Pi}]$ differs from its target value $\pi = \mathbb{E}_{\mathbb{Q}^*}[\Pi]$. The bias can obviously be reduced by increasing the number of time discretization steps N. In our example the bias is already negligible for $\Delta t = 1/12$ so that we can assume that $\mathbb{E}_{\mathbb{Q}^*}[\overline{\Pi}] \approx \pi$.

Second, the central limit theorem asserts that as the number of replications K increases, the simulation estimation error $\overline{\Pi} - \pi$ is approximately normal distributed with mean zero and approximate standard deviation of

$$s_\pi = \sqrt{\frac{\sum_{j=1}^{K}(\Pi^{(j)} - \overline{\Pi})}{K(K-1)}}.$$

The number s_π is also called the standard error of the Monte Carlo simulation. It means that $\overline{\Pi} \pm s_\pi$ is an asymptotically (as $K \to \infty$) valid 68% confidence interval for the true value π.

Third, there is an obvious trade-off between bias and variance for a given computing capacity, which has to be carefully balanced in general. A more thorough treatment of Monte Carlo is beyond the scope of this book. The interested reader is referred to [79]. This reference also treats bias and variance reduction techniques.

---

[2]Following Glasserman [79] we note that for the discrete approximation it makes a difference whether we define η right- or left-continuous, see (11.7). Indeed, if $t_{i-1} = T_l$ then $\eta(t_{i-1}) = l + 1$ and the sum in $\alpha_m(t_{i-1})$ starts at $k = l+1$. For the left-continuous specification of η the sum would start at $k = l$ and we would thus have an additional term in $\alpha_m(t_{i-1})$. Glasserman and Zhao [81] and Sidenius [150] both find that taking η right-continuous results in a smaller discretization error.

11.7 Volatility Structure and Calibration

So far, we have taken the volatility factors $\lambda(t, T_m)$ as given deterministic functions without indicating how they might be specified. In practice, these factors are chosen to calibrate the model to market prices of liquidly traded derivatives, such as caps and swaptions, or to historical time series. Note that the model is automatically calibrated to the initial bond prices via (11.3).

There are countless possible specifications of the volatility structure. Volatility estimation is a huge topic and a thorough discussion is beyond the scope of this book. The interested reader is referred to Brigo and Mercurio [27] and references therein. In this section, we will briefly discuss two approaches: historical volatility estimation via principal component analysis (PCA), and volatility calibration to market quotes of caps and swaptions.

11.7.1 Principal Component Analysis

The basic PCA approach, as outlined in Sect. 3.4, would roughly work as follows. Assume that $\lambda(t, T_m) = \lambda(T_m - t)$ is a function of time to maturity $T_m - t$, for all m. Suppose we have N observations $x(1), \ldots, x(N)$ of the random vectors

$$X(i) = (X_1(i), \ldots, X_M(i))^\top, \quad 1 \le i \le N,$$

where

$$X_m(i) = \log L(i\delta, (i+m-1)\delta) - \log L((i-1)\delta, (i+m-1)\delta), \quad 1 \le m \le M.$$

The Euler approximation (11.11) then yields

$$X_m(i) \approx \lambda(m\delta)\, Z(i)\, \sqrt{\delta},$$

where we neglected the drift term for simplicity.[3] Hence $X(i)$ are approximately independent identically distributed random vectors with zero mean. The PCA decomposition (3.15) of x then takes the form

$$x(i) = \hat{\mu} + \sum_{j=1}^{M} \hat{a}_j\, y_j(i) \approx \sum_{j=1}^{M} \hat{a}_j\, y_j(i)$$

with loadings $\hat{a}_j$ and principal components y_j in nonincreasing order

$$\mathrm{Var}[y_1] \ge \mathrm{Var}[y_2] \ge \cdots.$$

---

[3] Since the observations are made under the real-world measure $\mathbb{P}$, the drift term is of the order $\delta \, \|\lambda(m\delta)\| \times \max\{\|\lambda(m\delta)\|, \text{market price of risk}\}$.

We thus obtain the estimate for the functions $\lambda = (\lambda_1, \dots, \lambda_d)$

$$\lambda_j(m\delta) = \sqrt{\frac{\text{Var}[y_j]}{\delta}}\, \hat{a}_{jm}, \quad 1 \leq m \leq M.$$

It is a stylized fact that the first two to three principal components y_j are enough to explain most of the variance of x. The first three loadings $\hat{a}_j$, and thus the volatility curves $s \mapsto \lambda_j(s)$, are typically of the form as in Fig. 3.13: flat, upward (or downward) sloping, and hump-shaped.

11.7.2 Calibration to Market Quotes

As for volatility calibration to market quotes of caps and swaptions, we may adhere to the following facts. It becomes evident from Corollary 11.1 that calibrating to caplet prices constrains the norm $\|\lambda(t, T_m)\|$ of the d-vector $\lambda(t, T_m)$ only. There is no gain in flexibility in matching caplet implied volatilities by taking the number d of driving Brownian motions greater than one. The potential value of a multi-factor model lies in capturing correlations between forward LIBOR rates of different maturities. From the Euler approximation (11.11), we see that the instantaneous correlation between the increments of $\log L(t, T_m)$ and $\log L(t, T_n)$ is approximately

$$\rho_{mn}(t) = \frac{\lambda(t, T_m)\lambda(t, T_n)^{\top}}{\|\lambda(t, T_m)\|\|\lambda(t, T_n)\|}.$$

These correlations are often chosen to match market quotes of swaptions, which are sensitive to correlation as we have seen in Sect. 11.5, or historical correlations as indicated in the PCA above.

We formalize this dual aspect of volatility vs. correlation by writing

$$\lambda(t, T_m) = \sigma_m(t)\, \ell_m(t)$$

where $\sigma_m(t) = \|\lambda(t, T_m)\|$ is the volatility of $L(t, T_m)$, which is calibrated to caplet prices, and the row vector $\ell_m(t) = \lambda(t, T_m)/\|\lambda(t, T_m)\|$ captures the correlation between the different rates: $\rho_{mn}(t) = \ell_m(t)\, \ell_n(t)^{\top}$.

For further illustration, suppose that we are given the market quotes of all caplets $Cpl(T_{n-1}, T_n) = Cpl(0; T_{n-1}, T_n)$ in Corollary 11.1 in terms of their respective implied volatilities $\sigma_{Cpl(T_{n-1}, T_n)}$. We thus obtain

$$\int_0^{T_n} \sigma_n(t)^2\, dt = \sigma^2_{Cpl(T_n, T_{n+1})}\, T_n, \quad 1 \leq n \leq M - 1.$$

If we assume as in the PCA above that $\sigma_m(t) = \sigma(T_m - t)$ is a function of time to maturity, for all m, we infer

$$\int_{T_{n-1}}^{T_n} \sigma(t)^2\, dt = \begin{cases} \sigma^2_{Cpl(T_1, T_2)}\, T_1, & n = 1, \\ \sigma^2_{Cpl(T_n, T_{n+1})}\, T_n - \sigma^2_{Cpl(T_{n-1}, T_n)}\, T_{n-1}, & 2 \leq n \leq M - 1. \end{cases}$$

However, this specification bears some inconsistencies as it requires $\sigma^2_{Cpl(T_n,T_{n+1})} T_n$
$\geq \sigma^2_{Cpl(T_{n-1},T_n)} T_{n-1}$, which is not satisfied by the caplet data in general.

A simple and consistent alternative specification is to let $\sigma_m(t) \equiv \sigma_m$ be independent of t, for all m. In this case, the calibration is easy as

$$\sigma_n = \sigma_{Cpl(T_n,T_{n+1})}, \quad 1 \leq n \leq M - 1$$

However, this specification stipulates that the volatility does not change over time, which is not plausible for long-matured forward LIBOR rates.

A reasonable and tractable alternative are parametric forms. For illustration, we consider[4]

$$\sigma_m(t) = v_m e^{-\beta(T_m-t)} \tag{11.12}$$

for some common exponent β and individual factors v_m. Strictly speaking, this specification is under-determined as the system

$$v_n^2 \frac{1 - e^{-2\beta T_n}}{2\beta} = \sigma^2_{Cpl(T_n,T_{n+1})} T_n, \quad 1 \leq n \leq M - 1, \tag{11.13}$$

leaves us with one degree of freedom, which has to be fixed by some additional data point. For $\beta = 0$ we obtain back the constant volatility case from above. For a more systematic classification of admissible volatility specifications the reader is referred to [27, Sect. 6.3].

When it comes to the calibration of $\ell_m(t)$, one often assumes that $\ell_m(t) \equiv \ell_m$, and thus the correlation matrix $\rho_{mn} = \ell_m \ell_n^\top$, does not depend on t. For instance this is the framework in [27, Sect. 6.3]. The analytic approximation formula (11.9) for the implied swaption volatility σ^2_{swp} then reads

$$\sigma^2_{swp} T_\mu \approx \sum_{k,l=\mu}^{\nu-1} \frac{w_{k+1}(0) w_{l+1}(0) L(0, T_k) L(0, T_l) \int_0^{T_\mu} \sigma_k(t) \sigma_l(t)\, dt}{R^2_{swap}(0)} \rho_{k,l}. \tag{11.14}$$

This approximation may be used to estimate the correlation matrix ρ if we take the initial term-structure $L(0, T_k)$ as given and σ_m as calibrated to the caplet quotes.[5] In the absence of closed-form formulas or approximations such as (11.9), calibration is an iterative procedure that requires repeated Monte Carlo simulations at various parameter values until the model price matches the market.

---

[4]This volatility specification is underlying the empirical study by De Jong et al. [51]. Their estimates for β are always positive.

[5]Note that there may be some subtle differences between cap and swaption markets to be taken into account when combining the cap and swaption calibrations. For instance, in the euro zone caplets are written on semiannual LIBOR ($\delta = 1/2$), while swaps pay annual coupons ($\delta = 1$). See the example following in the text below.

Fig. 11.1 Perfectly
correlated case: trajectories of
$L(\cdot, T)$ for $T = 2, 4.5, 7, 9.5$

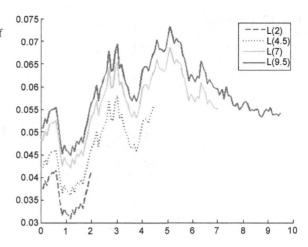

The specification of ℓ_m goes together with the choice of the dimension d of the driving Brownian motion. For the sake of illustration, we will confine ourselves to the following two extreme cases in what follows.[6]

- Specification I: one extreme case is $d = 1$ and $\ell_m = 1$, for all m. In this case, the instantaneous correlation between the increments of $\log L(t, T_m)$ and $\log L(t, T_n)$ is one, for all m, n. This situation is illustrated in Fig. 11.1.
- Specification II: the other extreme case is $d = M - 1$, where we have as many driving Brownian motions as forward LIBOR rates, and $\ell_m = e_m^\top$, for all m. In this case, ρ is the unit matrix, and the system of stochastic differential equations (11.10) becomes decoupled:

$$
dH_m(t) = \left(\frac{\delta e^{H_m(t)}}{1 + \delta e^{H_m(t)}} \sigma_m(t)^2 - \frac{1}{2}\sigma_m(t)^2 \right) dt + \sigma_m(t)\, dW_m(t), \quad t \le T_m,
$$

for all $1 \le m \le M - 1$. It is evident that the forward LIBOR rates are now independent. This is illustrated in Fig. 11.2.

A word on caplet quotes. The usual market quotes are on caps and floors, such as shown in Tables 11.1 and 11.2.[7] From these the caplet and floorlet volatilities have to be stripped. We illustrate a bootstrapping method analogous to the one in Sect. 3.1 for the term-structure estimation. The tenor is $T_i = i/2$, $i = 0, \ldots, 20$, where $T_1 = 1/2$ is the first caplet reset date and $T_{20} = 10$ the maturity of the last cap. The prevailing initial forward LIBOR curve is given in Table 11.3.

---

[6]In view of the stylized fact from Sect. 3.4.4, a reasonable specification of ρ_{mn} is exponentially decaying in $|T_m - T_n|$. That is, $\rho_{mn} = e^{-\gamma |T_m - T_n|}$ for some $\gamma \ge 0$. The extreme specifications I and II then correspond to $\gamma = 0$ and $\gamma = \infty$, respectively. See also Exercise 11.7.

[7]Tables 11.1 and 11.2 show prices in basis points. Often these prices have to be inferred from quoted implied volatilities, such as discussed in Sect. 2.6.

Fig. 11.2 Independent case: trajectories of $L(\cdot, T)$ for $T = 2, 4.5, 7, 9.5$

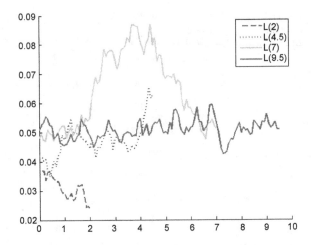

Table 11.1 Euro cap prices (in basis points) on 18 November 2008

T–K	3.5%	4%	4.5%	5%	5.5%	6%	6.5%	7%	7.5%
2	25.0	11.0	5.0	2.5	1.5	1.0	0.5	0.0	0.0
3	77.0	40.5	21.5	12.0	7.0	4.0	2.5	1.5	1.5
4	148.5	86.0	48.5	27.0	16.0	10.0	6.5	4.5	4.0
5	230.5	140.5	82.0	47.5	28.5	17.5	11.5	8.0	7.5
6	325.5	206.0	125.5	74.5	45.5	29.0	19.0	13.5	12.5
7	431.5	283.5	178.0	109.0	68.0	44.5	29.5	21.0	20.5
8	545.5	368.5	238.0	149.5	95.0	62.5	42.5	30.0	29.0
9	664.0	459.0	304.5	196.5	127.0	85.0	58.5	42.0	40.0
10	786.0	554.5	376.5	248.5	164.0	111.0	77.0	56.0	53.0

Table 11.2 Euro floor prices (in basis points) on 18 November 2008

T–K	3%	2.75%	2.5%	2.25%	2%	1.75%	1.5%	1.25%	1%
2	69.5	50.0	34.0	23.0	14.5	9.0	5.5	3.5	1.5
3	92.0	66.5	47.0	32.0	21.5	14.0	9.0	5.0	2.5
4	110.0	80.5	58.0	40.5	28.5	19.5	13.0	8.0	4.0
5	127.0	94.0	68.5	49.0	35.0	25.0	17.0	11.0	5.5
6	142.0	107.0	78.5	58.0	42.5	31.0	21.5	13.5	7.5
7	157.5	119.5	89.5	67.0	50.0	37.0	26.5	16.5	9.0
8	172.5	132.0	101.0	76.5	58.5	43.5	31.0	20.0	10.5
9	187.5	145.0	112.0	86.5	66.5	50.0	36.0	23.5	13.0
10	201.5	157.5	122.5	95.5	74.0	56.5	41.0	27.5	15.5

Table 11.3 Forward LIBOR curve (in %) on 18 November 2008

T_i	0	0.5	1	1.5	2	2.5	3	3.5	4	4.5
$L(0, T_i)$	4.228	2.791	3.067	3.067	3.728	3.728	4.051	4.051	4.199	4.199
T_i	5	5.5	6	6.5	7	7.5	8	8.5	9	9.5
$L(0, T_i)$	4.450	4.450	4.626	4.626	4.816	4.816	4.960	4.960	5.088	5.088

Table 11.4 Implied volatilities (in %) for caplets $Cpl(T_{i-1}, T_i)$ at strike rate 3.5%

i	1	2	3	4	5	6	7	8	9	10
$\sigma_{Cpl(T_{i-1}, T_i)}$	n/a	29.3	29.3	29.3	20.8	20.8	18.3	18.3	17.8	17.8
i	11	12	13	14	15	16	17	18	19	20
$\sigma_{Cpl(T_{i-1}, T_i)}$	16.3	16.3	16.7	16.7	16.1	16.1	15.7	15.7	15.7	15.7

Now consider, for instance, the cap at strike rate $K = 3.5\%$ and maturity in $T_4 = 2$ years. It is composed of the first three consecutive caplets $Cpl(T_{i-1}, T_i)$ with reset and settlement dates T_{i-1} and T_i, respectively, and strike rate 3.5%:

$$Cp(T_4) = Cpl(T_1, T_2) + Cpl(T_2, T_3) + Cpl(T_3, T_4).$$

From this we uniquely infer the caplet volatility

$$\sigma_{Cpl(T_1, T_2)} = \sigma_{Cpl(T_2, T_3)} = \sigma_{Cpl(T_3, T_4)} = 29.3\%$$

by inverting Black's formula. The next cap matures in $T_6 = 3$ years. We thus have

$$Cp(T_6) - Cp(T_4) = Cpl(T_4, T_5) + Cpl(T_5, T_6),$$

and again we can uniquely infer the implied caplet volatility

$$\sigma_{Cpl(T_4, T_5)} = \sigma_{Cpl(T_5, T_6)} = 20.8\%,$$

without altering the previous ones. Proceeding this way we arrive at the implied volatilities for all caplets at strike rate 3.5% shown in Table 11.4 ($\to$ Exercise 11.6).

As an application, we now calibrate the lognormal LIBOR market model to the 3.5% strike caplet data in Table 11.4 using the parametric specification (11.12) ($\to$ Exercise 11.7). From (11.13) we thus obtain the volatility parameters $v_1, \ldots, v_{19}$ as functions on β. Implementing the Monte Carlo algorithm described in Sect. 11.6, we recapture the original 3.5% cap prices from Table 11.1, independently of the choice of the correlation specification. In fact, we consider the two extreme specifications I ($d = 1$ and $\ell_m = 1$) and II ($d = M - 1$ and $\ell_m = e_m^\top$)

Fig. 11.3 The swaption price as function of β. The *straight horizontal line* indicates the real market quote of 248 bp. The *upper curves* are for the correlation specification I, the *lower curves* are for specification II. The *solid lines* show the Monte Carlo simulation based prices with standard errors indicated by the *dotted lines*. The *dashed lines* show the respective prices based on the approximation formula (11.14)

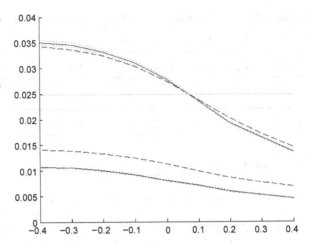

from above. We then price the at-the-money 4×6-swaption with maturity in 4 years and whose underlying swap is 6 years long. Its tenor is as follows: first reset date $T_8 = 4$, and annual(!) coupon payments at $T_{10} = 5, \ldots, T_{20} = 10$. As should have become clear by now, this price depends both on β and on the correlation specification. For comparison, we also compute the respective swaption prices based on the analytic approximation formula (11.14) for the implied swaption volatility. Figure 11.3 shows the results.

For correlation specification II there is an obvious diversification effect between the underlying LIBOR rates, which is due to their independence. This results in a lower aggregate volatility and thus a lower swaption price. It is interesting to note that the real market quote for this swaption was 248 bp. For correlation specification I, we obtain an estimate for β of approximately 0.07. But specification II could not calibrate to this data point. It seems that the lognormal LIBOR market model with specification II underprices swaption prices systematically. We also see that the approximation differs from the true values by order of 7 to 10 bp (specification I) and 20 to 35 bp (specification II), respectively. Of course, this simple example is merely of an indicative nature. More systematic tests for quality of the analytical approximation can be found in [27, Chap. 8].

Finally a word on the above caplet volatility calibration. The snag is that the implied volatilities in Table 11.4 will depend on the strike rate in general. Hence we can calibrate our lognormal LIBOR market model only to match the caplet prices at any maturity T_i for one strike rate at a time, no matter how many driving Brownian motions we use. Thus the situation is very much like in the Black–Scholes stock market model, which is incapable of fitting the market option prices across all strikes.[8] One way out is to let $\lambda(t, T_m) = \lambda(\omega, t, T_m)$ be a progressive process, analogous to Heston's generalization of the Black–Scholes model, and/or to replace

---

[8]Note that the Heston stochastic volatility model can produce volatility skews, see Fig. 10.4, and so does the simple affine stock model from Exercise 10.11.

the driving Brownian motion by some more general Lévy process. Much research has been done in this direction over the last ten years to achieve a good, possibly exact, fitting of market option data. The interested reader is referred to Brigo and Mercurio [27, Part IV] for a detailed overview.

11.8 Continuous-Tenor Case

In this section, we specify the continuum of all forward LIBOR rates $L(t, T)$, for $T \in [0, T_{M-1}]$. Given the discrete-tenor skeleton constructed in Sect. 11.2, it is enough to fill the gaps between the T_js. Each forward LIBOR rate $L(t, T)$ will follow a lognormal process under the forward measure for the date $T + \delta$.

The stochastic basis is the same as before, except that we now assume that $\mathcal{F}_t = \mathcal{F}_t^{W^{T_M}}$ is the filtration generated by the d-dimensional Brownian motion W^{T_M}, so that we can apply the representation theorem 4.8 for $\mathbb{Q}^{T_M}$-martingales. In addition, we now need a continuum of initial dates:

- for every $T \in [0, T_{M-1}]$, an $\mathbb{R}^d$-valued deterministic bounded measurable function $\lambda(t, T)$, $t \in [0, T]$, which represents the volatility of $L(t, T)$;
- a positive and nonincreasing initial term-structure

$$P(0, T), \quad T \in [0, T_M],$$

and hence a nonnegative initial forward LIBOR curve

$$L(0, T) = \frac{1}{\delta} \left(\frac{P(0, T)}{P(0, T + \delta)} - 1 \right), \quad T \in [0, T_{M-1}].$$

First, we construct a discrete-tenor model for $L(t, T_m)$, $m = 0, \ldots, M - 1$, as in the previous section.

Second, we focus on the forward measures for dates $T \in [T_{M-1}, T_M]$. We do not have to take into account forward LIBOR rates for these dates, since they are not defined there. However, we are given the values of the implied money-market account $B^*(T_{M-1})$ and $B^*(T_M)$ and the probability measure $\mathbb{Q}^*$. It satisfies

$$P(0, T_m) = \mathbb{E}_{\mathbb{Q}^*} \left[\frac{1}{B^*(T_m)} \right], \quad m \leq M.$$

By the monotonicity of $P(0, T)$, there exists a unique deterministic nondecreasing function

$$\alpha : [T_{M-1}, T_M] \to [0, 1]$$

with $\alpha(T_{M-1}) = 0$ and $\alpha(T_M) = 1$, such that

$$\log B^*(T) = (1 - \alpha(T)) \log B^*(T_{M-1}) + \alpha(T) \log B^*(T_M)$$

satisfies

$$P(0, T) = \mathbb{E}_{\mathbb{Q}^*}\left[\frac{1}{B^*(T)}\right], \quad T \in [T_{M-1}, T_M].$$

Let $T \in [T_{M-1}, T_M]$. Since $B^*(T)$ is $\mathcal{F}_T$-measurable and positive, and satisfies

$$\mathbb{E}_{\mathbb{Q}^*}\left[\frac{1}{B^*(T)P(0, T)}\right] = 1,$$

we can define the T-forward measure $\mathbb{Q}^T \sim \mathbb{Q}^*$ on $\mathcal{F}_T$ by

$$\frac{d\mathbb{Q}^T}{d\mathbb{Q}^*} = \frac{1}{B^*(T)P(0, T)}.$$

Then we have

$$\frac{d\mathbb{Q}^T}{d\mathbb{Q}^{T_M}} = \frac{d\mathbb{Q}^T}{d\mathbb{Q}^*}\frac{d\mathbb{Q}^*}{d\mathbb{Q}^{T_M}} = \frac{B^*(T_M)P(0, T_M)}{B^*(T)P(0, T)}.$$

By the representation theorem 4.8 for $\mathbb{Q}^{T_M}$-martingales there exists a unique $\sigma_{T,T_M} \in \mathcal{L}$ such that

$$\left.\frac{d\mathbb{Q}^T}{d\mathbb{Q}^{T_M}}\right|_{\mathcal{F}_t} = \mathbb{E}_{\mathbb{Q}^{T_M}}\left[\frac{B^*(T_M)P(0, T_M)}{B^*(T)P(0, T)}\bigg|\mathcal{F}_t\right] = \mathcal{E}_t\left(\sigma_{T,T_M} \bullet W^{T_M}\right),$$

for $t \in [0, T]$. Girsanov's theorem then tells us that

$$W^T(t) = W^{T_M}(t) - \int_0^t \sigma_{T,T_M}(s)^\top ds, \quad t \in [0, T],$$

is a $\mathbb{Q}^T$-Brownian motion.

Third, since $T \in [T_{M-1}, T_M]$ was arbitrary, we can now define the forward LIBOR process $L(t, T)$ for any $T \in [T_{M-2}, T_{M-1}]$ as

$$dL(t, T) = L(t, T)\lambda(t, T)\,dW^{T+\delta}(t),$$

$$L(0, T) = \frac{1}{\delta}\left(\frac{P(0, T)}{P(0, T + \delta)} - 1\right).$$

This in turn defines the bounded progressive process

$$\sigma_{T,T+\delta}(t) = \frac{\delta L(t, T)}{\delta L(t, T) + 1}\lambda(t, T), \quad t \in [0, T],$$

for any $T \in [T_{M-2}, T_{M-1}]$. The forward measures for $T \in [T_{M-2}, T_{M-1}]$ are now given by

$$\frac{d\mathbb{Q}^T}{d\mathbb{Q}^{T+\delta}} = \mathcal{E}_T\left(\sigma_{T,T+\delta} \bullet W^{T+\delta}\right).$$

Hence we have

$$\frac{d\mathbb{Q}^T}{d\mathbb{Q}^{T_M}}\bigg|_{\mathcal{F}_t} = \frac{d\mathbb{Q}^T}{d\mathbb{Q}^{T+\delta}}\bigg|_{\mathcal{F}_t} \frac{d\mathbb{Q}^{T+\delta}}{d\mathbb{Q}^{T_M}}\bigg|_{\mathcal{F}_t}$$

$$= \mathcal{E}_t\left(\sigma_{T,T+\delta} \bullet W^{T+\delta}\right)\mathcal{E}_t\left(\sigma_{T+\delta,T_M} \bullet W^{T_M}\right)$$

$$= \mathcal{E}_t\left(\sigma_{T,T_M} \bullet W^{T_M}\right), \quad t \in [0,T],$$

for any $T \in [T_{M-2}, T_{M-1}]$, where

$$\sigma_{T,T_M} = \sigma_{T,T+\delta} + \sigma_{T+\delta,T_M}.$$

Proceeding by backward induction yields the forward measure $\mathbb{Q}^T$ and the corresponding $\mathbb{Q}^T$-Brownian motion W^T for all $T \in [0, T_M]$, and forward LIBOR rates $L(t,T)$ for all $T \in [0, T_{M-1}]$.

This way, we obtain the zero-coupon bond prices for all maturities $0 \le T \le S \le T_M$. Indeed, in view of Lemma 7.1, it is reasonable to define the forward price process

$$\frac{P(t,S)}{P(t,T)} = \frac{P(0,S)}{P(0,T)}\frac{d\mathbb{Q}^S}{d\mathbb{Q}^T}\bigg|_{\mathcal{F}_t}$$

$$= \frac{P(0,S)}{P(0,T)}\frac{d\mathbb{Q}^S}{d\mathbb{Q}^{T_M}}\bigg|_{\mathcal{F}_t}\frac{d\mathbb{Q}^{T_M}}{d\mathbb{Q}^T}\bigg|_{\mathcal{F}_t}$$

$$= \frac{P(0,S)}{P(0,T)}\mathcal{E}_t\left(-\sigma_{T,S} \bullet W^T\right), \quad t \in [0,T], \qquad (11.15)$$

where we set ($\rightarrow$ Exercise 11.8)

$$\sigma_{T,S} = \sigma_{T,T_M} - \sigma_{S,T_M}.$$

In particular, for $t = T$ we get

$$P(T,S) = \frac{P(0,S)}{P(0,T)}\mathcal{E}_T\left(-\sigma_{T,S} \bullet W^T\right).$$

Notice that now $P(T,S)$ may be greater than 1, unless $S - T = m\delta$ for some integer m. Hence even though all δ-period forward LIBOR rates $L(t,T)$ are nonnegative, there may be negative interest rates for other than δ periods.

11.9 Exercises

Exercise 11.1 Derive (11.2), by showing that

$$\frac{\delta L(t,T)}{\delta L(t,T)+1} = 1 - e^{-\int_T^{T+\delta} f(t,u)\,du}.$$

Exercise 11.2 Give a full proof of Lemma 11.1.

Exercise 11.3 Finish the proof of Lemma 11.2.

Exercise 11.4 Prove Lemma 11.4.

Exercise 11.5 Prove Lemma 11.5.

Exercise 11.6 Derive the caplet implied volatilities in Table 11.4 by the described bootstrapping method.

Exercise 11.7 Write a code for the Monte Carlo simulation of the LIBOR market model as outlined in Sect. 11.6, both for the risk-neutral measure $\mathbb{Q}^*$ as well as the forward measure $\mathbb{Q}^{T_M}$, and with tenor structure $T_i = i/2$ for $0 \le i \le M = 20$. If not mentioned otherwise, implement the two extremal correlation specifications I ($d = 1$ and $\ell_m = 1$) and II ($d = M - 1$ and $\ell_m = e_m^\top$).

(a) Calibrate the parametric specification (11.12) to the 3.5% strike caplet data in Table 11.4 as a function of β.
(b) Compute the 3.5% cap prices with maturities $2, \ldots, 10$ and compare your result to the original quotes in Table 11.1. Convince yourself that the computed cap prices are the same in both cases I and II.
(c) Compute the at-the-money 4×6-swaption price via Monte Carlo simulation as function of β and the correlation specification, as shown in Fig. 11.3. Note that the underlying swap has annual coupon payments.
(d) Compute this swaption price using the analytic approximation formula (11.14) for the implied swaption volatility and Black's swaption formula. Compare these results to the findings in (c).
(e) Compute this swaption price for a intermediary correlation matrix specified by $\rho_{mn} = e^{-\gamma |T_i - T_j|}$ for various values of $\gamma > 0$ (hint: use $d = M - 1$ and find the corresponding ℓ_m via Cholesky factorization).
(f) Run the Monte Carlo algorithm under both the risk-neutral $\mathbb{Q}^*$ as well as the terminal forward measure $\mathbb{Q}^{T_{20}}$. Verify that the results are the same under both measures.

Exercise 11.8 Consider the setup of Sect. 11.8, and let $0 \le T \le S \le T_M$.

(a) Let $k \in \mathbb{N}_0$ be such that $T_{M-1} \le T + k\delta < T_M$. Prove that

$$\left. \frac{d\mathbb{Q}^T}{d\mathbb{Q}^{T_M}} \right|_{\mathcal{F}_t} = \mathcal{E}_t \left(\sigma_{T,T_M} \bullet W^{T_M} \right), \quad t \in [0, T],$$

where $\sigma_{T,T_M} = \sigma_{T,T+\delta} + \sigma_{T+\delta,T+2\delta} + \cdots + \sigma_{T+(k-1)\delta,T+k\delta} + \sigma_{T+k\delta,T_M}$.

(a) Prove that the forward price process defined in (11.15) satisfies

$$\frac{P(t,S)}{P(t,T)} = \frac{P(0,S)}{P(0,T)} \mathcal{E}_t \left(-\sigma_{T,S} \bullet W^T \right), \quad t \in [0, T]$$

where $\sigma_{T,S} = \sigma_{T,T_M} - \sigma_{S,T_M}$. Compare this result with Lemma 7.1.

(c) Use (b) to prove that

$$\frac{P(t, T)}{P(t, T + \delta)} = \frac{P(0, T)}{P(0, T + \delta)} \mathcal{E}_t \left(\sigma_{T,T+\delta} \bullet W^{T+\delta}\right), \quad t \in [0, T].$$

(d) Deduce from (c) that actually we have

$$\frac{P(t, T)}{P(t, T + \delta)} = 1 + \delta L(t, T), \quad t \in [0, T].$$

(e) Use (d) to show that $P(T, S) \le 1$ if $S - T = k\delta$ for some $k \in \mathbb{N}$.

(f) Verify that $P(t, T)/B^*(t), t \le T$, is a $\mathbb{Q}^*$-martingale.

11.10 Notes

Apart from what has been said in the main text, an overview of who contributed to the development of LIBOR market models is given in [127, Sect. 12.4]. The backward induction approach from Sects. 11.2 and 11.8 was developed by Musiela and Rutkowski [126]. An analytic approximation for swaption pricing in an affine framework in the spirit of Sect. 11.5.2 ("freezing the coefficients") has been provided by Schrager and Pelsser [145], see also references therein. Section 11.6 is partly based on Glasserman [79, Sect. 7.3]. Some of the data material in Sect. 11.7.2 has been prepared jointly by Antoon Pelsser and the author for a joint course at the WU (Vienna University of Economics and Business Administration) Executive Academy in December 2008.

This book gives an introduction to the calibration to and pricing of European style standard products, such as caps, floors and swaptions. Path dependent, such as American or Bermudan style, options have become popular and require more sophisticated valuation methods than the one presented here. Computing continuation values, or conditional expectations in general, with Monte Carlo becomes cumbersome since it requires nested simulation. Several authors, especially Carrière [37], Longstaff and Schwartz [118] and Tsitsiklis and Van Roy [158, 159], have proposed the use of regression to estimate continuation values from simulated path and thus to price American options by simulation. This approach has by now become standard. The interested reader is referred to Glasserman [79, Sect. 8.6] for a thorough introduction. An alternative to the Monte Carlo approach to path dependent option valuation is given by the so-called Markov-functional interest rate models, first introduced by Hunt, Kennedy and Pelsser [97]. The interested reader is also referred to Pelsser [131, Chap. 9].

Chapter 12
Default Risk

So far bond price processes $P(t, T)$ had the property that $P(T, T) = 1$. That is, the payoff was certain, there was no risk of default of the issuer. This may be the case for treasury bonds. Corporate bonds, however, may bear a substantial risk of default. Investors should be adequately compensated by a risk premium, which is reflected by a higher yield on the bond.

In this chapter, we will briefly review the two most common approaches to credit risk modeling: the structural and the intensity-based approach. The structural approach models the value of a firm's assets. Default is when this value hits a certain lower bound. This approach goes back to Merton's [123] seminal corporate debt model. In the intensity-based approach, default is specified exogenously by a stopping time with given intensity process. This approach can be traced back to work of Jarrow, Lando and Turnbull in the early 1990s.

12.1 Default and Transition Probabilities

Rating agencies aim at providing timely, objective information and credit analysis of obligors. Usually they operate without government mandate and are independent of any investment banking firm or similar organization. Among the biggest agencies are Moody's Investors Service (Moody's), Standard&Poor's (S&P), and Fitch Ratings.

Rating agencies assign a credit rating that reflects the creditworthiness of an obligor. After issuance and assignment of the initial obligor's rating, the rating agency regularly checks and adjusts the rating. If there is a tendency observable that may affect the rating, the obligor is set on the Rating Review List (Moody's) or the Credit Watch List (S&P). The interpretation of the letter ratings by S&P's and Moody's are summarized in Table 12.1.

For the quantitative assessment of credit risk we thus have not only to consider default probabilities, but also the probabilities for transitions between credit ratings.[1] Note that the rating is based on objective probabilities, while for the pricing we need the corresponding risk-neutral probabilities. The equivalent change of measure will be discussed in Sect. 12.3.3 below.

A stylized formal definition of objective default and transition rates is given in [142, Chap. 2] as follows:

---

[1] Another important risk element are the recovery rates. That is, the proportion of value delivered after default has occurred. See Sect. 12.3.3 below.

D. Filipović, *Term-Structure Models,*
Springer Finance,
DOI 10.1007/978-3-540-68015-4_12, © Springer-Verlag Berlin Heidelberg 2009

Table 12.1 Rating symbols

S&P	Moody's	Interpretation
Investment-grade ratings		
AAA	Aaa	Highest quality, extremely strong
AA+	Aa1	
AA	Aa2	High quality
AA−	Aa3	
A+	A1	
A	A2	Strong payment capacity
A−	A3	
BBB+	Baa1	
BBB	Baa2	Adequate payment capacity
BBB−	Baa3	
Speculative-grade ratings		
BB+	Ba1	Likely to fulfill obligations,
BB	Ba2	ongoing uncertainty
BB−	Ba3	
B+	B1	
B	B2	High-risk obligations
B−	B3	
CCC+	Caa1	
CCC	Caa2	Current vulnerability to default
CCC−	Caa3	
CC		
C	Ca	In bankruptcy or default
D		or other marked shortcoming

Definition 12.1

(a) The historical one-year *default rate*, based on the time frame $[Y_0, Y_1]$, for an
   R-rated issuer is

$$d_R = \frac{\sum_{y=Y_0}^{Y_1} M_R(y)}{\sum_{y=Y_0}^{Y_1} N_R(y)},$$

where $N_R(y)$ is the number of issuers with rating R at beginning of year y,
and $M_R(y)$ is the number of issuers with rating R at beginning of year y which
defaulted in that year.

Table 12.2 S&P's one-year transition and default rates, based on the time frame [1981, 2007]. N.R. stands for not rated

Initial rating (R)	Rating at end of year (R')								
	AAA	AA	A	BBB	BB	B	CCC	D	N.R.
AAA	86.05	13.95	0.00	0.00	0.00	0.00	0.00	0.00	0.00
AA	0.73	90.73	6.10	0.24	0.24	0.00	0.00	0.00	1.95
A	0.00	1.80	87.57	6.13	0.54	0.00	0.00	0.18	3.78
BBB	0.00	0.12	1.60	88.07	3.81	0.12	0.00	0.49	5.78
BB	0.00	0.00	0.00	4.12	79.02	5.69	0.20	0.59	10.39
B	0.00	0.00	0.00	0.31	4.33	73.37	3.41	6.81	11.76
CCC/C	0.00	0.00	0.00	0.00	0.00	10.00	35.00	40.00	15.00

(b) The historical one-year *transition rate* from rating R to R', based on the time frame $[Y_0, Y_1]$, is

$$tr_{R,R'} = \frac{\sum_{y=Y_0}^{Y_1} M_{R,R'}(y)}{\sum_{y=Y_0}^{Y_1} N_R(y)},$$

where $N_R(y)$ is as above, and $M_{R,R'}(y)$ is the number of issuers with rating R at beginning of year y and R' at the end of that year.

Transition rates are gathered in a *transition matrix* as shown in Table 12.2. Note that the actual estimation methods for default and transition probabilities used by rating agencies, such as S&P's, are more sophisticated than appears from the Definition 12.1. For instance, the above statistics have to be adjusted for issuers that changed to not rated (N.R.) during the underlying time frame. See the report [153].

The rating-based default and transition probabilities bear some shortcomings. Rating agencies appear to be too slow to change ratings.[2] This may result in a systematic overestimation of $tr_{R,R}$ and d_R, and hence underestimation of $tr_{R,R'}$, at least for some ratings $R \neq R'$. Finally, note that Definition 12.1 neglects the default rate volatility. Transition and default probabilities are dynamic and vary over time, depending on economic conditions. In the following sections we will consider two different dynamic model approaches.

12.2 Structural Approach

Merton [123] proposed a simple capital structure of a firm consisting of equity and one type of zero-coupon debt with promised terminal constant payoff $X > 0$ at maturity T. The firm defaults by T if the total market value of its assets $V(T)$ at T is

---

[2]For instance, rating agencies have been subject to criticism in the wake of large losses beginning in 2007 in the collateralized debt obligation (CDO) market that occurred despite being assigned top ratings.

less than its liabilities X. Thus the probability of default by time T conditional on the information available at $t \leq T$ is

$$p_d(t, T) = \mathbb{P}[V(T) < X \mid \mathcal{F}_t],$$

with respect to some stochastic basis $(\Omega, \mathcal{F}, (\mathcal{F}_t)_{t \in [0,T]}, \mathbb{P})$. The dynamics of $V(t)$ is modeled as geometric Brownian motion

$$\frac{dV(t)}{V(t)} = \mu \, dt + \sigma \, dW(t), \quad t \in [0, T],$$

that is

$$V(T) = V(t) \exp\left(\sigma(W(T) - W(t)) + \left(\mu - \frac{1}{2}\sigma^2 \right)(T - t) \right), \quad t \in [0, T].$$

Then we have

$$p_d(t, T) = \Phi\left(\frac{\log(\frac{X}{V(t)}) - (\mu - \frac{1}{2}\sigma^2)(T - t)}{\sigma\sqrt{T - t}} \right), \quad t \in [0, T]. \tag{12.1}$$

If the firm value process $V(t)$ is continuous, as in the Merton model, the instantaneous default intensity $\partial_T^+ p_d(t, T)|_{T=t}$ is zero ($\to$ Exercise 12.1). To include "unexpected" defaults one has to consider firm value processes with jumps. Zhou [164] models $V(t)$ as a jump-diffusion process

$$V(T) = V(t) \left(\prod_{j=N(t)+1}^{N(T)} e^{Z_j} \right) e^{(\mu - \frac{\sigma^2}{2})(T-t) + \sigma(W(T)-W(t))},$$

where $N(t)$ is a Poisson process with intensity λ and $Z_1, Z_2, \ldots$ is a sequence of i.i.d. Gaussian $\mathcal{N}(m, \rho^2)$ distributed random variables. It is assumed that W, N and Z_j are mutually independent. A dynamic description of V is

$$V(t) = V(0) + \int_0^t V(s)(\mu \, ds + \sigma \, dW(s)) + \sum_{j=1}^{N(t)} V(\tau_j-)\left(e^{Z_j} - 1 \right),$$

where $\tau_1, \tau_2, \ldots$ are the jump times of N.

It is clear that the distribution of $\log V(T)$ conditional on $\mathcal{F}_t$ and $N(T) - N(t) = n$ is Gaussian with mean

$$\log V(t) + mn + \left(\mu - \frac{\sigma^2}{2} \right)(T - t)$$

and variance

$$n\rho^2 + \sigma^2(T - t).$$

Hence the conditional default probability is

$$p_d(t, T) = \mathbb{P}\left[\log V(T) < \log X \mid \mathcal{F}_t\right]$$

$$= \sum_{n=0}^{\infty} \mathbb{P}\left[\log V(T) < \log X \mid \mathcal{F}_t, \, N(T) - N(t) = n\right] \mathbb{P}\left[N(T) - N(t) = n\right]$$

$$= \sum_{n=0}^{\infty} \Phi\left(\frac{\log(\frac{X}{V(t)}) - mn - (\mu - \frac{\sigma^2}{2})(T - t)}{\sqrt{n\rho^2 + \sigma^2(T - t)}}\right) e^{-\lambda(T-t)} \frac{(\lambda(T - t))^n}{n!},$$

$$(12.2)$$

so that now the instantaneous default intensity $\partial_T^+ p_d(t, T)|_{T=t}$ is positive on $\{V(t) \geq X\}$ ($\to$ Exercise 12.1).

One drawback of the above approach is that the event "default by T" is defined as $\{V(T) < X\}$, which does not depend on the asset values $V(t)$ prior to T. First passage time models make this approach more realistic by admitting default at any time $T_d \in [0, T]$, and not just at maturity T. That means, bankruptcy occurs if the firm value $V(t)$ crosses a specified stochastic boundary $X(t)$, such that

$$T_d = \inf\{t \mid V(t) < X(t)\}.$$

In this case, "default by T" means $\{T_d \leq T\}$, and thus the conditional default probability is

$$p_d(t, T) = \mathbb{P}[T_d \leq T \mid \mathcal{F}_t], \quad t \in [0, T],$$

which has to be determined by Monte Carlo simulation in general. We will now present an approach where the default time T_d is directly modeled via its intensity.

12.3 Intensity-Based Approach

Default is often a complicated event. The precise conditions under which it must occur (such as hitting a barrier) are easily misspecified. The above structural approach has the additional deficiency that it is usually difficult to determine and trace a firm's value process.

In this section we focus directly on describing the evolution of the default probabilities $p_d(t, T)$ without defining the exact default event. Formally, we fix a probability space $(\Omega, \mathcal{F}, \mathbb{P})$. The flow of the complete market information is represented by a filtration $(\mathcal{F}_t)$ satisfying the usual conditions. The default time T_d is assumed to be an $(\mathcal{F}_t)$-stopping time, hence the right-continuous default process

$$H(t) = 1_{\{T_d \leq t\}}$$

is $(\mathcal{F}_t)$-adapted. The $\mathcal{F}_t$-conditional default probability is now

$$p_d(t, T) = \mathbb{E}[H(T) \mid \mathcal{F}_t], \quad t \in [0, T].$$

Obviously, H is a uniformly integrable submartingale. By the Doob–Meyer decomposition ([106, Theorem 1.4.10]) there exists a unique $(\mathcal{F}_t)$-predictable[3] non-decreasing process $A(t) = A(t \wedge T_d)$ with $A(0) = 0$ and such that

$$M(t) = H(t) - A(t) \tag{12.3}$$

is a martingale. Hence

$$p_d(t, T) = 1_{\{T_d \le t\}} + \mathbb{E}\left[A(T) - A(t) \mid \mathcal{F}_t\right].$$

This formula is the best we can hope for in general. We next proceed in several steps towards an explicit expression for $p_d(t, T)$ by imposing more and more restrictive conditions.

Throughout, we will assume that there exists a sub-filtration $(\mathcal{G}_t) \subset (\mathcal{F}_t)$ (partial market information) such that

$$\mathcal{F}_t = \mathcal{G}_t \vee \mathcal{H}_t,$$

where $\mathcal{H}_t = \sigma(H(s) \mid s \le t)$ and $\mathcal{G}_t \vee \mathcal{H}_t$ stands for the smallest σ-algebra containing $\mathcal{G}_t$ and $\mathcal{H}_t$. Intuitively speaking, events in $\mathcal{F}_t$ are $\mathcal{G}_t$-observable given that $T_d > t$. The formal statement is as follows:

Lemma 12.1 *Let $t \in \mathbb{R}_+$. For every $A \in \mathcal{F}_t$ there exists $B \in \mathcal{G}_t$ such that*

$$A \cap \{T_d > t\} = B \cap \{T_d > t\}. \tag{12.4}$$

Proof Let

$$\mathcal{F}_t^* = \{A \in \mathcal{F}_t \mid \exists B \in \mathcal{G}_t \text{ with property } (12.4)\}.$$

The inclusion $\mathcal{G}_t \subset \mathcal{F}_t^*$ is obvious. Simply take $B = A$. Moreover $\mathcal{H}_t \subset \mathcal{F}_t^*$. Indeed, for every $A \in \mathcal{H}_t$ the intersection $A \cap \{T_d > t\}$ is either $\emptyset$ or $\{T_d > t\}$, so we can take for B either $\emptyset$ or Ω.

Since $\mathcal{F}_t^*$ is a σ-algebra and $\mathcal{F}_t$ is defined to be the smallest σ-algebra containing $\mathcal{G}_t$ and $\mathcal{H}_t$, we conclude that $\mathcal{F}_t \subset \mathcal{F}_t^*$. This proves the lemma. $\qquad\square$

We next elaborate on the following assumption:

(D1) there exists a nonnegative $(\mathcal{G}_t)$-progressive process λ such that

$$\mathbb{P}[T_d > t \mid \mathcal{G}_t] = e^{-\int_0^t \lambda(s)\,ds}.$$

Hence the default probability by t as seen by a $\mathcal{G}_t$-informed observer satisfies $\mathbb{P}[T_d \le t \mid \mathcal{G}_t] < 1$. In particular, the inclusion $\mathcal{G}_t \subset \mathcal{F}_t$ is strict: a market participant

---

[3] The $(\mathcal{F}_t)$-predictable σ-algebra on $\Omega \times \mathbb{R}_+$ is generated by all left-continuous $(\mathcal{F}_t)$-adapted processes; or equivalently, by the sets $B \times \{0\}$ where $B \in \mathcal{F}_0$ and $B \times (s, t]$ where $s < t$ and $B \in \mathcal{F}_s$. A predictable process is always progressive.

with access to the partial market information $\mathcal{G}_t$ cannot observe whether default has occurred by time t ($T_d \leq t$) or not ($T_d > t$). In other words, T_d is not a stopping time for ($\mathcal{G}_t$). This nicely reflects the aforementioned difficulties to determine the exact default event in practice. We will give an interpretation of λ after Lemma 12.3 below.

A consequence of the following lemma is that for any $\mathcal{F}_t$-measurable random variable Y there exists an $\mathcal{G}_t$-measurable random variable $\tilde{Y}$ such that $Y = \tilde{Y}$ on $\{T_d > t\}$.

Lemma 12.2 *Assume* (**D1**), *and let Y be a nonnegative random variable. Then*

$$\mathbb{E}\left[1_{\{T_d > t\}} Y \mid \mathcal{F}_t\right] = 1_{\{T_d > t\}} e^{\int_0^t \lambda(s)\,ds} \mathbb{E}\left[1_{\{T_d > t\}} Y \mid \mathcal{G}_t\right] \qquad (12.5)$$

for all t.

Proof Let $A \in \mathcal{F}_t$. By Lemma 12.1 there exists a $B \in \mathcal{G}_t$ with (12.4), that is, $1_A 1_{\{T_d > t\}} = 1_B 1_{\{T_d > t\}}$. Hence, by the very definition of the $\mathcal{G}_t$-conditional expectation,

$$\int_A 1_{\{T_d > t\}} Y \mathbb{P}[T_d > t \mid \mathcal{G}_t]\,d\mathbb{P} = \int_B 1_{\{T_d > t\}} Y \mathbb{P}[T_d > t \mid \mathcal{G}_t]\,d\mathbb{P}$$

$$= \int_B \mathbb{E}\left[1_{\{T_d > t\}} Y \mid \mathcal{G}_t\right] \mathbb{P}[T_d > t \mid \mathcal{G}_t]\,d\mathbb{P}$$

$$= \int_B 1_{\{T_d > t\}} \mathbb{E}\left[1_{\{T_d > t\}} Y \mid \mathcal{G}_t\right]\,d\mathbb{P}$$

$$= \int_A 1_{\{T_d > t\}} \mathbb{E}\left[1_{\{T_d > t\}} Y \mid \mathcal{G}_t\right]\,d\mathbb{P}.$$

This implies

$$\mathbb{E}\left[1_{\{T_d > t\}} Y \mathbb{P}[T_d > t \mid \mathcal{G}_t] \mid \mathcal{F}_t\right] = 1_{\{T_d > t\}} \mathbb{E}\left[1_{\{T_d > t\}} Y \mid \mathcal{G}_t\right],$$

which proves the lemma. □

We have now the following expression for the conditional default probabilities.

Lemma 12.3 *Assume* (**D1**). *For any $t \leq T$ we have*

$$\mathbb{P}[T_d > T \mid \mathcal{F}_t] = 1_{\{T_d > t\}} \mathbb{E}\left[e^{-\int_t^T \lambda(s)\,ds} \mid \mathcal{G}_t\right], \qquad (12.6)$$

$$\mathbb{P}[t < T_d \leq T \mid \mathcal{F}_t] = 1_{\{T_d > t\}} \mathbb{E}\left[1 - e^{-\int_t^T \lambda(s)\,ds} \mid \mathcal{G}_t\right]. \qquad (12.7)$$

Moreover, the processes

$$L(t) = 1_{\{T_d > t\}} e^{\int_0^t \lambda(s)\,ds} = (1 - H(t)) e^{\int_0^t \lambda(s)\,ds}$$

is an ($\mathcal{F}_t$)-*martingale.*

Proof Let $t \leq T$. Then $1_{\{T_d > T\}} = 1_{\{T_d > t\}} 1_{\{T_d > T\}}$. Using this and (12.5) we derive

$$\mathbb{P}[T_d > T \mid \mathcal{F}_t] = \mathbb{E}\left[1_{\{T_d > t\}} 1_{\{T_d > T\}} \mid \mathcal{F}_t\right]$$

$$= 1_{\{T_d > t\}} e^{\int_0^t \lambda(s)\,ds} \mathbb{E}\left[1_{\{T_d > T\}} \mid \mathcal{G}_t\right]$$

$$= 1_{\{T_d > t\}} e^{\int_0^t \lambda(s)\,ds} \mathbb{E}\left[\mathbb{E}\left[1_{\{T_d > T\}} \mid \mathcal{G}_T\right] \mid \mathcal{G}_t\right]$$

$$= 1_{\{T_d > t\}} e^{\int_0^t \lambda(s)\,ds} \mathbb{E}\left[e^{-\int_0^T \lambda(s)\,ds} \mid \mathcal{G}_t\right],$$

which proves (12.6). Equation (12.7) follows since

$$1_{\{t < T_d \leq T\}} = 1_{\{T_d > t\}} - 1_{\{T_d > T\}}.$$

For the second statement it is enough to consider

$$\mathbb{E}[L(T) \mid \mathcal{F}_t] = \mathbb{E}\left[1_{\{T_d > t\}} 1_{\{T_d > T\}} e^{\int_0^T \lambda(s)\,ds} \mid \mathcal{F}_t\right]$$

$$= 1_{\{T_d > t\}} e^{\int_0^t \lambda(s)\,ds} \mathbb{E}\left[1_{\{T_d > T\}} e^{\int_0^T \lambda(s)\,ds} \mid \mathcal{G}_t\right] = L(t),$$

since by **(D1)**

$$\mathbb{E}\left[1_{\{T_d > T\}} e^{\int_0^T \lambda(s)\,ds} \mid \mathcal{G}_t\right] = \mathbb{E}\left[\mathbb{E}\left[1_{\{T_d > T\}} \mid \mathcal{G}_T\right] e^{\int_0^T \lambda(s)\,ds} \mid \mathcal{G}_t\right] = 1. \qquad \square$$

Replacing T by $t + \Delta t$ in (12.7) gives, in first order in Δt,

$$\mathbb{P}[t < T_d \leq t + \Delta t \mid \mathcal{F}_t] \approx 1_{\{T_d > t\}} \lambda(t) \Delta t,$$

so $\lambda(t)\Delta t$ is approximately the conditional probability of a default in a small time interval after t given survival up to and including t. Whence we refer to $\lambda(t)$ as default intensity prevailing at time t.

Here is a rather surprising and important identification result for the Doob–Meyer decomposition (12.3) in terms of λ.

Lemma 12.4 *Assume* **(D1)**. *Then the process*

$$N(t) = H(t) - \int_0^t \lambda(s) 1_{\{T_d > s\}}\,ds$$

is an $(\mathcal{F}_t)$-*martingale. Hence, by the uniqueness of the predictable Doob–Meyer decomposition (12.3), we have*

$$A(t) = \int_0^t \lambda(s) 1_{\{T_d > s\}}\,ds.$$

Proof Let $t \leq T$. In view of (12.6) and (12.5) we have

$$\mathbb{E}[N(T) \mid \mathcal{F}_t] = 1 - \mathbb{E}\left[1_{\{T_d > T\}} \mid \mathcal{F}_t\right] - \int_0^t \lambda(s) 1_{\{T_d > s\}} \, ds$$

$$- \int_t^T \mathbb{E}\left[\lambda(s) 1_{\{T_d > s\}} \mid \mathcal{F}_t\right] ds$$

$$= 1 - 1_{\{T_d > t\}} \mathbb{E}\left[e^{-\int_t^T \lambda(u) \, du} \mid \mathcal{G}_t\right] - \int_0^t \lambda(s) 1_{\{T_d > s\}} \, ds$$

$$- \underbrace{\int_t^T 1_{\{T_d > t\}} e^{\int_0^t \lambda(u) \, du} \mathbb{E}\left[\lambda(s) 1_{\{T_d > s\}} \mid \mathcal{G}_t\right] ds}_{=:I}.$$

We have further

$$I = \int_t^T 1_{\{T_d > t\}} e^{\int_0^t \lambda(u) \, du} \mathbb{E}\left[\lambda(s) \mathbb{E}\left[1_{\{T_d > s\}} \mid \mathcal{G}_s\right] \mid \mathcal{G}_t\right] ds$$

$$= 1_{\{T_d > t\}} \mathbb{E}\left[\int_t^T \lambda(s) e^{-\int_t^s \lambda(u) \, du} \, ds \mid \mathcal{G}_t\right]$$

$$= 1_{\{T_d > t\}} \mathbb{E}\left[1 - e^{-\int_t^T \lambda(u) \, du} \mid \mathcal{G}_t\right],$$

hence

$$\mathbb{E}[N(T) \mid \mathcal{F}_t] = 1 - 1_{\{T_d > t\}} - \int_0^t \lambda(s) 1_{\{T_d > s\}} \, ds = N(t). \qquad \square$$

The next assumption leads the way to implement a default risk model:

(D2) $\mathbb{P}[T_d > t \mid \mathcal{G}_\infty] = \mathbb{P}[T_d > t \mid \mathcal{G}_t]$, $\quad t \geq 0$.

Stopping times which satisfy **(D1)** and **(D2)** are called $(\mathcal{G}_t)$-doubly stochastic, see e.g. [24, Sect. II.1].

Here are two lemmas which put **(D2)** in context.

Lemma 12.5 *The following properties are equivalent:*

(a) **(D2)** *holds.*
(b) *Every bounded $\mathcal{G}_\infty$-measurable X satisfies $\mathbb{E}[X \mid \mathcal{F}_t] = \mathbb{E}[X \mid \mathcal{G}_t]$.*
(c) *Every $(\mathcal{G}_t)$-martingale is an $(\mathcal{F}_t)$-martingale.*

Property (c) is known in the literature as "hypothesis H", see [25, 64].

Proof (a) $\Leftrightarrow$ (b): For $A \in \mathcal{G}_t$, $u \leq t$ and some bounded $\mathcal{G}_\infty$-measurable X, define

$$I_1 = \int_{A \cap \{T_d > u\}} X \, d\mathbb{P} = \int_A X \mathbb{E}[1_{\{T_d > u\}} \mid \mathcal{G}_\infty] \, d\mathbb{P},$$

$$I_2 = \int_{A \cap \{T_d > u\}} \mathbb{E}[X \mid \mathcal{G}_t] \, d\mathbb{P} = \int_A X \mathbb{E}[1_{\{T_d > u\}} \mid \mathcal{G}_t] \, d\mathbb{P}.$$

Since $\mathcal{F}_t = \mathcal{G}_t \vee \mathcal{H}_t$ is generated by sets of the form $A \cap \{T_d > u\}$, it is clear that both (a) and (b) are equivalent to $I_1 = I_2$.

(b) $\Leftrightarrow$ (c): $\rightarrow$ Exercise 12.2. □

Lemma 12.6 *The following properties are equivalent*:

(a) $M(t) = H(t) - \int_0^t \ell(s) 1_{\{T_d > s\}} \, ds$ *is a $(\mathcal{G}_\infty \vee \mathcal{H}_t)$-martingale for some nonnegative $(\mathcal{G}_t)$-progressive process ℓ.*
(b) **(D1)** *and* **(D2)** *hold for* $\lambda = \ell$.

Proof (a) $\Rightarrow$ (b): The function $\phi(T) = \mathbb{E}[1_{\{T_d > T\}} \mid \mathcal{G}_\infty \vee \mathcal{H}_t]$ for $T \geq t$ satisfies

$$\phi(T) = \mathbb{E}\left[1 - M(T) - \int_0^T \ell(s) 1_{\{T_d > s\}} \, ds \mid \mathcal{G}_\infty \vee \mathcal{H}_t\right]$$

$$= 1 - M(t) - \int_0^t \ell(s) 1_{\{T_d > s\}} \, ds - \int_t^T \mathbb{E}\left[\ell(s) 1_{\{T_d > s\}} \, ds \mid \mathcal{G}_\infty \vee \mathcal{H}_t\right]$$

$$= 1_{\{T_d > t\}} - \int_t^T \ell(s) \phi(s) \, ds. \tag{12.8}$$

This property is equivalent to

$$\mathbb{E}\left[1_{\{T_d > T\}} \mid \mathcal{G}_\infty \vee \mathcal{H}_t\right] = 1_{\{T_d > t\}} e^{-\int_t^T \ell(s) \, ds}. \tag{12.9}$$

For $t = 0$, we obtain

$$\mathbb{E}\left[1_{\{T_d > T\}} \mid \mathcal{G}_\infty\right] = e^{-\int_0^T \ell(s) \, ds}.$$

Since the right-hand side is $\mathcal{G}_T$-measurable, conditioning on $\mathcal{G}_T$ yields **(D1)** and **(D2)** for $\lambda = \ell$.

(b) $\Rightarrow$ (a): A straightforward modification of the proofs of Lemmas 12.1 and 12.2 shows that ($\rightarrow$ Exercise 12.3)

$$\mathbb{E}\left[1_{\{T_d > t\}} Y \mid \mathcal{G}_\infty \vee \mathcal{H}_t\right] = 1_{\{T_d > t\}} e^{\int_0^t \lambda(s) \, ds} \mathbb{E}\left[1_{\{T_d > t\}} Y \mid \mathcal{G}_\infty\right] \tag{12.10}$$

for every random variable Y. Combining this and **(D2)**, we obtain

$$\mathbb{E}\left[1_{\{T_d > T\}} \mid \mathcal{G}_\infty \vee \mathcal{H}_t\right] = 1_{\{T_d > t\}} e^{\int_0^t \lambda(s) \, ds} \mathbb{E}\left[1_{\{T_d > T\}} \mid \mathcal{G}_\infty\right]$$

$$= 1_{\{T_d > t\}} e^{\int_0^t \lambda(s) \, ds} e^{-\int_0^T \lambda(s) \, ds},$$

and thus (12.9), which again is equivalent to (12.8), for $\ell = \lambda$. Hence M is a $(\mathcal{G}_\infty \vee \mathcal{H}_t)$-martingale. □

The next lemma gives the key idea for how to construct an intensity-based model.

Lemma 12.7 *Suppose, in addition to* **(D1)** *and* **(D2)**, *that* $\int_0^\infty \lambda(s)\,ds = \infty$. *Then* $\int_0^{T_d} \lambda(s)\,ds$ *is an exponential random variable with parameter 1 and is independent of* $\mathcal{G}_\infty$.

Proof Define $\Lambda(t) = \int_0^t \lambda(s)\,ds$. Then $\Lambda(t)$ is nondecreasing and continuous, and satisfies $\Lambda(\mathbb{R}_+) = \mathbb{R}_+$. We can define its right inverse

$$C(s) = \inf\{t \mid \Lambda(t) > s\},$$

which is $\mathcal{G}_\infty$-measurable. Then $\Lambda(t) > s$ if and only if $t > C(s)$. Moreover, $\Lambda(C(s)) = s$ for all $s \ge 0$, and so

$$\mathbb{P}[\Lambda(T_d) > s \mid \mathcal{G}_\infty] = \mathbb{P}[T_d > C(s) \mid \mathcal{G}_\infty] = e^{-\Lambda(C(s))} = e^{-s}.$$

This proves that $\Lambda(T_d)$ is an exponential random variable with parameter 1 and independent of $\mathcal{G}_\infty$. $\qquad\square$

12.3.1 Construction of Doubly Stochastic Intensity-Based Models

The construction of a model that satisfies **(D1)** and **(D2)** is now straightforward by reversion of the above approach. We start with a filtration $(\mathcal{G}_t)$ satisfying the usual conditions and

$$\mathcal{G}_\infty = \sigma(\mathcal{G}_t \mid t \in \mathbb{R}_+) \subset \mathcal{F}.$$

Let $\lambda(t)$ be a nonnegative $(\mathcal{G}_t)$-progressive process with the property

$$\int_0^t \lambda(s)\,ds < \infty \quad \text{a.s. for all } t \in \mathbb{R}_+.$$

Motivated by Lemma 12.7, we then fix an exponential random variable ϕ with parameter 1 and independent of $\mathcal{G}_\infty$, and define the random time

$$T_d = \inf\left\{t \mid \int_0^t \lambda(s)\,ds \ge \phi\right\}$$

with values in $(0, \infty]$. Note that $\int_0^\infty \lambda(s)\,ds$ may be finite, so that we cannot necessarily reconstruct ϕ from T_d and λ as in Lemma 12.7. Nevertheless, by the independence of ϕ and $\mathcal{G}_\infty$, we obtain

$$\mathbb{P}[T_d > t \mid \mathcal{G}_\infty] = \mathbb{P}\left[\phi > \int_0^t \lambda(u)\,du \mid \mathcal{G}_\infty\right] = e^{-\int_0^t \lambda(u)\,du}.$$

Conditioning both sides on $\mathcal{G}_t$ yields

$$\mathbb{P}[T_d > t \mid \mathcal{G}_t] = e^{-\int_0^t \lambda(u)\,du}.$$

Hence **(D1)** and **(D2)** hold. We finally define $\mathcal{F}_t = \mathcal{G}_t \vee \mathcal{H}_t$, as above.

12.3.2 Computation of Default Probabilities

When it comes to computations of the default probabilities (12.6) we need a tractable model for the intensity process λ. But the right-hand side of (12.6) looks just like what we had for the risk-neutral valuation of zero-coupon bonds in terms of a given short-rate process (Chap. 5). Notice that $\lambda \geq 0$ is essential. An obvious and popular choice for λ is thus an affine process. So let W be a $(\mathcal{G}_t)$-Brownian motion, $b \geq 0$, $\beta \in \mathbb{R}$ and $\sigma > 0$ some constants, and let

$$d\lambda(t) = (b + \beta\lambda(t))\, dt + \sigma\sqrt{\lambda(t)}\, dW(t), \quad \lambda(0) \geq 0. \qquad (12.11)$$

Now construct a doubly stochastic model as outlined in Sect. 12.3.1. The proof of the following lemma is left as an exercise.

Lemma 12.8 *For the intensity process* (12.11) *the conditional default probability is*

$$p_d(t, T) = \mathbb{P}[T_d \leq T \mid \mathcal{F}_t] = \begin{cases} 1 - e^{-A(T-t) - B(T-t)\lambda(t)}, & \text{if } T_d > t, \\ 1, & \text{else}, \end{cases}$$

where

$$A(u) = -\frac{2b}{\sigma^2} \log\left(\frac{2\gamma e^{(\gamma-\beta)u/2}}{(\gamma - \beta)(e^{\gamma u} - 1) + 2\gamma}\right),$$

$$B(u) = \frac{2(e^{\gamma u} - 1)}{(\gamma - \beta)(e^{\gamma u} - 1) + 2\gamma},$$

$$\gamma = \sqrt{\beta^2 + 2\sigma^2}.$$

Proof $\rightarrow$ Exercise 12.4. $\square$

12.3.3 Pricing Default Risk

We suppose now that we are given a risk-neutral probability measure $\mathbb{Q} \sim \mathbb{P}$ and a $(\mathcal{G}_t)$-progressive short-rate process $r(t)$. We also assume that there exists a nonnegative $(\mathcal{G}_t)$-progressive process $\lambda_{\mathbb{Q}}$ such that

$$\int_0^t \left(|r(s)| + \lambda_{\mathbb{Q}}(s)\right) ds < \infty \quad \text{for all } t \in \mathbb{R}_+,$$

and properties **(D1)** and **(D2)** are satisfied[4] for $\mathbb{Q}$.

---

[4]Properties **(D1)** and **(D2)** are not necessarily preserved under an equivalent change of measure in general, see Sect. 12.3.4 below.

We will determine the price $C(t, T)$ of a corporate zero-coupon bond with maturity T, which may default. As for the recovery we fix a constant recovery rate $\delta \in (0, 1)$ and distinguish three cases:

- Zero recovery: the cash flow at T is $1_{\{T_d > T\}}$.
- Partial recovery at maturity: the cash flow at T is $1_{\{T_d > T\}} + \delta 1_{\{T_d \le T\}}$.
- Partial recovery at default: the cash flow is

$$
\begin{cases}
1 \text{ at } T & \text{if } T_d > T, \\
\delta \text{ at } T_d & \text{if } T_d \le T.
\end{cases}
$$

12.3.3.1 Zero Recovery

The arbitrage price of $C(t, T)$ is

$$
C(t, T) = \mathbb{E}_{\mathbb{Q}} \left[e^{-\int_t^T r(s)\,ds} 1_{\{T_d > T\}} \mid \mathcal{F}_t \right].
$$

In view of Lemma 12.2 this is

$$
C(t, T) = 1_{\{T_d > t\}} e^{\int_0^t \lambda_{\mathbb{Q}}(s)\,ds} \mathbb{E}_{\mathbb{Q}} \left[e^{-\int_t^T r(s)\,ds} 1_{\{T_d > T\}} \mid \mathcal{G}_t \right]
$$

$$
= 1_{\{T_d > t\}} e^{\int_0^t \lambda_{\mathbb{Q}}(s)\,ds} \mathbb{E}_{\mathbb{Q}} \left[e^{-\int_t^T r(s)\,ds} \mathbb{E}_{\mathbb{Q}} \left[1_{\{T_d > T\}} \mid \mathcal{G}_T \right] \mid \mathcal{G}_t \right]
$$

$$
= 1_{\{T_d > t\}} \mathbb{E}_{\mathbb{Q}} \left[e^{-\int_t^T (r(s) + \lambda_{\mathbb{Q}}(s))\,ds} \mid \mathcal{G}_t \right]. \tag{12.12}
$$

Note that this is a very nice formula: pricing a corporate bond boils down to the pricing of a non-defaultable zero-coupon bond with the short-rate process replaced by

$$
r(s) + \lambda_{\mathbb{Q}}(s) \ge r(s).
$$

A tractable (e.g. affine) doubly stochastic model, based on the construction in Sect. 12.3.1, is easily found. For the short rates we choose CIR: let W^* be a $(\mathbb{Q}, \mathcal{G}_t)$-Brownian motion, $b \ge 0$, $\beta \in \mathbb{R}$, $\sigma > 0$ constant parameters and

$$
dr(t) = (b + \beta r(t))\,dt + \sigma \sqrt{r(t)}\,dW^*(t), \quad r(0) \ge 0. \tag{12.13}
$$

For the intensity process we choose the affine combination

$$
\lambda_{\mathbb{Q}}(t) = c_0 + c_1 r(t), \tag{12.14}
$$

for two constants $c_0, c_1 \ge 0$.

Lemma 12.9 *For the above affine model* (12.13)–(12.14) *we have*

$$
C(t, T) = 1_{\{T_d > t\}} e^{-A(T-t) - B(T-t) r(t)},
$$

where

$$A(u) = c_0 u - \frac{2b(1+c_1)}{\sigma^2} \log\left(\frac{2\gamma e^{(\gamma-\beta)u/2}}{(\gamma-\beta)(e^{\gamma u}-1)+2\gamma}\right),$$

$$B(u) = \frac{2(e^{\gamma u}-1)}{(\gamma-\beta)(e^{\gamma u}-1)+2\gamma}(1+c_1),$$

$$\gamma = \sqrt{\beta^2 + 2(1+c_1)\sigma^2}.$$

Proof → Exercise 12.5.   □

A special case is $c_1 = 0$ (constant intensity). Here we have

$$C(t, T) = 1_{\{T_d > t\}} e^{-c_0(T-t)} P(t, T),$$

where $P(t, T)$ is the CIR price of a default-free zero-coupon bond.

12.3.3.2 Partial Recovery at Maturity

This is an easy modification of the preceding case since

$$1_{\{T_d > T\}} + \delta 1_{\{T_d \leq T\}} = (1-\delta)1_{\{T_d > T\}} + \delta.$$

We thus obtain for the corporate bond price with partial recovery at maturity

$$C(t, T) = (1-\delta) C_0(t, T) + \delta P(t, T),$$

where $C_0(t, T)$ is the bond price with zero recovery, and $P(t, T)$ denotes the price of the default-free zero-coupon bond.

12.3.3.3 Partial Recovery at Default

The price of the corporate bond with partial recovery at default is

$$C(t, T) = C_0(t, T) + \delta \Pi(t, T),$$

where $C_0(t, T)$ denotes the bond price with zero recovery, and

$$\Pi(t, T) = \mathbb{E}_{\mathbb{Q}}\left[e^{-\int_t^{T_d} r(s)\, ds} 1_{\{t < T_d \leq T\}} \mid \mathcal{F}_t\right]$$

is the unit price of the recovery at default given that $t < T_d \leq T$.

We now further develop $\Pi(t, T)$. From (12.10) we obtain for $t \leq u$

$$\mathbb{Q}[t < T_d \leq u \mid \mathcal{G}_\infty \vee \mathcal{H}_t] = 1_{\{T_d > t\}} e^{\int_0^t \lambda_\mathbb{Q}(s)\,ds} \mathbb{E}_\mathbb{Q}\left[1_{\{t < T_d \leq u\}} \mid \mathcal{G}_\infty \right]$$

$$= 1_{\{T_d > t\}} e^{\int_0^t \lambda_\mathbb{Q}(s)\,ds} \left(e^{-\int_0^t \lambda_\mathbb{Q}(s)\,ds} - e^{-\int_0^u \lambda_\mathbb{Q}(s)\,ds} \right)$$

$$= 1_{\{T_d > t\}} \left(1 - e^{-\int_t^u \lambda_\mathbb{Q}(s)\,ds} \right),$$

which is the regular $\mathcal{G}_\infty \vee \mathcal{H}_t$-conditional distribution of T_d given $\{T_d > t\}$.[5] Differentiation in with respect to u yields its density function

$$1_{\{T_d > t\}} \lambda_\mathbb{Q}(u) e^{-\int_t^u \lambda_\mathbb{Q}(s)\,ds} 1_{\{t \leq u\}}.$$

We thus obtain by disintegration

$$\Pi(t, T) = \mathbb{E}_\mathbb{Q}\left[e^{-\int_t^{T_d} r(s)\,ds} 1_{\{t < T_d \leq T\}} \mid \mathcal{F}_t \right]$$

$$= \mathbb{E}_\mathbb{Q}\left[\mathbb{E}_\mathbb{Q}\left[e^{-\int_t^{T_d} r(s)\,ds} 1_{\{t < T_d \leq T\}} \mid \mathcal{G}_\infty \vee \mathcal{H}_t \right] \mid \mathcal{F}_t \right]$$

$$= 1_{\{T_d > t\}} \mathbb{E}_\mathbb{Q}\left[\int_t^T e^{-\int_t^u r(s)\,ds} \lambda_\mathbb{Q}(u) e^{-\int_t^u \lambda_\mathbb{Q}(s)\,ds}\,du \mid \mathcal{F}_t \right]$$

$$= 1_{\{T_d > t\}} \int_t^T \mathbb{E}_\mathbb{Q}\left[\lambda_\mathbb{Q}(u) e^{-\int_t^u (r(s) + \lambda_\mathbb{Q}(s))\,ds} \mid \mathcal{G}_t \right] du.$$

The replacement of $\mathcal{F}_t$ by $\mathcal{G}_t$ in the last equality is justified by Lemma 12.5. This can be made more explicit:

Lemma 12.10 *For the above affine model* (12.13)–(12.14) *we have*

$$\Pi(t, T) = 1_{\{T_d > t\}} \left(\frac{c_0}{1 + c_1} \int_0^{T-t} e^{-A(u) - B(u)r(t)}\,du \right.$$

$$\left. + \frac{c_1}{1 + c_1} \left(1 - e^{-A(T-t) - B(T-t)r(t)} \right) \right),$$

for A and B as in Lemma 12.9.

Proof This follows from the identity

$$\mathbb{E}_\mathbb{Q}\left[\lambda_\mathbb{Q}(u) e^{-\int_t^u (r(s) + \lambda_\mathbb{Q}(s))\,ds} \mid \mathcal{G}_t \right] = \frac{c_0}{1 + c_1} \mathbb{E}_\mathbb{Q}\left[e^{-\int_t^u (r(s) + \lambda_\mathbb{Q}(s))\,ds} \mid \mathcal{G}_t \right]$$

$$- \frac{c_1}{1 + c_1} \frac{d}{du} \mathbb{E}_\mathbb{Q}\left[e^{-\int_t^u (r(s) + \lambda_\mathbb{Q}(s))\,ds} \mid \mathcal{G}_t \right]$$

($\rightarrow$ Exercise 12.6). $\qquad\qquad\qquad\qquad\qquad\qquad\qquad\qquad\qquad\qquad\qquad\qquad\quad\square$

---

[5] See the footnote on regular conditional distributions on page 64.

The above calculations and an extension to stochastic recovery go back to Lando [111].

12.3.4 Measure Change

We consider an equivalent change of measure and derive the behavior of the compensator process A in the Doob–Meyer decomposition (12.3) for the stopping time T_d. Again, we take the above stochastic setup and let **(D1)** and **(D2)** hold. So that

$$M(t) = H(t) - \int_0^t \lambda(s) 1_{\{T_d > s\}} \, ds$$

is a $(\mathbb{P}, \mathcal{G}_\infty \vee \mathcal{H}_t)$-martingale, see Lemma 12.6. Let μ be a positive $(\mathcal{G}_t)$-predictable process such that $\lambda_\mathbb{Q} = \mu\lambda$ satisfies

$$\int_0^t \lambda_\mathbb{Q}(s) \, ds < \infty \quad \text{for all } t \in \mathbb{R}_+. \tag{12.15}$$

We will now construct an equivalent probability measure $\mathbb{Q} \sim \mathbb{P}$ such that **(D1)** and **(D2)** hold under $\mathbb{Q}$ for $\lambda_\mathbb{Q}$.

The following analysis involves stochastic calculus for càdlàg processes of finite variation (FV), which in a sense is simpler than for Brownian motion since it is a pathwise calculus. The reader is referred to Protter [132] or Rogers and Williams [137] for an introduction. We recall the integration by parts formula for two right-continuous FV functions[6] f and g

$$f(t)g(t) = f(0)g(0) + \int_0^t f(s-) \, dg(s) + \int_0^t g(s-) \, df(s) + [f, g](t),$$

where we denote the covariation of f and g by

$$[f, g](t) = \sum_{0 < s \le t} \Delta f(s) \Delta g(s), \quad \text{and write} \quad \Delta f(s) = f(s) - f(s-).$$

Lemma 12.11 *The process*

$$D(t) = C(t)V(t)$$

with

$$C(t) = e^{\int_0^t (1 - \mu(s))\lambda(s) 1_{\{T_d > s\}} \, ds},$$

---

[6]See Protter [132, Corollary 2 and Theorem 26 in Chap. II] or Rogers and Williams [137, Sect. IV.3.18].

$$V(t) = \left(1_{\{T_d > t\}} + \mu(T_d)1_{\{T_d \le t\}}\right) = \begin{cases} 1, & t < T_d, \\ \mu(T_d), & t \ge T_d \end{cases}$$

satisfies

$$D(t) = 1 + \int_0^t D(s-)(\mu(s) - 1)\,dM(s)$$

and is thus a positive $(\mathbb{P}, \mathcal{G}_\infty \vee \mathcal{H}_t)$*-local martingale.*

Proof Notice that $[C, V] = 0$ and

$$V(t) = 1 + \int_0^t (\mu(s) - 1)\,dH(s) = 1 + \int_0^t V(s-)(\mu(s) - 1)\,dH(s).$$

Hence

$$D(t) = 1 + \int_0^t C(s-)\,dV(s) + \int_0^t V(s-)\,dC(s)$$

$$= 1 + \int_0^t C(s-)V(s-)(\mu(s) - 1)\,dH(s)$$

$$+ \int_0^t C(s)V(s-)(1 - \mu(s))\lambda(s)1_{\{T_d > s\}}\,ds$$

$$= 1 + \int_0^t D(s-)(\mu(s) - 1)\,dM(s).$$

Since M is a locally bounded $(\mathbb{P}, \mathcal{G}_\infty \vee \mathcal{H}_t)$-martingale, and since $D(s-)$ is locally bounded and by (12.15) we conclude by Protter [132, Theorems 17 and 20 in Chap. II] that D is a $(\mathbb{P}, \mathcal{G}_\infty \vee \mathcal{H}_t)$-local martingale. $\qquad \square$

Lemma 12.12 *Let* $T \in \mathbb{R}_+$*. Suppose* $\mathbb{E}[D(T)] = 1$*, so that we can define an equivalent probability measure* $\mathbb{Q} \sim \mathbb{P}$ *on* $\mathcal{G}_\infty \vee \mathcal{H}_T$ *by*

$$\frac{d\mathbb{Q}}{d\mathbb{P}} = D(T).$$

Then the process

$$M_{\mathbb{Q}}(t) = H(t) - \int_0^t \lambda_{\mathbb{Q}}1_{\{T_d > s\}}\,ds, \quad t \in [0, T], \tag{12.16}$$

is a $(\mathbb{Q}, \mathcal{G}_\infty \vee \mathcal{H}_t)$*-martingale, and thus* **(D1)** *and* **(D2)** *hold under* $\mathbb{Q}$ *for* $\lambda_{\mathbb{Q}}$*.*

Proof It is enough to show that $M_{\mathbb{Q}}$ is a $(\mathbb{Q}, \mathcal{G}_\infty \vee \mathcal{H}_t)$-local martingale. Indeed, (12.16) is the unique Doob–Meyer decomposition of H under $\mathbb{Q}$. Since H is uniformly integrable, so is $M_{\mathbb{Q}}$ ([106, Theorem 1.4.10]), and thus martingality follows from local martingality.

From Bayes' rule we know that $M_\mathbb{Q}$ is a $(\mathbb{Q}, \mathcal{G}_\infty \vee \mathcal{H}_t)$-local martingale if and only if $DM_\mathbb{Q}$ is a $(\mathbb{P}, \mathcal{G}_\infty \vee \mathcal{H}_t)$-local martingale. Notice that

$$[D, M_\mathbb{Q}](t) = \Delta D(T_d) 1_{\{T_d \geq t\}} = D(T_d-)(\mu(T_d) - 1) 1_{\{T_d \geq t\}}$$

$$= \int_0^t D(s-)(\mu(s) - 1) dH(s).$$

Integration by parts gives

$$DM_\mathbb{Q}(t) = \int_0^t D(s-) dM_\mathbb{Q}(s) + \int_0^t M_\mathbb{Q}(s-) dD(s) + [D, M_\mathbb{Q}](t)$$

$$= \int_0^t D(s-) dH(s) - \int_0^t D(s-)\mu(s)\lambda(s) 1_{\{T_d > s\}} ds$$

$$+ \int_0^t M_\mathbb{Q}(s-) dD(s) + \int_0^t D(s-)(\mu(s) - 1) dH(s)$$

$$= \int_0^t M_\mathbb{Q}(s-) dD(s) + \int_0^t D(s-)\mu(s) dM(s),$$

which proves the claim. The last statement follows from Lemma 12.6.   □

We finally remark that the conclusion of Lemma 12.12 does not necessarily hold under any equivalent probability measure $\mathbb{Q} \sim \mathbb{P}$. Indeed, we chose the density process D above such that the compensator process A in the Doob–Meyer decomposition (12.3) for $\mathbb{Q}$ is of the form $A(t) = \int_0^t \lambda_\mathbb{Q}(s) 1_{\{T_d > s\}} ds$ for some nonnegative $(\mathcal{G}_t)$-adapted process $\lambda_\mathbb{Q}$. In general, $\lambda_\mathbb{Q}(t)$ may depend on $\{T_d > t\} \in \mathcal{H}_t$, even though the corresponding λ under $\mathbb{P}$ does not. In that case, the density process D, and thus $M_\mathbb{Q}$, need not be a $(\mathbb{P}, \mathcal{G}_\infty \vee \mathcal{H}_t)$-martingale. A counterexample, involving more than one default times, has been constructed by Kusuoka [108], see also [10, Sect. 7.3]. Under these circumstances, the pricing approach from Sect. 12.3.3 does not apply, and one has to switch to other methods. A detailed analysis can be found in [10, Sect. 8.3].

12.4 Exercises

Exercise 12.1 Using elementary calculus and the fact that $\mathbb{P}[V(t) = X] = 0$, show that:

(a) $\lim_{T \downarrow t} p_d(t, T) = 1_{\{V(t) < X\}}$ a.s. both for Merton's and Zhou's default probabilities (12.1) and (12.2).

(b) $\lim_{T \downarrow t} \partial_T^+ p_d(t, T) = 0$ a.s. for Merton's default probability (12.1)

(c) $\lim_{T \downarrow t} \partial_T^+ p_d(t, T) = \lambda \Phi(d) 1_{\{V(t) \geq X\}} - \lambda \Phi(-d) 1_{\{V(t) < X\}}$ a.s. where $d = \frac{\log(X/V(t)) - m}{\rho}$ for Zhou's default probability (12.2).

Exercise 12.2 Consider Lemma 12.5.

(a) Finish the proof of (b) $\Leftrightarrow$ (c).
(b) Show that **(D2)** holds if and only if $\mathcal{G}_\infty$ and $\mathcal{F}_t$ are conditionally independent given $\mathcal{G}_t$; that is,

$$\mathbb{E}[XY \mid \mathcal{G}_t] = \mathbb{E}[X \mid \mathcal{G}_t]\mathbb{E}[Y \mid \mathcal{G}_t]$$

for all bounded $\mathcal{G}_\infty$-measurable X and bounded $\mathcal{F}_t$-measurable Y.

Exercise 12.3 Show that Lemma 12.1 holds for $\mathcal{F}_t$ replaced by $\mathcal{G}_T \vee \mathcal{H}_t$ and $\mathcal{G}_t$ replaced by $\mathcal{G}_T$, for every $t \leq T \leq \infty$. Use this to prove (12.10).

Exercise 12.4 Prove Lemma 12.8.

Exercise 12.5 Prove Lemma 12.9.

Exercise 12.6 Complete the proof of Lemma 12.10.

12.5 Notes

Sections 12.1 and 12.2 are based on [142, Chap. 2]. Table 12.2 is taken from [153, Table 6]. Section 12.3 is a substantially simplified blend of Bielecki and Rutkowski's textbook [10, Chaps. 5 to 8]. Further recommended textbooks include Duffie and Singleton [59], Schönbucher [143], McNeil, Frey and Embrechts [122, Chaps. 8 and 9], Lando [112].

Default risk has been an area of active research in mathematical finance since the early 1990s. However, the development and application of option pricing techniques to the study of corporate liabilities is where the modeling of credit risk has its foundations. That can be traced back to Black and Scholes' and Merton's milestone papers [18, 123].

The introduction to default risk in this book focused on the single name case only. An current area of active research is the modeling of portfolio credit risk, which involves several default times. One of the main problems there is the dependence modeling of the default times. Recent references include Bennani [9], Chen and Filipović [38], Cont et al. [45, 46], Ehlers and Schönbucher [63, 64], Filipović, Overbeck and Schmidt [75], Giesecke and Goldberg [78], Laurent and Gregory [113], Schönbucher [144], and Sidenius, Piterbarg and Anderson [151].

References

1. Ahn, D.H., Dittmar, R.F., Gallant, A.R.: Quadratic term structure models: Theory and evidence. Rev. Financ. Stud. **15**, 243–288 (2002)
2. Aït-Sahalia, Y.: Non-parametric pricing of interest rate-derivative securities. Econometrica **64**(3), 527–560 (1996)
3. Amann, H.: Ordinary Differential Equations. de Gruyter Studies in Mathematics, vol. 13, p. 458. de Gruyter, Berlin (1990). An introduction to nonlinear analysis, Translated from the German by Gerhard Metzen. ISBN 3-11-011515-8
4. Andersen, L.B.G., Piterbarg, V.V.: Moment explosions in stochastic volatility models. Finance Stoch. **11**(1), 29–50 (2007)
5. Avellaneda, M., Laurence, P.: Quantitative Modelling of Derivate Securities: From Theory to Practice. Chapman & Hall/CRC, London/Boca Raton (2000)
6. Bachelier, L.: Théorie de la spéculation. Ann. Sci. École Norm. Sup. (3) **17**, 21–86 (1900)
7. Bauer, H.: Wahrscheinlichkeitstheorie, 5th edn. de Gruyter Lehrbuch. [de Gruyter Textbook], p. 520. de Gruyter, Berlin (2002). ISBN 3-11-017236-4
8. Becker, P.A., Bouwman, K.E.: Arbitrage smoothing in fitting a sequence of yield curves. Preprint (2007)
9. Bennani, N.: The forward loss model: A dynamic term structure approach for the pricing of portfolio credit derivatives. Working Paper (2005)
10. Bielecki, T.R., Rutkowski, M.: Credit Risk: Modelling, Valuation and Hedging. Springer Finance, p. 500. Springer, Berlin (2002). ISBN 3-540-67593-0
11. BIS: Zero-coupon yield curves. Technical Documentation, Bank for International Settlements, Basle (1999)
12. Björk, T.: A geometric view of interest rate theory. In: Option Pricing, Interest Rates and Risk Management. Handb. Math. Finance, pp. 241–277. Cambridge Univ. Press, Cambridge (2001)
13. Björk, T.: Arbitrage Theory in Continuous Time, 2nd edn. Oxford University Press, London (2004)
14. Björk, T., Christensen, B.J.: Interest rate dynamics and consistent forward rate curves. Math. Finance **9**(4), 323–348 (1999)
15. Björk, T., Svensson, L.: On the existence of finite-dimensional realizations for nonlinear forward rate models. Math. Finance **11**(2), 205–243 (2001)
16. Björk, T., Kabanov, Y., Runggaldier, W.: Bond market structure in the presence of marked point processes. Math. Finance **7**(2), 211–239 (1997)
17. Black, F., Karasinski, P.: Bond and option pricing when short rates are lognormal. Financ. Anal. J. **47**, 52–59 (1991)
18. Black, F., Scholes, M.: The pricing of options and corporate liabilities. J. Polit. Econ. **81**, 637–654 (1973)
19. Black, F., Derman, E., Toy, W.: A one-factor model of interest rates and its application to treasury bond options. Financ. Anal. J. **46**, 33–39 (1990)
20. Bouchaud, A., Cont, R., Karoui, N.E., Potters, M., Sagna, N.: Phenomenology of the interest rate curve. Appl. Math. Finance **6**, 209–232 (1999)
21. Boyarchenko, N., Levendorskiĭ, S.: The eigenfunction expansion method in multi-factor quadratic term structure models. Math. Finance **17**(4), 503–539 (2007)
22. Boyle, P., Tian, W.: Quadratic interest rate models as approximations to effective rate models. J. Fixed Income **9**, 69–81 (1999)

D. Filipović, *Term-Structure Models*,
Springer Finance,
DOI 10.1007/978-3-540-68015-4, © Springer-Verlag Berlin Heidelberg 2009

23. Brace, A., Gątarek, D., Musiela, M.: The market model of interest rate dynamics. Math. Finance **7**(2), 127–155 (1997)
24. Brémaud, P.: Point Processes and Queues. Springer Series in Statistics, p. 354. Springer, New York (1981). Martingale dynamics. ISBN 0-387-90536-7
25. Brémaud, P., Yor, M.: Changes of filtrations and of probability measures. Z. Wahrsch. Verw. Geb. **45**(4), 269–295 (1978)
26. Brezis, H.: Analyse fonctionnelle. Collection Mathématiques Appliquées pour la Maîtrise. [Collection of Applied Mathematics for the Master's Degree], p. 234. Masson, Paris (1983). Théorie et applications [Theory and applications]. ISBN 2-225-77198-7
27. Brigo, D., Mercurio, F.: Interest Rate Models—Theory and Practice, 2nd edn. Springer Finance, p. 981. Springer, Berlin (2006). With smile, inflation and credit. ISBN 978-3-540-22149-4; 3-540-22149-2
28. Brown, R.H., Schaefer, S.M.: Why do long term forward rates (almost always) slope downwards? London Business School Working Paper, November 1994
29. Brown, R.H., Schaefer, S.M., Rogers, L.C.G., Mehta, S., Pezier, J.: Interest rate volatility and the shape of the term structure [and discussion]. Philos. Trans.: Phys. Sci. Eng. **347**(1684), 563–576 (1994). http://www.jstor.org/stable/54367
30. Bru, M.F.: Wishart processes. J. Theor. Probab. **4**(4), 725–751 (1991)
31. Buehler, H.: Consistent variance curve models. Finance Stoch. **10**(2), 178–203 (2006)
32. Buraschi, B., Porchia, P., Trojani, F.: Correlation risk and optimal portfolio choice. J. Finance (2009, forthcoming)
33. Cairns, A.J.G.: Interest Rate Models. Princeton University Press, Princeton (2004)
34. Carleton, W.T., Cooper, I.A.: Estimation and uses of the term structure of interest rates. J. Finance **31**, 1067–1083 (1976)
35. Carmona, R.A., Tehranchi, M.R.: Interest Rate Models: An Infinite Dimensional Stochastic Analysis Perspective. Springer Finance, p. 235. Springer, Berlin (2006). ISBN 978-3-540-27065-2; 3-540-27065-5
36. Carr, P., Madan, D.: Option valuation using the fast Fourier transform. J. Comput. Finance **2**, 61–73 (1999)
37. Carriere, J.F.: Valuation of the early-exercise price for options using simulations and non-parametric regression. Insurance Math. Econom. **19**(1), 19–30 (1996)
38. Chen, L., Filipović, D.: Credit derivatives in an affine framework. Asia-Pac. Financ. Mark. **14**, 123–140 (2007)
39. Chen, L., Poor, H.V.: Parametric estimation of quadratic models for the term structure of interest rate. Working Paper, Princeton University (2002)
40. Chen, L., Filipović, D., Poor, H.V.: Quadratic term structure models for risk-free and defaultable rates. Math. Finance **14**(4), 515–536 (2004)
41. Cheng, P., Scaillet, O.: Linear-quadratic jump-diffusion modeling. Math. Finance **17**(4), 575–598 (2007)
42. Cheridito, P., Filipović, D., Yor, M.: Equivalent and absolutely continuous measure changes for jump-diffusion processes. Ann. Appl. Probab. **15**(3), 1713–1732 (2005)
43. Cheridito, P., Filipović, D., Kimmel, R.L.: A note on the Dai–Singleton canonical representation of affine term structure models. Math. Finance (2009, forthcoming)
44. Collin-Dufresne, P., Goldstein, R.S., Jones, C.S.: Identification of maximal affine term structure models. J. Finance **63**(2), 743–795 (2008)
45. Cont, R., Minca, A.: Recovering portfolio default intensities implied by CDO quotes. Financial Engineering Report 2008-01, Columbia University Center for Financial Engineering (2008)
46. Cont, R., Savescu, I.: Forward equations for portfolio credit derivatives. In: Cont, R. (ed.) Frontiers in Quantitative Finance: Volatility and Credit Risk Modeling. Wiley Finance Series, pp. 269–293. Wiley, New York (2009), Chap. 11
47. Cox, J.C., Ingersoll, J.E., Ross, S.A.: A theory of the term structure of interest rates. Econometrica **53**(2), 385–407 (1985)
48. Cuchiero, C., Filipović, D., Mayerhofer, E., Teichmann, J.: Affine processes on positive semi-definite matrices. Working Paper (2009)

49. Da Fonseca, J., Grasselli, M., Tebaldi, C.: A multifactor volatility Heston model. J. Quant. Finance **8**(6), 591–604 (2008)
50. Dai, Q., Singleton, K.J.: Specification analysis of affine term structure models. J. Finance **55**(5), 1943–1978 (2000)
51. De Jong, F., Driessen, J., Pelsser, A.: Libor market models versus swap market models for pricing interest rate derivatives: An empirical analysis. Eur. Finance Rev. **5**, 201–237 (2001)
52. Delbaen, F.: Representing martingale measures when asset prices are continuous and bounded. Math. Finance **2**, 107–130 (1992)
53. Delbaen, F., Schachermayer, W.: A general version of the fundamental theorem of asset pricing. Math. Ann. **300**(3), 463–520 (1994)
54. Delbaen, F., Schachermayer, W.: The no-arbitrage property under a change of numéraire. Stoch. Stoch. Rep. **53**(3–4), 213–226 (1995)
55. Dieudonné, J.: Foundations of Modern Analysis. Pure and Applied Mathematics, vol. X, p. 361. Academic Press, San Diego (1960)
56. Dothan, M.: On the term structure of interest rates. J. Financ. Econ. **6**, 59–69 (1978)
57. Duffie, D., Huang, C.F.: Multiperiod security markets with differential information: martingales and resolution times. J. Math. Econ. **15**(3), 283–303 (1986)
58. Duffie, D., Kan, R.: A yield-factor model of interest rates. Math. Finance **6**(4), 379–406 (1996)
59. Duffie, D., Singleton, K.J.: Credit Risk: Pricing, Measurement, and Management. Princeton University Press, Princeton (2003)
60. Duffie, D., Pan, J., Singleton, K.: Transform analysis and asset pricing for affine jump-diffusions. Econometrica **68**(6), 1343–1376 (2000)
61. Duffie, D., Filipović, D., Schachermayer, W.: Affine processes and applications in finance. Ann. Appl. Probab. **13**(3), 984–1053 (2003)
62. Dybvig, P., Ingersoll, J., Ross, S.: Long forward and zero coupon rates can never fall. J. Bus. **69**, 1–25 (1996)
63. Ehlers, P., Schönbucher, P.J.: Pricing interest rate-sensitive credit portfolio derivatives. Working Paper, ETH Zurich (2006)
64. Ehlers, P., Schönbucher, P.J.: Background filtrations and canonical loss processes for top-down models of portfolio credit risk. Finance Stoch. **13**, 79–103 (2009)
65. Eksi, Z.: A Black-Scholes like model with Vasiček interest rates. Vienna Institute of Finance Working Paper No. 1, www.vif.ac.at/papers (2007)
66. Filipović, D.: A note on the Nelson-Siegel family. Math. Finance **9**(4), 349–359 (1999)
67. Filipović, D.: Exponential-polynomial families and the term structure of interest rates. Bernoulli **6**(6), 1081–1107 (2000)
68. Filipović, D.: Consistency Problems for Heath-Jarrow-Morton Interest Rate Models. Lecture Notes in Mathematics, vol. 1760, p. 134. Springer, Berlin (2001). ISBN 3-540-41493-2
69. Filipović, D.: A general characterization of one factor affine term structure models. Finance Stoch. **5**(3), 389–412 (2001)
70. Filipović, D.: Separable term structures and the maximal degree problem. Math. Finance **12**(4), 341–349 (2002)
71. Filipović, D.: Time-inhomogeneous affine processes. Stoch. Process. Appl. **115**(4), 639–659 (2005)
72. Filipović, D., Mayerhofer, E.: Affine diffusion processes: Theory and applications. In: Advanced Financial Modelling. Radon Ser. Comput. Appl. Math., vol. 8. de Gruyter, Berlin (2009)
73. Filipović, D., Teichmann, J.: Existence of invariant manifolds for stochastic equations in infinite dimension. J. Funct. Anal. **197**(2), 398–432 (2003)
74. Filipović, D., Teichmann, J.: On the geometry of the term structure of interest rates. Proc. R. Soc. Lond. Ser. A Math. Phys. Eng. Sci. **460**(2041), 129–167 (2004). Stochastic analysis with applications to mathematical finance
75. Filipović, D., Overbeck, L., Schmidt, T.: Dynamic CDO term structure modelling. Math. Finance (2009, forthcoming)

76. Fischer, S.: Call option pricing when the exercise price is uncertain, and the valuation of index bonds. J. Finance **33**, 169–176 (1978)
77. Geman, H., El Karoui, N., Rochet, J.C.: Changes of numéraire, changes of probability measure and option pricing. J. Appl. Probab. **32**(2), 443–458 (1995)
78. Giesecke, K., Goldberg, L.: A top down approach to multi-name credit. Working Paper, Stanford University (2007)
79. Glasserman, P.: Monte Carlo Methods in Financial Engineering. Applications of Mathematics (New York), vol. 53, p. 596. Springer, New York (2004). Stochastic Modelling and Applied Probability. ISBN 0-387-00451-3
80. Glasserman, P., Kim, K.K.: Moment explosions and stationary distributions in affine diffusion models. Math. Finance (2008/2009, to appear)
81. Glasserman, P., Zhao, X.: Arbitrage-free discretization of lognormal forward Libor and swap rate models. Finance Stoch. **4**(1), 35–68 (2000)
82. Goldammer, V., Schmock, U.: Generalization of the Dybvig-Ingersoll-Ross theorem and asymptotic minimality. Working Paper, Vienna University of Technology (2008)
83. Gombani, A., Runggaldier, W.J.: A filtering approach to pricing in multifactor term structure models. Int. J. Theor. Appl. Finance **4**(2), 303–320 (2001). Information modeling in finance (Évry, 2000)
84. Gourieroux, C., Sufana, R.: Wishart quadratic term structure models. Working Paper, CREF HRC Montreal (2003)
85. Gourieroux, C., Sufana, R.: A classification of two factor affine diffusion term structure models. J. Financ. Econom. **4**(1), 31–52 (2006)
86. Grasseli, M., Tebaldi, C.: Solvable affine term structure models. Math. Finance **18**(1), 135–153 (2008)
87. Hamilton, J.D.: The daily market for federal funds. J. Polit. Econ. **104**(1), 26–56 (1996). http://ideas.repec.org/a/ucp/jpolec/v104y1996i1p26-56.html
88. Harrison, J.M., Pliska, S.R.: Martingales and stochastic integrals in the theory of continuous trading. Stoch. Process. Appl. **11**(3), 215–260 (1981)
89. Harrison, M., Kreps, J.: Martingales and arbitrage in multiperiod securities markets. J. Econ. Theory **20**(3), 381–408 (1979)
90. Heath, D., Jarrow, R., Morton, A.: Bond pricing and the term structure of interest rates: A new methodology for contingent claims valuation. Econometrica **60**, 77–105 (1992)
91. Heston, S.: A closed-form solution for options with stochastic volatility with applications to bond and currency options. Rev. Financ. Stud. **6**, 327–344 (1993)
92. Ho, T.S.Y., Lee, S.B.: Term structure movements and pricing interest rate contingent claims. J. Finance **41**, 1011–1029 (1986)
93. Hubalek, F., Klein, I., Teichmann, J.: A general proof of the Dybvig–Ingersoll–Ross theorem: long forward rates can never fall. Math. Finance **12**(4), 447–451 (2002)
94. Hubalek, F., Kallsen, J., Krawczyk, L.: Variance-optimal hedging for processes with stationary independent increments. Ann. Appl. Probab. **16**(2), 853–885 (2006)
95. Hull, J.C.: Options, Futures, and Other Derivatives, 4th edn. Prentice-Hall International, Englewood Cliffs (2000)
96. Hull, J.C., White, A.: Pricing interest rate derivative securities. Rev. Financ. Stud. **3**(4), 573–592 (1990)
97. Hunt, P., Kennedy, J., Pelsser, A.: Markov-functional interest rate models. Finance Stoch. **4**(4), 391–408 (2000)
98. Hurd, T., Zhou, Z.: A Fourier transform method for spread option pricing. Working Paper, McMaster University (2009)
99. Ikeda, N., Watanabe, S.: Stochastic Differential Equations and Diffusion Processes. North-Holland Mathematical Library, vol. 24, p. 464. North-Holland, Amsterdam (1981). ISBN 0-444-86172-6
100. James, J., Webber, N.: Interest Rate Modelling. Wiley, New York (2000)
101. Jamshidian, R.: An exact bond option formula. J. Finance **44**(1), 205–209 (1989)
102. Jamshidian, R.: Libor and swap market models and measures. Finance Stoch. **1**(4), 290–330 (1997)

103. Jarrow, R.: Modelling Fixed Income Securities and Interest Rate Options. McGraw-Hill, New York (1996)
104. Johnson, N.L., Kotz, S., Balakrishnan, N.: Continuous Univariate Distributions. Vol. 2, 2nd edn. Wiley Series in Probability and Mathematical Statistics: Applied Probability and Statistics, p. 719. Wiley, New York (1995). A Wiley-Interscience Publication. ISBN 0-471-58494-0
105. Joslin, S.: Can unspanned stochastic volatility models explain the cross section of bond volatilities? Working Paper, Stanford University (2006)
106. Karatzas, I., Shreve, S.E.: Brownian Motion and Stochastic Calculus, 2nd edn. Graduate Texts in Mathematics, vol. 113, p. 470. Springer, New York (1991). ISBN 0-387-97655-8
107. Kardaras, C., Platen, E.: On the Dybvig–Ingersoll–Ross theorem. Working Paper, University of Technology Sydney (2009)
108. Kusuoka, S.: A remark on default risk models. In: Advances in Mathematical Economics, Tokyo, 1997. Adv. Math. Econ., vol. 1, pp. 69–82. Springer, Berlin (1999)
109. Lakshmikantham, V., Shahzad, N., Walter, W.: Convex dependence of solutions of differential equations in a Banach space relative to initial data. Nonlinear Anal. 27(12), 1351–1354 (1996)
110. Lamberton, D., Lapeyre, B.: Introduction to Stochastic Calculus Applied to Finance, p. 185. Chapman & Hall, London (1996). Translated from the 1991 French original by Nicolas Rabeau and François Mantion. ISBN 0-412-71800-6
111. Lando, D.: On Cox processes and credit-risky securities. Rev. Deriv. Res. 2, 99–120 (1998)
112. Lando, D.: Credit Risk Modeling. Princeton University Press, Princeton (2004)
113. Laurent, J.P., Gregory, J.: Basket default swaps, cdos and factor copulas. J. Risk 7, 103–122 (2005)
114. Lee, R.W.: The moment formula for implied volatility at extreme strikes. Math. Finance 14(3), 469–480 (2004)
115. Leippold, M., Wu, L.: Asset pricing under the quadratic class. J. Financ. Quant. Anal. 37, 271–295 (2002)
116. Leippold, M., Wu, L.: Design and estimation of quadratic term structure models. Eur. Finance Rev. 7, 47–73 (2003)
117. Litterman, R., Scheinkman, J.K.: Common factors affecting bond returns. J. Fixed Income 1, 54–61 (1991)
118. Longstaff, F., Schwartz, E.: Valuing American options by simulation: a simple least-squares approach. Rev. Financ. Stud. 14, 113–147 (2001)
119. Lorimier, S.: Interest rate term structure estimation based on the optimal degree of smoothness of the forward rate curve. Ph.D. Thesis, University of Antwerp (1995)
120. Lukacs, E.: Characteristic Functions, 2nd edn., revised and enlarged, p. 350. Hafner, New York (1970)
121. Margrabe, W.: The value of an option to exchange one asset for another. J. Finance 33, 177–186 (1978)
122. McNeil, A.J., Frey, R., Embrechts, P.: Quantitative Risk Management. Princeton University Press, Princeton (2005)
123. Merton, R.C.: On the pricing for corporate debt: the risk structure of interest rates. J. Finance 29(2), 449–470 (1974)
124. Miltersen, K., Sandmann, K., Sondermann, D.: Closed form solutions for term structure derivatives with lognormal interest rates. J. Finance 52, 409–430 (1997)
125. Morton, A.J.: Arbitrage and martingales. Technical Report 821, School of Operations Research and Industrial Engineering, Cornell University (1988)
126. Musiela, M., Rutkowski, M.: Continuous-time term structure models: forward measure approach. Finance Stoch. 1, 261–291 (1997)
127. Musiela, M., Rutkowski, M.: Martingale Methods in Financial Modelling, 2nd edn. Stochastic Modelling and Applied Probability, vol. 36, p. 636. Springer, Berlin (2005). ISBN 3-540-20966-2
128. Nelson, C.R., Siegel, A.F.: Parsimonious modeling of yield curves. J. Bus. 60(4), 473–489 (1987)

129. Neumaier, A.: Introduction to Numerical Analysis, p. 356. Cambridge University Press, Cambridge (2001). ISBN 0-521-33323 7; 0-521-33610-4
130. Nielsen, L.T.: Pricing and Hedging of Derivative Securities, p. 439. Oxford University Press, London (1999)
131. Pelsser, A.: Efficient Methods for Valuing Interest Rate Derivatives. Springer Finance, p. 172. Springer, Berlin (2000). ISBN 1-85233-304-9
132. Protter, P.E.: Stochastic Integration and Differential Equations, 2nd edn. Applications of Mathematics (New York), vol. 21, p. 415. Springer, Berlin (2004). Stochastic Modelling and Applied Probability. ISBN 3-540-00313-4
133. Rabinovitch, R.: Pricing stock and bond options when the default-free rate is stochastic. JFQA **24**(4), 447–457 (1989)
134. Rebonato, R.: Interest-Rate Option Models: Understanding, Analysing and Using Models for Exotic Interest-Rate Options, 2nd edn. Wiley Series in Financial Engineering, p. 521. Wiley, New York (1998)
135. Revuz, D., Yor, M.: Continuous Martingales and Brownian Motion, 3rd edn. Grundlehren der Mathematischen Wissenschaften [Fundamental Principles of Mathematical Sciences], vol. 293, p. 602. Springer, Berlin (1999). ISBN 3-540-64325-7
136. Rockafellar, R.T.: Convex Analysis. Princeton Landmarks in Mathematics, p. 451. Princeton University Press, Princeton (1997). Reprint of the 1970 original, Princeton Paperbacks. ISBN 0-691-01586-4
137. Rogers, L.C.G., Williams, D.: Diffusions, Markov Processes, and Martingales. Vol. 2. Cambridge Mathematical Library, p. 480. Cambridge University Press, Cambridge (2000). Itô calculus, Reprint of the second (1994) edition. ISBN 0-521-77593-0
138. Rudin, W.: Real and Complex Analysis, 3rd edn., p. 416. McGraw-Hill, New York (1987). ISBN 0-07-054234-1
139. Sandmann, K., Sondermann, D.: A note on the stability of lognormal interest rate models and the pricing of Eurodollar futures. Math. Finance **7**, 119–128 (1997)
140. Schachermayer, W.: Martingale measures for discrete-time processes with infinite horizon. Math. Finance **4**(1), 25–55 (1994)
141. Schaefer, S.: The problem with redemption yields. Financ. Anal. J. (**July/August**), 59–67 (1977)
142. Schmid, B.: Pricing Credit Linked Financial Instruments. Lecture Notes in Economics and Mathematical Systems, vol. 516, p. 246. Springer, Berlin (2002). Theory and empirical evidence. ISBN 3-540-43195-0
143. Schönbucher, P.J.: Credit Derivatives Pricing Models. Wiley, New York (2003)
144. Schönbucher, P.J.: Portfolio losses and the term structure of loss transition rates: A new methodology for the pricing of portfolio credit derivatives. Working Paper, ETH Zürich (2005)
145. Schrager, D.F., Pelsser, A.A.J.: Pricing swaptions and coupon bond options in affine term structure models. Math. Finance **16**(4), 673–694 (2006)
146. Sharef, E.: Quantitative evaluation of consistent forward rate processes. An empirical study. Senior's Thesis, Princeton University (2003)
147. Sharef, E., Filipović, D.: Conditions for consistent exponential-polynomial forward rate processes with multiple nontrivial factors. Int. J. Theor. Appl. Finance **7**(6), 685–700 (2004)
148. Shkolnikov, M.: Consistent diffusions for the Lorimier forward rate model. Private communication (2008)
149. Shreve, S.E.: Stochastic Calculus for Finance. II. Springer Finance, p. 550. Springer, New York (2004). Continuous-time models. ISBN 0-387-40101-6
150. Sidenius, J.: Libor market models in practice. J. Comput. Finance **3**(2), 5–26 (2000)
151. Sidenius, J., Piterbarg, V., Andersen, L.: A new framework for dynamic credit portfolio loss modelling. Int. J. Theor. Appl. Finance **11**(2), 163–197 (2008)
152. Siegel, A.F.: The noncentral chi-squared distribution with zero degrees of freedom and testing for uniformity. Biometrika **66**(2), 381–386 (1979)

153. Standard and Poor's: Default, transition, and recovery: Canadian ratings performance 2007: Benign credit conditions end abruptly. Ratings Direct, URL: http://www2.standardandpoors. com/spf/pdf/media/_default_study_2007_pub_04_08.pdf, April 2008
154. Steele, J.M.: Stochastic Calculus and Financial Applications. Applications of Mathematics (New York), vol. 45, p. 300. Springer, New York (2001). ISBN 0-387-95016-8
155. Steeley, J.: Estimating the Gilt-edged term structure: basis splines and confidence intervals. J. Bus. Finance Account. **18**, 513–530 (1991)
156. Stein, E.M., Weiss, G.: Introduction to Fourier Analysis on Euclidean Spaces. Princeton Mathematical Series, vol. 32, p. 297. Princeton University Press, Princeton (1971)
157. Svensson, L.E.O.: Estimating and interpreting forward interest rates: Sweden 1992–1994. Technical Report 114, International Monetary Fund, Washington, DC (1994)
158. Tsitsiklis, J.N., Van Roy, B.: Optimal stopping of Markov processes: Hilbert space theory, approximation algorithms, and an application to pr icing high-dimensional financial derivatives. IEEE Trans. Automat. Contr. **44**(10), 1840–1851 (1999)
159. Tsitsiklis, J.N., Van Roy, B.: Regression methods for pricing complex American-style options. IEEE Trans. Neural Netw. **12**, 694–703 (2001)
160. Vasiček, O.: An equilibrium characterization of the term structure. J. Financ. Econ. **5**, 177–188 (1977)
161. Williams, D.: Probability with Martingales. Cambridge Mathematical Textbooks, p. 251. Cambridge University Press, Cambridge (1991). ISBN 0-521-40455-X; 0-521-40605-6
162. Yamada, T., Watanabe, S.: On the uniqueness of solutions of stochastic differential equations. J. Math. Kyoto Univ. **11**, 155–167 (1971)
163. Zagst, R.: Interest-Rate Management. Springer Finance, p. 341. Springer, Berlin (2002). ISBN 3-540-67594-9
164. Zhou, C.: A jump-diffusion approach to modeling credit risk and valuing defaultable securities. Working Paper, Federal Reserve Board, Washington, DC (1996)

Index

A

Accrued interest, 19
Adapted, 59
Affine
  (diffusion) process, 143
    admissible parameters, 147
    canonical state space, 146
    existence and uniqueness, 172
    term-structure (ATS), 84, 127, 151
Appreciation rate, 65
Arbitrage, 1, 9, 67
  -free, 67
  price, 74

B

Bank account, 10
Basis point, 8
Bayes' rule, 77
Bessel function, 164
Beta function, 160
Bill, 18
Black–Scholes
  (implied) volatility, 168, 190
  model, 76
    with stochastic short rates, 110
  option price formula, 113
Black's formula, 21, 24, 198, 204, 208
Bond, 5
  callable, 23
  (fixed) coupon, 11, 18
  T-, 5
  zero-coupon, 5
Bootstrapping, 29, 215

C

Cap, 20
  at-the-money (ATM), 21
  Black's formula, 21, 198, 204
  (implied) volatility, 21
  in-the-money (ITM), 21
  out-of-the-money (OTM), 21
  rate, 20
Caplet, 20
Characteristic function, 156, 174

Cholesky factorization, 172
Claim
  attainable, 71
  contingent, 71
  T-, 71
Clean price, 19
Complete market, 72
Consistency condition
    for parametrizations, 126
Contingent claim, 71
Convexity, 17
Convexity adjustment, 33, 121
Covariation, 61
Credit risk, 225

D

Day-count convention, 17
Default
  event, 229
  intensity, 228, 232
  probability, 228, 229
  rate, 226
  risk, 225
    intensity-based approach, 229
    structural approach, 227
  time, 229
Diffusion, 63
  matrix, 63
Dirty price, 19
Discount curve, 5, 34
Discount factor, 10
Doob–Meyer decomposition, 230
Drift, 63
Duration, 16
Dybvig–Ingersoll–Ross theorem, 108

E

Equivalent (local) martingale measure
    (E(L)MM), 68
Estimation
  non-parametric, 34
  parametric, 38
Euler approximation, 211
Exchange option, 158

D. Filipović, *Term-Structure Models,*
Springer Finance,
DOI 10.1007/978-3-540-68015-4, © Springer-Verlag Berlin Heidelberg 2009

Expectation hypothesis, 107
Exponential–polynomial family, 50, 134

F
Feynman–Kac formula, 81
First passage time model, 229
Floating rate note, 12
Floor, 20
    at-the-money (ATM), 21
    Black's formula, 21
    (implied) volatility, 21
    in-the-money (ITM), 21
    out-of-the-money (OTM), 21
Floorlet, 20
Forward
    contract, 117
    price, 117
Forward curve, 7
Forward measure, 105, 199
Forward rate
    continuously compounded, 7
    instantaneous, 7
    simple, 6
Forward rate agreement (FRA), 6
Forward swap measure, 207
Forward swap rate, 13
Fourier
    inversion formula, 156
    transform, 156
        examples, 157
Fubini's theorem for stochastic integrals, 99
Fundamental theorem of asset pricing
    first, 70
    second, 72
Futures
    contract, 118
    Eurodollar, 119
    interest rate, 30, 119
    price, 118
    rate, 30, 120

G
Girsanov's change of measure theorem, 68
Gronwall's inequality, 150

H
Heath–Jarrow–Morton (HJM)
    drift condition, 95
    framework, 93
    model, 98
        Gaussian, 109
        with proportional volatility, 99
    short-rate dynamics, 97

Hedge, 71
Heston stochastic volatility model, 166
Hypothesis H, 233

I
Itô process, 61
Itô's formula, 62

L
Lévy's characterization theorem, 62
LIBOR, 8, 197
    market model, 199
        calibration, 213
        Monte Carlo simulation, 210
Loading, 52
    empirical, 53
London Interbank Offered Rate, 8
Lorimier family, 140

M
Macaulay duration, 16
Margrabe option, 158
Market model, 197
    LIBOR, 199
    swap, 208
Market price of risk, 69
Marking to market, 118
Markov property, 64
Maximal degree problem, 129–132
Merton model, 227
Money account, 10
Money-market account, 10, 65
Monte Carlo simulation, 211
    standard error, 211
Multi-factor model, 123

N
Nelson–Siegel family, 49, 134
Noncentral χ^2-distribution, 164
    characteristic function of, 164, 188
Note, 18
Novikov's condition, 69
Numeraire, 66
    change of, 106

P
Par swap rate, 13
Polynomial term-structure (PTS), 128
Portfolio, 65
    admissible, 70
    arbitrage, 67
    self-financing, 66
Principal component, 52

empirical, 53
Principal component analysis (PCA), 51, 212
  of the forward curve, 53
Progressive, 59
Progressively measurable, 59

Q
Quadratic term-structure (QTS), 129
Quadratic variation, 62

R
Rating agency, 225
  Fitch, 225
  Moody's, 225
  S&P, 225
Rebonato's formula, 210
Recovery
  partial at default, 237, 238
  partial at maturity, 237, 238
  zero, 237
Recovery rate, 225
Representation theorem, 72
Resettlement, 118
Riccati equation, 144
  solution of, 180
Risk-free
  asset, 10, 65
  rate of return, 10
Risk-neutral measure, 106, 151, 199, 205
Risky asset, 65

S
Savings account, 10
Self-financing, 66
Short rate, 7
Short-rate model, 79
  affine, 151, 169
    canonical representation, 169
  affine term-structure (ATS), 84
  Black–Derman–Toy, 83, 88
  Black–Karasinski, 83, 88
  Cox–Ingersoll–Ross (CIR), 83, 87, 114,
    163, 237
  diffusion, 81
  Dothan, 83, 88
  Ho–Lee, 83, 89, 97, 115
  Hull–White, 83, 90, 103, 135
  lognormal, 88
  Vasiček, 82, 85, 110, 113, 114, 162
Spline
  B-, 39
  cubic, 38

kth-order, 38
  smoothing, 43
Spot LIBOR measure, 205
Spot rate
  continuously compounded, 7
  simple, 7
Spread option, 159
State-price density, 75
Stochastic differential equation, 63
  solution, 63
    strong, 63
    weak, 173
Stochastic exponential, 64
Stochastic integral, 60
Stochastic invariance, 177
Stopping time, 59
  doubly stochastic, 233
    construction of, 235
Strategy, 65
  admissible, 70
  arbitrage, 67
  self-financing, 66
STRIPS, 18
Svensson family, 49, 135
Swap
  payer, 12
  receiver, 13
Swap market model, 208
Swaption, 22, 206
  at-the-money (ATM), 23
  Black's formula, 24, 208
  (implied) volatility, 24
  in-the-money (ITM), 23
  out-of-the-money (OTM), 23
  payer, 22
  receiver, 22
  tenor, 22
  $x \times y$-, 23

T
Term-structure, 1
  affine (ATS), 84, 127, 151
  of zero-coupon bond prices, 5, 34
  polynomial (PTS), 128
  quadratic (QTS), 129
Term-structure equation, 82
Transition
  matrix, 227
  rate, 227

U
Usual conditions, 59

V
Value process, 66
Volatility, 65
    row vector, 65

Y
Yield
    -to-maturity, 16, 19

zero-coupon, 15
Yield curve, 15

Z
Zero-coupon
    bond, 5
    yield, 15